ALL ABOUT SCIENCE
PHILOSOPHY, HISTORY, SOCIOLOGY & COMMUNICATION

Science Matters Series

Lui Lam
Founder and Editor

Scimat (Science Matters) is the new discipline that treats all human-dependent matters as part of science, wherein, humans (the material system of *Homo sapiens*) are studied scientifically from the perspective of complex systems. That "Everything in Nature is Part of Science" was well recognized by Aristotle and da Vinci and many others. Yet, it is only recently, with the advent of modern science and experiences gathered in the study of evolutionary and cognitive sciences, neuroscience, statistical physics, complex systems and other disciplines, that we know how the human-related disciplines can be studied scientifically. Science Matters Series covers new developments in all the topics in the humanities and social sciences from the scimat perspective, with emphasis on the humanities.

Published

1. *Science Matters: Humanities as Complex Systems*
   M. Burguete & L. Lam, editors

2. *Arts: A Science Matter*
   M. Burguete & L. Lam, editors

3. *All About Science: Philosophy, History, Sociology & Communication*
   M. Burguete & L. Lam, editors

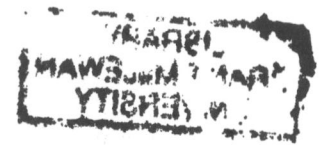

Science Matters Series No. 3

ALL ABOUT SCIENCE
PHILOSOPHY, HISTORY, SOCIOLOGY & COMMUNICATION

Maria Burguete
Bento da Rocha Cabral Institute for Scientific Research, Portugal

Lui Lam
San Jose State University, USA

Editors

 World Scientific

NEW JERSEY • LONDON • SINGAPORE • BEIJING • SHANGHAI • HONG KONG • TAIPEI • CHENNAI

Published by

World Scientific Publishing Co. Pte. Ltd.
5 Toh Tuck Link, Singapore 596224
USA office: 27 Warren Street, Suite 401-402, Hackensack, NJ 07601
UK office: 57 Shelton Street, Covent Garden, London WC2H 9HE

Library of Congress Cataloging-in-Publication Data
All about science : philosophy, history, sociology & communication / edited by Maria Burguete (Scientific Research Institute Bento da Rocha Cabral, Portugal) and Lui Lam (San Jose State University, USA).
   pages cm. -- (Science matters series ; v. 3)
   Includes bibliographical references and index.
   ISBN 978-9814472920 (hardcover : alk. paper) -- ISBN 978-9814508193 (pbk : alk. paper)
   1. Science. 2. Science--Philosophy. I. Burguete, Maria, editor. II. Lam, Lui, editor.
   Q158.5.A445 2014
   500--dc23
                                                                    2014027465

British Library Cataloguing-in-Publication Data
A catalogue record for this book is available from the British Library.

Cover design: Lui Lam

Artwork: Cover: *Two Linked Animals—A Metaphor for Science* (Lui Lam, 2014; see Fig. 2.1)
         Parts I-V: *Disputed Islands* (Charlene Lam, 2014)

Copyright © 2014 by World Scientific Publishing Co. Pte. Ltd.

All rights reserved. This book, or parts thereof, may not be reproduced in any form or by any means, electronic or mechanical, including photocopying, recording or any information storage and retrieval system now known or to be invented, without written permission from the publisher.

For photocopying of material in this volume, please pay a copying fee through the Copyright Clearance Center, Inc., 222 Rosewood Drive, Danvers, MA 01923, USA. In this case permission to photocopy is not required from the publisher.

Printed in Singapore by Mainland Press Pte Ltd.

Preface

The importance of science goes without saying. Yet there is a lot of confusion and misconception concerning Science. The nature and contents of science is an unsettled problem. For example, Thales of 2,600 years ago is recognized as the father of science but the word science was introduced only in the 14th century; the definition of science is often avoided in books about philosophy of science. This book aims to clear up all these confusions and present new developments in the philosophy, history, sociology and communication of science.

In fact, through a careful examination of the historical development it is not too hard to recognize that science is a subset of human activities aiming to understand how Nature—consisting of the human system and all nonhuman systems—works *without* bringing in God or any supernatural. In other words, what characterizes science is its secularity. This simple definition of Science—historically correct but missed by many people—is expounded by Lui Lam in Chapter 1. Also included there is an elaboration of the new discipline *Scimat* (Science Matters) which treats all human-dependent matters as part of science, with its immediate goal of setting up scimat centers around the world.

The nature and development of science are analyzed by the three interrelated disciplines, Philosophy of Science, History of Science and Sociology of Science while Science Communication, which depends heavily on the other three, is the discipline that connects the public to science. These four important disciplines are very young, emerged within the last century, and are part of the humanities. Chapter 2 by Lam examines the four disciplines with new insights, from the perspective of scimat, and urges the expansion of their scope to include more complex systems, humans in particular. It should be pointed out that this book is probably the first one ever which treats the four disciplines collectively together.

In China, these four disciplines are grouped under the umbrella term "scientific culture" since the 1980s. Top scholars from China were invited to present their newest works here. The lack of scientific culture in ancient China is explained by Hong-Sheng Wang (Renmin University of China) in Chapter 4. The history of its development in contemporary China is summarized by Bing Liu (Tsinghua University) and Mei-Fang Zhang (University of Science and Technology Beijing) in Chapter 13. Guo-Sheng Wu (Peking University) writes on his favorable topic, the phenomenological philosophy of science (Chapter 3) while Jin-Yang Liu (Remin University of China) discusses in detail his idea of holism (Chapter 6). New insights on science in Victorian Era, on the "mistake" of Friedrick Engel and Mitsutomo Yuasa, are provided by Dun Liu (Chinese Academy of Sciences) in Chapter 8. Moreover, in Chapter 15, a thorough description of popular-science writings in early modern China, which played a crucial role in the introduction of science from the West, is written by Lin Yin (China Research Institute for Science Popularization).

Of course, the scientific culture originated in the West and has been widely covered in numerous books and articles. Still, presented in this book are four important articles: A summary of the three waves of science studies (Chapter 11) by sociologist Harry Collins, a unique insight on what scientists really know (Chapter 5) by physicist Nigel Sanitt, a historical description of medical studies in Portugal around 1911 (Chapter 9) by historian Maria Burguete, and a history and review on science communication (Chapter 14) by the expert Peter Broks.

Two more important articles on the history and sociology of science are written by Lam. Chapter 10 is his detailed, personal account of the why and how as well as the background and crucial events in establishing the International Liquid Crystal Society, a story never told before. It is written for those working in or interested in science, liquid crystals in particular, and for science historians. Chapter 12 is his personal recollection of the six years of working in China, starting from the "Science Spring" of 1978, the year China's reform-and-opening up *revolution* began. In this chapter, the development of soliton research and political climate in China experienced by the author is revealed for the first time.

Science consists of two parts: the scientific process and the resulting scientific knowledge. An example of these two parts in action is nicely demonstrated by Robin Warren's description of his Nobel Prize-winning work on *Helicobacter* (Chapter 7).

Science, according to the definition in Chapter 1, consists of not just "natural science" but also the humanities and social science. We are thus very happy to be given the chance to showcase three articles in this book to illustrate this point. Kajsa Berg's Chapter 16 describes neuroarthistory, a relatively new discipline, from Socrates to the "contextual brain", a concept invented by the author. Ting-Ting Wang (Peking University), in Chapter 17, presents her in-depth analysis of online spy video games from the narrative and cultural perspectives, on how a game's text is constructed and how the player's pleasure is generated. Finally, the physicist-turned-historian Dietrich Stauffer's Chapter 18 provides an easy-to-understand tutorial on statistical physics for humanists, with a step-to-step description of the simple but useful Ising model, finished with interesting applications in social science and a Fortran program. On top of that, a list of history titles that the author, and hopefully the reader, finds interesting is included.

The book's 18 chapters are organized into five parts: Part I: Philosophy of Science; Part II: History of Science; Part III: Sociology of Science; Part IV: Science Communication; Part V: Other Science Matters. Hopefully, research scholars and laypeople will both find this book enlightening and useful.

Rio Maior, Portugal   Maria Burguete
San Jose, California   Lui Lam

A Note on Chinese Names

There is no perfect way to write Chinese names in English. The spelling and ordering conventions of a Chinese name's characters are different in different geological areas—mainland China, Hong Kong, Taiwan and United States. The conventions adopted in this book are as follows.

1. A contributor's Chinese name after the chapter's title always appears with family name *last* and the first name's characters (if more than one) connected by a hyphen. For example, Guo-Sheng Wu, a contributor in this book, corresponds to Wu Guosheng in mainland China.

2. All Chinese names in text and references are written with family name *first*, with first name's characters connected by a hyphen.

3. All Chinese names from mainland China are spelled out in pinyin.

4. For those who made their career in the US, whether they settled later in mainland China or not, their name's old spelling is adopted, i.e., *not* in pinyin. For example, Yang Chen-Ning in this book is Chen Ning Yang in the US (which would be Yang Zhengning if he made his career in mainland China but not in the US).

5. Lui Lam made his career in both places, outside and inside China. The name Lui Lam appears the same as a contributor and in text while his pinyin name Lin Lei appears also in text and reference list. (His family name, Lin in pinyin, is Lam in Cantonese.)

Contents Summary

Preface		**v**
1	About Science 1: Basics—Knowledge, Nature, Science and Scimat *Lui Lam*	1
2	About Science 2: Philosophy, History, Sociology and Communication *Lui Lam*	50
PART I PHILOSOPHY OF SCIENCE		**101**
3	Towards a Phenomenological Philosophy of Science *Guo-Sheng Wu*	103
4	The Predicament of Scientific Culture in Ancient China *Hong-Sheng Wang*	116
5	What Do Scientists Know! *Nigel Sanitt*	137
6	How to Deal with the Whole: Two Kinds of Holism in Methodology *Jin-Yang Liu*	147
PART II HISTORY OF SCIENCE		**175**
7	*Helicobacter*: The Ease and Difficulty of a New Discovery *Robin Warren*	177

8	Science in Victorian Era: New Observations on Two Old Theses *Dun Liu*	185
9	Medical Studies in Portugal around 1911 *Maria Burguete*	193
10	The Founding of the International Liquid Crystal Society *Lui Lam*	209

PART III SOCIOLOGY OF SCIENCE — **241**

11	Three Waves of Science Studies *Harry Collins*	243
12	Solitons and Revolution in China: 1978–1983 *Lui Lam*	253
13	Scientific Culture in Contemporary China *Bing Liu and Mei-Fang Zhang*	290

PART IV COMMUNICATION OF SCIENCE — **305**

14	Science Communication: A History and Review *Peter Broks*	307
15	Popular-Science Writings in Early Modern China *Lin Yin*	330

PART V OTHER SCIENCE MATTERS — **347**

16	Understanding Art through Science: From Socrates to the Contextual Brain *Kajsa Berg*	349
17	Spy Video Games after 9/11: Narrative and Pleasure *Ting-Ting Wang*	371
18	Statistical Physics for Humanists: A Tutorial *Dietrich Stauffer*	383

Acknowledgments	**407**
Contributors	**409**
Index	**415**

Contents

Preface v

1 About Science 1: Basics—Knowledge, Nature,
   Science and Scimat 1
   Lui Lam

   1.1 Introduction 1
   1.2 Human Knowledge and the Knowscape 3
   1.3 Scimat 1: The Humanities 5
   1.4 Religion and Philosophy 7
   1.5 Nature and Science 11
       1.5.1 The Idea of Nature 11
       1.5.2 The Idea of Science 12
   1.6 Scimat 2: Science, Scientist and the Science Room 16
       1.6.1 Science Defined 16
       1.6.2 Scientist Defined 18
       1.6.3 The Science Room 19
   1.7 How Science Is Done 21
       1.7.1 Simple Systems 21
       1.7.2 Complex Systems 26
   1.8 The Essence of Science 34
   1.9 Scimat 3: Q & A and Ramifications 36
       1.9.1 Q & A 36
       1.9.2 Ramifications 39
   1.10 Discussion and Conclusion 41
   References 44

2 About Science 2: Philosophy, History, Sociology and Communication   50
Lui Lam

2.1	Introduction	50
2.2	Science in a Nutshell	52
2.3	Philosophy of Science	54
	2.3.1 Ernst Mach (1838–1916)	55
	2.3.2 Karl Popper (1902–1994)	56
	2.3.3 Thomas Kuhn (1922–1996)	59
	2.3.4 Paul Feyerabend (1924–1994)	61
2.4	History of Science	62
	2.4.1 Scope of History of Science	63
	2.4.2 How Much Detail?	64
	2.4.3 Beyond Narrative	65
	2.4.4 An Open Problem	66
2.5	Sociology of Science	66
	2.5.1 The Scientific Process	67
	2.5.2 The Scientific Results and the Book Drop Test	67
	2.5.3 The Generalization Trap	68
	2.5.4 The Outsider Problem	69
	2.5.5 The Laboratory Visits	71
	2.5.6 The Role of Theory	73
2.6	Science Communication	74
	2.6.1 Two Brief Histories: United Kingdom and China	76
	2.6.2 Two Modes of Operation: United States and China	78
	2.6.3 Why Communicating Science Is Difficult: A Few Examples	79
2.7	What Happened and What to Do	83
	2.7.1 The Beginning	84
	2.7.2 Ancient Time	84
	2.7.3 Modern Time	86
	2.7.4 Last Century	87
	2.7.5 Near Future	94

2.8	Conclusion: An Old but New Frontier	95
References		95

PART I PHILOSOPHY OF SCIENCE

3 Towards a Phenomenological Philosophy of Science 103
Guo-Sheng Wu

3.1	Introduction	103
3.2	Phenomenology as Reverse Thinking (Reflection)	105
3.3	Philosophy of Science from "Positive Thinking" to "Reverse Thinking"	106
References		115

4 The Predicament of Scientific Culture in Ancient China 116
Hong-Sheng Wang

4.1	Introduction	116
4.2	Scientific Knowledge Was Plentiful in Ancient Chinese Civilization but Was Not Treasured Up	118
4.3	Reasons for the Absence of Scientific Spirit in Ancient China	124
4.4	Political and Social Influences on Scholastic Culture	128
	4.4.1 Ancient Period	128
	4.4.2 Modern Period	133
4.5	Conclusion	134
References		135

5 What Do Scientists Know! 137
Nigel Sanitt

5.1	Introduction	137
5.2	Main Strands	138
5.3	Questions	143
5.4	Conclusion	145
References		146

6 How to Deal with the Whole: Two Kinds of Holism in Methodology   147
Jin-Yang Liu

6.1	Introduction		147
6.2	Methodological Shift in Holism		149
	6.2.1	A Hidden Assumption	149
	6.2.2	Basic Conditions of a Whole in Methodology	151
6.3	Constitutive Holism		155
	6.3.1	Constitutive Whole	155
	6.3.2	An Example: Synthetic Microanalysis	157
6.4	Generative Holism		158
	6.4.1	An Example: Cellular Automaton	159
	6.4.2	Philosophical Discussion on "Generation"	163
	6.4.3	Generative Whole	164
6.5	Comparison between the Two Approaches		166
	6.5.1	Constitutive Holism: Carve Nature at Its Joint	166
	6.5.2	Generative Holism: How the Mechanism Works	168
6.6	Conclusion		171
References			172

PART II HISTORY OF SCIENCE

7 *Helicobacter*: The Ease and Difficulty of a New Discovery   177
Robin Warren

7.1	Introduction	177
7.2	Before 1979	178
7.3	The Breakthrough	179
7.4	Marshall and I	181
7.5	Conclusion	184
References		184

8	**Science in Victorian Era: New Observations on Two Old Theses**		**185**
	Dun Liu		
	8.1	Introduction	185
	8.2	What Engels Missed Out	186
	8.3	Questioning Yuasa's Thesis	188
	8.4	Conclusion	191
	References		192
9	**Medical Studies in Portugal around 1911**		**193**
	Maria Burguete		
	9.1	Introduction	193
	9.2	First European Universities	194
	9.3	Emergence of Laboratory Teaching at Coimbra University	195
	9.4	Faculty of Medicine at Coimbra University (1863–1892)	196
		9.4.1 Faculty of Medicine (1863–1872)	197
		9.4.2 Medical Laboratories and Scientific Travellers	197
		9.4.3 The Berlin School of Medicine	200
		9.4.4 The Dawn of the Natural Science Era	200
		9.4.5 Coimbra Microscopes Collection (1748–1872)	202
		9.4.6 Faculty of Medicine (1872–1892)	202
	9.5	Medical Science after 1911: The Lisbon Case	204
	9.6	Important Achievements	205
	9.7	Conclusion	207
	References		208
10	**The Founding of the International Liquid Crystal Society**		**209**
	Lui Lam		
	10.1	Introduction	209
	10.2	My Involvement in Liquid Crystals	213
		10.2.1 In the West (1972–1977)	213

10.2.2	In China (1978–1983)	215
10.2.3	In New York (1984–1987)	221
10.3	Founding the International Liquid Crystal Society (1987–1990)	224
10.4	After 1990	234
References		237

PART III SOCIOLOGY OF SCIENCE

11 Three Waves of Science Studies — 243
Harry Collins

11.1	Introduction	243
11.2	The First Wave of Science Studies	244
11.3	The Second Wave of Science Studies	245
11.4	The Third Wave of Science Studies	248
References		252

12 Solitons and Revolution in China: 1978–1983 — 253
Lui Lam

12.1	Introduction		253
12.2	Returning to China		254
12.3	Arriving Beijing and the Early History of Institute of Physics		259
12.4	Life at Institute of Physics		261
	12.4.1	Living in Zhongguancun	262
	12.4.2	Science Spring 1978	265
	12.4.3	My 1979 Trip to America as a Visiting Scholar	270
	12.4.4	Other Things	272
12.5	Solitons in China		275
12.6	Leaving China		282
12.7	Conclusion: The Missing Link		284
Appendix 12.1: A Brief History of Chinese Students Going Abroad and the Returnees			284
References			286

13	**Scientific Culture in Contemporary China**	**290**
	Bing Liu and Mei-Fang Zhang	
	13.1 A Brief History	290
	13.2 Definitions of Scientific Culture	294
	13.3 Scientific Culture Studies in Recent Years	297
	13.4 Academic Journals in Scientific Culture Studies	300
	13.5 Conclusion	302
	References	303

PART IV COMMUNICATION OF SCIENCE

14	**Science Communication: A History and Review**	**307**
	Peter Broks	
	14.1 Nineteenth-Century Origins of "Popular Science"	307
	14.1.1 Early-Nineteenth Century: Republic of Science	308
	14.1.2 Late-Nineteenth Century: The Rise of the Expert	310
	14.1.3 Popular Science Redefined	311
	14.2 Two Moments	312
	14.2.1 Sputnik and Fears about Science Literacy	313
	14.2.2 Bodmer and the Public Understanding of Science	315
	14.3 New Challenges and New Models	317
	14.3.1 PUS: Problems and Politics	318
	14.3.2 From PUS to PEST	320
	14.3.3 Meanings and Trust	324
	14.4 Conclusion: Scientific Literacy	325
	References	328
15	**Popular-Science Writings in Early Modern China**	**330**
	Lin Yin	
	15.1 Introduction	330
	15.2 Science Writings before Modern Time	331
	15.2.1 Chinese Classics Concerning Science	332

		15.2.2	Translation of Foreign Scientific Literature	333

- 15.3 Science Writings at the Beginning of Modern Time (1840–1860)   335
  - 15.3.1 Science Writing Activities of the Western Missionaries   335
  - 15.3.2 Representative Science Writings by Chinese Authors   336
- 15.4 Popular-Science Writings during the Westernization Movement (1861–1895)   337
  - 15.4.1 Newspapers and Magazines   338
  - 15.4.2 Translation and Publication of Science and Technology Books   340
- 15.5 Popular-Science Writings from After the Westernization Movement to the Beginning of the Twentieth Century   341
  - 15.5.1 Springing Up of Local Newspapers and Magazines   341
  - 15.5.2 Two New Forms of Popular-Science Writing   343
- 15.6 Conclusion   344
- References   345

PART V OTHER SCIENCE MATTERS

16 Understanding Art through Science: From Socrates to the Contextual Brain   349
Kajsa Berg

- 16.1 Introduction   349
- 16.2 Empathic Responses to Art from Socrates to Gombrich   352
  - 16.2.1 Ancient Greece and Rome   352
  - 16.2.2 The Renaissance   353
  - 16.2.3 Charles Le Brun   354
  - 16.2.4 Nineteenth-Century Empathy Theory   355
  - 16.2.5 Ekman, Gombrich and Bryson   356
  - 16.2.6 Baxandall and Bryson (Revised)   357

	16.3	Mirror Neurons	360
	16.4	Neuroarthistory	361
	16.5	The Contextual Brain, Empathy and Caravaggio	363
	16.6	Conclusion	366
	References		368

17 Spy Video Games after 9/11: Narrative and Pleasure — 371
Ting-Ting Wang

	17.1	Introduction	371
	17.2	Narrative and Pleasure	372
	17.3	Narratives in Post-9/11 Spy Games	376
	17.4	Games in Reality and Reality in Games	379
	17.5	Conclusion	381
	References		382

18 Statistical Physics for Humanists: A Tutorial — 383
Dietrich Stauffer

	18.1	Introduction		383
	18.2	Zipf Plots and Random Walks		384
		18.2.1	Zipf Plots	384
		18.2.2	Random Walks	385
	18.3	Model Building		386
		18.3.1	What Is a Model?	386
		18.3.2	Binary versus More Complicated Models	386
		18.3.3	Humans Are Neither Spins Nor Atoms	387
		18.3.4	Deterministic or Statistical?	388
	18.4	Statistical Physics and the Ising Model		389
		18.4.1	Boltzmann Distribution	389
		18.4.2	Ising Model	389
	18.5	Applications		391
		18.5.1	Schelling Model for Social Segregation	392
		18.5.2	Sociophysics and Networks	394
	Appendix 18.1: How to Program the Ising Model			395
	Appendix 18.2: Some Formulas for Boltzmann Distributions			402

Appendix 18.3: History 403
References 403

Acknowledgments **407**
Contributors **409**
Index **415**

1

About Science 1: Basics—Knowledge, Nature, Science and Scimat

Lui Lam

There is a lot of confusion and misconception concerning Science. The nature and contents of science is an unsettled problem. For example, Thales of 2,600 years ago is recognized as the "Father of Science" but the word science was introduced only in the 14th century, and so it is obvious wrong if science is understood as modern science only, which started with Galileo about 400 years ago. If science is mainly about nonliving systems, then social science cannot be part of science. And if social science is part of science, then why the humanities, which are also about humans, are not part of science? All these confusions and dilemmas concerning science could be traced to the historical evolution of the word and concept of Science and the many misconceptions perpetuated by various philosophers and historians of science, due to the lack of an agreed upon definition of science. This chapter aims to clear up all these confusions by retracing the historical development of science—the word, concept and practice. The nature of knowledge, Nature, religion and philosophy are covered. A simple definition of science according to scimat, the new discipline that treats all human-dependent matters as part of science, is provided. Three important lessons learned about science, including the required Reality Check (which differentiates science from other forms of knowledge) are given. Important ramifications from this definition concerning antiscience and pseudoscience in particular are discussed.

1.1 Introduction

Science is one of the three pillars that support an advanced civilization, East and West. While the other two pillars, ethics/religion and arts, have an extremely long history of at least one million years [Lam, 2011] science, counting from the days of Thales (c. 624-c. 546 BC)—the Father of Science—has a "short" history of only about 2,600 years. Short as it is, it is long compared to the span of modern science, a mere 400 years or so since Galileo (1564-1642).

While the tremendous success of modern science did lead to positive results (and important applications like the cell phone), unfortunately, it also led to all sorts of confusion among the philosophers, historians, sociologists and communicators whose works are related to science; e.g., the definition of science is often avoided in philosophy of science books [Oldroyd, 1986; Godfrey-Smith, 2003]. The confusions concern three aspects of science: (1) the contents of science, (2) the existence and nature of the different stages in the development of science, and (3) the scientific research process.

The crux of the problem is that the concept and practice of science are not constants but have evolved over time, with many twists and turns, contributing to the *lack of an agreed-upon definition of science*. Thus, most statements made on science more than 56 years ago (the year 1957, see Section 1.7.1) by scholars and even by scientists turn out to be no longer valid. To clear up the confusion, the historical development of science over the last 2,600 years since Thales is retraced. And since the essence of science is to gather knowledge about Nature, the idea of Nature and the nature of knowledge are reexamined, too. It turns out that, based on our current scientific knowledge and the historical record, two simple conclusions are reached: (1) Humans are part of Nature, and (2) the aim of science was never to challenge the existence of God but merely to see how far humans can go in understanding what is around them by reasoning without appealing to God. These two recognitions, missed by many others, form the premise of *scimat* (Science Matters), the new discipline initiated in 2007/2008.

In Section 1.2 below, two kinds of human knowledge and the "knowscape", a metaphor for the landscape of knowledge, are introduced. The rationale behind scimat, which treats all human-related matters (covered in the humanities and social science) as part of science and the importance of the humanities are presented in Section 1.3. And since science first appeared (without the name) as part of philosophy in early Greek time, which was humans' first attempt to break away from superstitions (if you exclude God from superstitions), the connection between religion and philosophy is examined in Section 1.4. The historical developments of the idea of Nature and of the idea of science (a very recent concept) follow (Section 1.5).

In Section 1.6, the very definition of science and of scientist according to scimat are presented, which are simple, clear and historically correct. Science Room, the metaphor for science, is also included. Section 1.7 outlines how science was actually done, for both simple (mostly inanimate) systems and complex systems (which include human and nonhuman animate systems). The contents of this Section influence heavily our discussion on the philosophy, history, sociology and communication of science, presented in [Lam, 2014]. Three important lessons learned about science are given in Section 1.8. A summary of the spirit and essence of scimat (in the format of Q & A), and the ramifications from the basics about science are given in Section 1.9. Finally, Section 1.10 concludes with discussion and a take-home message.

1.2 Human Knowledge and the Knowscape

In spite of the many different opinions, we could probably agree that "Science is to understand Nature."[1] Then what is Nature? The present understanding is: Nature consists of all material systems including humans and (living and nonliving) nonhumans. That humans are part of Nature is a relatively new recognition. It follows from Darwin's evolution theory [1859] that humans are a living system like the chimpanzees and evolved from the fishes, say. Humans are thus one of the many kinds of animals[2] and a material system and, like all material systems, are made up of atoms. Note that the existence of atoms was

---

[1] Some people prefer "Science is to understand the universe". But since Nature includes everything in the universe (see Sections 1.3 and 1.5.1) the two statements are equivalent to each other. Here, "understand" means to understand without appealing to supernatural or God. See Sections 1.5.2 and 1.6.1 for a historical discussion of this position, which is adopted in this chapter and in scimat.
[2] Plato seems to recognize this when he defines humans as featherless, bipedal animals with broad nails [Läertius, 2011].

established only about 100 years ago due to Einstein's work on Brownian motion [1905]. Consequently, most discussions on the contents of science published 150 years ago are simply wrong or misleading, resulting in many misconceptions.

All knowledge about Nature accumulated through science by humans could be divided into two parts: A human-independent part[3] and a human-dependent part [Lam, 2008a]. An example of the former is the law of gravity, which could be discovered by aliens, too, if they exist. The latter includes knowledge of human-made materials such as semiconductors and all the topics in social science and the humanities.[4]

To facilitate discussion in this chapter, we introduce the *knowscape*, the landscape of knowledge. It is meant as a metaphor, and, as a metaphor it is far from perfect and has its limitations, e.g., not everything in the picture should be taken literally. In the knowscape (Fig. 1.1) there are hills/mountains, valleys and plateaus all linked to each other in a vast terrain, and some man-made lakes. Each hilltop represents a highlight in human knowledge; the height of the hill corresponds to the difficulty of reaching it. And, as in real mountain climbing, the explorer usually has to pass a lower hill before reaching a higher one. A researcher (whether you call her/him a scientist or not) is the explorer.

However, there are two types of mountains. One type is human independent which is out there whether humans exist or not (and could be found by aliens, say), represented by the upper curve in Fig. 1.1. The other type is human dependent, the knowledge found in the humanities and social sciences (lower curve). The isolated lakes represent artificial, nonhuman systems (such as semiconductor, computer and artificial life) which are human dependent, too (ellipses).

---

[3] The existence of human-independent knowledge is denied by some relativists. *Relativism* is a branch of philosophy started by the ancient Greeks in 5th century BC. In those times, the profession of lawyer did not exist and everyone has to argue for himself in court. The relativists played the role of legal advisors. Like in today, winning the case is the only aim of arguing, not reaching for the "truth" or "reality" [Buckingham et al, 2011, p. 42].

[4] See [Doren, 1991] for a concise review of human knowledge.

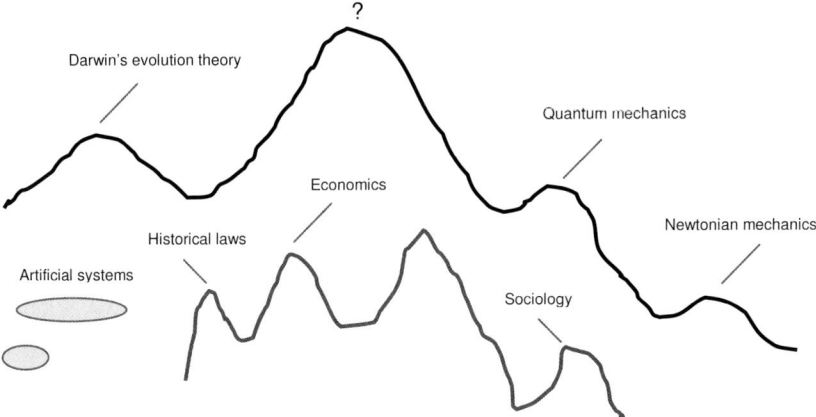

Fig. 1.1. The knowscape: Landscape of knowledge. The upper curve represents the human-independent part; lower curve, human-dependent part related to human matters; ellipse, human-dependent part related to artificial systems.

1.3 Scimat 1: The Humanities

And so we have these three recognitions.

1. Science is to understand Nature.
2. Nature includes all material systems.
3. Humans are a material system.

The logical conclusion derived from these three statements is that "all things related to humans are part of science". What are "all things"?

Let us consider bees which, like humans, are a kind of animal. When we say "all things related to bees" we mean the biological property of a single bee, the behavior of a single bee or a group of bees living together (how they communicate with each other; how they divide jobs among themselves; how their society is organized; etc.), the competition or cooperation between different groups of bees, and so on. Everything! The same goes for chimpanzees. And so, the same goes for humans.

Historically, the study of different aspects of humans was classified into medical science (including human biology), social science and the humanities. Medical science is about the biology of a single human or a group of humans; an example of the latter is epidemiology. Social science (e.g., economics and sociology) is the study of some, but not all, aspects of a group of humans. Cultural study, a branch of the humanities, is about a group of humans, too. It is thus clear that the division of social science and the humanities is not according to the number of humans under study. Rather, as pointed out before [Lam, 2008a, p. 13], it is due to the scientific level achieved in these two large group of studies. Yet, for some people, it is due to the belief that the humanities can *never* be a part of science either (1) because humans are so complex that it cannot be handled by science or (2) because humans are fundamentally different from the bees and chimpanzees. These issues form the core of this chapter and will be clarified later. At this point, let us note that this classification of human studies was established well before the times of Darwin and Einstein, and the separation of the humanities from science has been maintained, for many people, even today. As shown below, it is due mostly to the misconceptions about science.

Why is it crucial to recognize the humanities, as it should, as part of science? The answer lies in the importance of the humanities. This point could never be overstated even though it is usually overlooked in every country, perhaps with the exception of France where philosophy is a required examination for graduating high school students since the Napoleon days. Humanities' importance could be seen from these two considerations.

1. If all the present "science"[5] research projects were frozen or eliminated, the world would still be the same—chaos and tragedies would continue—because it is the humanities which include decision making, underdeveloped in the last 2,600 years since Plato (427-347 BC), that matter in human affairs.[6]

---

[5] In this chapter "science" with double-quotation marks means science in the conventional sense, which is the sum of natural science and social science but excludes the humanities (see Fig. 1.2).

[6] For example, according to Jean Ziegler [2013], the hunger and malnutrition suffered by nearly one billion people in the world results not from failure of agriculture, science or technology but from inhumane and shortsighted politics.

2. Apple company is successful because they put a humanist, Steven Jobs (1955-2011), in charge of the engineers—good for the economy.

Further discussion on the humanities is given in Section 1.10.

Scimat is the discipline initiated by Lam [2008a; 2008b] that treats all human-dependent matters, humanities in particular, as part of science. (See Sections 1.6 and 1.9 for more discussion of scimat.) The Scimat Program[7] is the latest concerted (international) effort in reviving the Aristotle tradition of treating human and nonhuman systems alike in the pursuit of knowledge. (See Fig. 1.3 and Section 1.10 for a discussion of the past efforts that failed.)

That such a simple and almost trivial conclusion that the humanities are part of science is not immediately and universally accepted by every learned person is at first puzzling. It turns out that the reasons lie in people's understanding of what science is and what the word science represents to them. A little bit of research shows that (1) not just laypersons but even some good "scientists" hold the wrong ideas about science, (2) the root of the problem is mostly historical, and (3) misconceptions about science could be traced to the inadequacies in the four disciplines concerning science, viz., the philosophy, history, sociology and communication of science. Below the historical root of the problem is discussed; each of these four disciplines is examined in the next chapter [Lam, 2014].

1.4 Religion and Philosophy

The chimp and human lineages split from each other six million years ago. Four million years later *Homo erectus* appeared and already possessed the ability of mimesis [Donald, 2006]. Then, 1.6 million year ago, fire and complex stone tools were invented. It is thus not hard to imagine that about a million years ago, our ancestors though primitive were sophisticated enough to think and wonder about things they saw

---

[7] The Scimat Program was started by Maria Burguete and Lui Lam in 2007 with the first international scimat conference in Portugal and the forming of the International Science Matters Committee. For more see: www.sjsu.edu/people/lui.lam/scimat.

and the happenings in their lives. They might ask: Why does the sun rises and disappears everyday? Why am I sick and recovered but not the one next to me? They might even ask: Who am I? This was likely to happen because, apart from hunting and mating, there was plenty of leisure time; there were no televisions or football games to watch [Lam, 2011]. Besides, curiosity helped in survival [Lam, 2004, p. 36].

We do not know what answers they came up with since there was no record to show; writing had not been invented yet. But we do know what answers their descendents, *Homo sapiens* who appeared in the scene 195,000 years ago, came up with. More precisely, we mean the later generations; their answer: Everything is due to "something out there", the supernatural.

This is understandable. In the absence of scientific knowledge at that time, it is natural to explain everything by using analogies and lessons learned in their daily lives. Consequently, the ascent and descent of the sun or the moon is governed by a human-like god, like the way a piece of rock could be moved by a human being. When one is sick, without the benefit of any medical knowledge, an easy explanation is that a certain god was offended. To get well again, the god's anger had to be removed, and that could be done by bribery—in the form of animal or human sacrifices—in the same way that it works with humans themselves.

What we call superstition was refined by the ancient Greeks in the form of mythology, with numerous gods with specific names. For example, Apollo takes care of the Sun while multitasking in light, knowledge, music, healing and the arts; Boreas, the north wind; Notus, the south wind; Zephyrus, the west wind; Apheliotes, the east wind. Similarly, human matters are governed by gods; e.g., Eros the god for love and sexual intercourse; Athena the goddess for intelligence, skill and wisdom [Buxton, 2004]. Two points about this "theory" of gods: (1) It is consistent with everything they know at that time.[8] (2) It treats human and nonhuman systems alike, by the same mechanism.

---

[8] In fact, even today's science has not proved that Apollo (or any of the Greek Gods) does not exist. We just do not need them or believe in them anymore because we have a simpler and better answer in explaining why the sun rises in the East and sets in the West.

This romantic theory of gods is clumsy and suffers from the lack of evidence and predictive power, even though predictive power is not the necessary quality of a new theory. In fact, the gods hypothesis is already being criticized in the 6th century BC by what is later called philosophers[9] who prefer to explain natural phenomena in terms of natural causes. Thales is the most important and famous one who claims the nature of things is water and is said to have predicted the eclipse of the sun happening in 585 BC [Cornford, 2004, p. 1]. But he also maintains that the universe is alive with soul in it, full of daemons or gods [Cornford, 2004, p. 127; Lloyd, 1970, p. 9]. Aristotle (384-322 BC) suggests that the inquiry into the causes of things begins with Thales [Lloyd, 1970, p. 1] who is the first to define general principles and set forth hypotheses, and has thus been dubbed the "Father of Science".[10] (The word science first appears in the 14th century; see Fig. 1.2.)

Meanwhile, the number of gods is reduced from many to one [Armstrong, 1993], reflecting humans' desire for simplicity. This principle of simplicity remains in the core of modern science and is called Occam's Razor: What can be done with fewer is done in vain with more, i.e., the simpler the better (as long as the simpler works). With one God, organized religion as we know it today emerges.

Philosophy, starting with Thales, is the effort to understand and explain things in the universe—the nonhuman systems and even the human system[11]—through reasoning without bringing in the supernatural or the gods/heroes. It has two parts: one part that God is not brought in explicitly and another part that God is purposively and explicitly put there [Buckingham et al, 2011]. Thales' water hypothesis is an example of the former; Aristotle's "unmoved movers" in *Metaphysics*, the latter

---

[9] The two words philosophy (meaning "love of wisdom") and philosopher are coined by Pythagoras (c. 570-c. 495 BC) [Bertman, 2010, p. 244]. The word 'philosophy' with the meaning of "all learning exclusive of technological concepts and practical arts", enters the English language in the 13th century [see: *Webster Ninth New Collegiate Dictionary* (Merriam-Webster, Springfield, MA, 1984)]. Unless otherwise specified, all the dates of words appearing in the rest of this chapter are from this Webster dictionary.
[10] en.wikipedia.org/wiki/Thales (July 27, 2013).
[11] The assertion of the historian Robin Collingwood (1889-1943) in *The Idea of Nature* [(1945) 1960, p. 3] that philosophy is exclusively concerned with the physical universe is wrong.

[Collingwood, 1960, p. 87]. It is the former (but not the latter) that is identified as science in modern time.

Philosophy is a transition from the "narrative" or "story telling" to "reasoning"—a big step forward. Note that what the ancient Greeks abandon are the primitive supernatural and the relatively simple gods/heroes; they do keep the more sophisticated God, which actually is a central part of their knowledge system. In fact, philosophy *never* challenges the existence of God; it just assumes *a priorily* that part of the universe could be understood through reasoning (which turns out to be correct as shown by later developments). That this is at all possible is quite trivial since no one knows how God runs the universe after he creates it. For example, God could lay down the rules or laws and go fishing and comes back occasionally to burn down a city if he finds a sizable portion of the residents' attitudes are God-incorrect, or he could be a CEO who manages minutely every happening, big or small. In either case, philosophers, if they so wish could easily claim that what they do is just reading the mind of God. And this explains why, e.g., Plato, Aristotle and Immanuel Kant (1724-1804) could do their philosophy while still believing in God.

Since philosophy comes after religion which includes mystic considerations, there exist in philosophy two lines of abstract speculations among pre-Socratic thinkers, leading to two traditions: scientific and mystical. The scientific trend puts the gods completely away in reasoning, reduce the Soul to material particles, and concentrate in inanimate systems, more in line with modern science. It starts with the Milesian school (Thales, Anaximander and Anaximenes) and leads to Democritus (c. 460-c. 370 BC) who proposes that everything is made of atoms. The mystical trend is "rooted in certain beliefs about the nature of the divine and the destiny of the human soul" and tries "to justify faith to reason." It is exemplified by Pythagoras and Plato [Cornford, 2004]. The two traditions coexist and influence the development of philosophy for a long, long time, and are still among us today.

1.5 Nature and Science

Both the concepts of Nature and of science evolved with time which, when ignored, resulted in many misconceptions and a lot of confusion among the scholars on science.

1.5.1 *The Idea of Nature*

The word "nature" has two meanings, same in ancient Greek time and modern time: (1) the sum total of natural things; (2) the principle (or source) governing natural things. It is mainly the first meaning, the contents of Nature, which concerns us here. In particular, (1) Do the natural things include humans? (2) If so, do humans differ from other animals? (3) And if so, could the distinctions be explained completely on a material basis?

First issue first. In the 6th and 5th centuries BC, philosophers of the Ionian school (of which the Milesian school is a subset) believe that (1) there is such a thing as "nature"; (2) nature is "one"; (3) the thing which in its relation to behavior is called nature is itself a substance or matter [Collingwood, (1945) 1960, p. 46].

For Aristotle of the 4th century BC, nature is the essence of things which have a source of movement in them [p. 81], which, therefore, should include humans. It is in later years that humans are excluded from the domain of natural things. This issue is settled with Darwin's discovery [1859] that humans, like other animals, evolve from other more primitive creatures and organisms; *humans are thus part of Nature.*[12]

The second issue of whether humans are distinct from other animals has two levels. At the first level, we all know that through adaptation and

---

[12] Of course, it took many years before Darwin's evolution theory was accepted by the mainstream, which, in fact, is still rejected by many laypersons on religious grounds. It is thus worthwhile to point out that in 1996, Pope John Paul II has declared that Darwin's evolution theory is indeed correct and covers humans, too, except that a human being, unlike other animals, is infused a soul by God when it becomes into being [Pope, 1996]. And this was four years after the same Pope admitted that Galileo was mistreated by the Vatican and apologized (www.vaticanobservatory.org/index.php/en/history-of-astronomy/197-the-galileo-affair, Nov. 29, 2013; New York Times, Nov. 1, 1992).

heredity humans have evolved to be quite distinct from other animals; e.g., our brains are larger; we have invented written language; etc.[13] [Suddendorf, 2013; Pollard, 2009]. The second level is more sophisticated: the existence of "soul", consciousness, and free will that are thought to be uniquely human. This is fine. We all have a vague idea of what these are since we could experience it ourselves.

The third issue of where do soul and free will come from has been investigated intensively by philosophers, past and present, and more recently, by neurobiologists. The majority opinion currently is that they are emergent phenomena/features derived from the neurons and their connections [Gazzaniga, 2012; Tse, 2013]. In other words, the mind-body problem will be solved by neuroscience; it also means that it is not yet a settled issue.

1.5.2 The Idea of Science

The word science first came into English in the 14th century and its present usage appeared even later, in 1867, soon after the Age of Revolution (1775-1848, such as the Industrial Revolution and the French Revolution). Science comes from Latin *scientia*, meaning "knowledge" or "the pursuit of knowledge", and especially "established theories" when it first appears [Williams, 1983; Ferris, 2010, p. 3].

Since Thales is called the Father of Science, naturally, science (no matter what it means) should already exist in 600 BC. What was it called then? It was called philosophy; more precisely, it is that part of philosophy that God is not mentioned explicitly (see Section 1.4).[14] For example, Aristotle's works on biology are science; his works on the human system such as ethics and arts, in our opinion, are also science (at the empirical level; see Section 1.7.1). Another example: the works of Archimedes (c. 287-c. 212 BC) on buoyancy, the Archimedes' Principle, is obviously science, even by the present definition of the word.

---

[13] On the other hand, there are more similarities between humans and other animals than we are ready to admit [Natterson-Horowitz & Bowers, 2013].

[14] Collingwood has remarked: What we call science, Aristotle called it philosophy [1922].

However, in the 14th century the term "natural philosophy" appears and philosophy is split into three parts: 'philosophy', theology (including natural theology) and natural philosophy (Fig. 1.2). Here, *'philosophy'* with single-quotation marks means philosophy in a restricted sense, the study of deep questions about humans (e.g., ethics and metaphysics). *Natural theology* is a branch of theology based on reason and ordinary experience, in contrast to revealed theology based on scripture and religious experience.[15] *Natural philosophy* is the study of (living and nonliving) nonhuman systems and non-religious issues, which is divided into two parts: one part without invoking God and another part with God invoked. The former is later absorbed into "science"; the latter falls into the "God of the gaps"[16] category and later becomes part of theology. A noted example of the latter is provided by Isaac Newton (1642-1727) who, knowing well that the stars always attract each other due to gravity, has invoked God's active intervention to prevent them from falling in on each others—a notion that was ridiculed by his competitor, Gottfried Leibniz (1646-1716) [Alexander, 1956, p. xvii].

Modern "science", beginning with Galileo Galilei (1564-1642) who died the year Newton was born, has blossomed rapidly in the last 400 years, especially in physics (see Section 1.7). After Galileo, a major player in the so-called Scientific Revolution, Newton's (deterministic) mechanics was so successful that people in the Enlightenment (1688-1789) wanted to make human matters a science, too [Porter, 2001]. They succeeded partially. Before the Enlightenment ended prematurely by the French Revolution (1789-1799), Adam Smith (1723-1790) published *The*

---

[15] http://en.wikipedia.org/wiki/Natural_ theology (August 4, 2013).
[16] The concept of "God of the gaps", due to the evangelist Henry Drummond (1851-1897), suggests that gaps in *scientific* knowledge should be taken as evidence or proof of God's existence (en.wikipedia.org/wiki/God_of_the_gaps, Nov. 30, 2013). Since new gaps keep on appearing while science closes the old gaps, this argument will never fail [Lam, 2004]. For instance, when Newton's argument that it is God's hand that keeps the stars from sticking to each other is no longer needed after the expansion of the universe is established and explained by the Big Bang theory, Vatican embraces the Big Bang theory in 1951 before many scientists are willing to do [Linder, 2004]. And that is 14 years before Big Bang's final confirmation by Arno Penzias and Robert Wilson's experimental discovery of the cosmic background radiation comes in. The reason: Origin of the Big Bang is the new gap in science but Vatican has the answer: God did it.

Wealth of Nations [(1776) 1977], ushering in Economics, the first discipline in Social Science (Fig. 1.3).

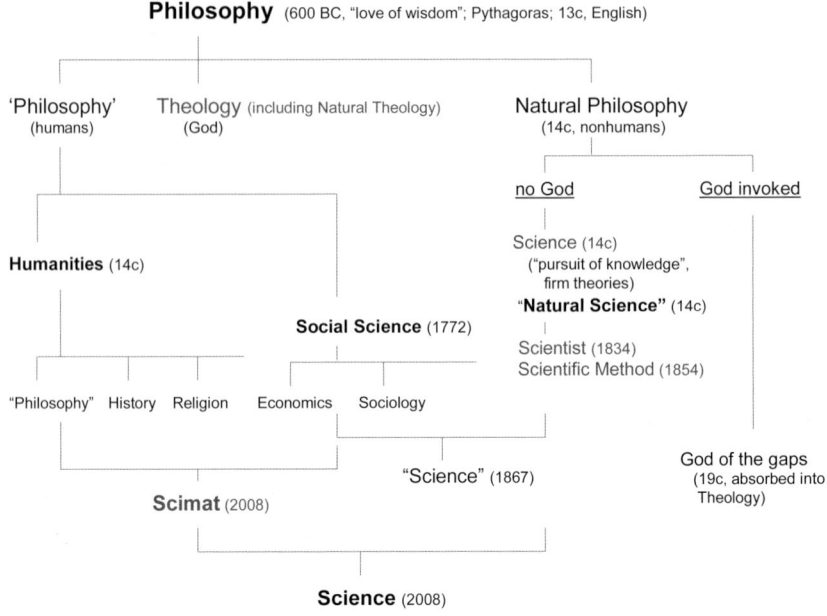

Fig. 1.2. A brief history of words, from philosophy to "science" and to scimat. "Philosophy" and "science", with double-quotation marks, correspond to philosophy and science, respectively, in their present, restrictive use (see text). The god-invoked part in natural philosophy is huge and frequently motivates the basic assumptions (such as Newton's absolute space) adopted by individuals working on the no-god part [Henry, 1997, pp. 73-85]. The year of 1834 for scientist comes from the Webster dictionary while 1840, attributed to William Whewell, appears in [Williams, 1983, p. 279].

By 1840, with enough number of people working full time in "science" the English scientist and theologian William Whewell (1794-1866) felt the need to coin the word "scientist" to describe these professionals; he was inspired by the word artist. After all, arts have been in existence long before science does, for at least 35,000 or perhaps a million years [Lam, 2011]. And the word scientist was not fully accepted until the early 20th century [Ross, 1962].

> 1687 Newton's *Mathematical Principles of Natural Philosophy* published
> 1688 Enlightenment begins
> 1762 Probability used in gambling and insurance business
> 1776 Adam Smith (**Economics**; *The Wealth of Nations*)—birth of **Social Science**
> 1789 Enlightenment ends
> 1812 Pierre-Simon Laplace (*Analytical Theory of Probabilities*)
> 1844 Auguste Comte (**Sociology**)
> 1859 Probability used in Darwin's evolutionary theory and Maxwell's kinetic gas theory—first time in science

Fig. 1.3. Enlightenment, Social Science and probability theory. The Enlightenment (1688-1789) aims to make a "Science of Man", which succeeds in creating Social Science but fails with the humanities. The reasons: (1) Human matters are not deterministic like in Newtonian mechanics but are probabilistic; (2) the necessary tools of a probabilistic science were not yet there [Lestienne, 1998].

Somehow, the word "science" in its present usage (written with double-quotation marks in this chapter) was not in place until April, 1867 [Harrison et al, 2011, p. 2] even though the word science, as remarked above, appeared already in the 14th century. "Science" means the sum of "Natural Science"[17] and Social Science, after 'philosophy' split into two parts in 1772: the humanities and social science. At this point, the assumed existence of God is confined entirely to Theology.

"*Natural Science*" differs from natural philosophy in that all theological and metaphysical considerations are excluded in the former but not necessarily in the latter. By these definitions, Newton was doing neither "science" nor "natural science" but natural philosophy.[18] However, we are so fond of Newton that we want to make him one of our own, a "scientist." The way to do that is to retain only parts of his book *Mathematical Principles of Natural Philosophy* (1687) in which

---

[17] In this chapter "natural science" with quotation marks means the science of nonhuman systems, the conventional definition. We prefer to call natural science the "science of all things in Nature"; i.e., science = natural science, by scimat's definition [Lam, 2008a].

[18] As confirmed by private documents and personal papers which became known since 1936, for Newton, religion and science were two inseparable parts of the same life-long quest to understand the universe. See PBS's NOVA program "Newton's dark secrets" (www.pbs.org/wgbh/nova/physics/newton-dark-secrets.html, Sept. 1, 2013), and also [Buchwald & Feingold, 2013].

God is not mentioned, notably, the mathematical description of the three laws of motion and the law of gravity, as we are doing it today in every physics textbook.

Presently, "natural science" consists of physics (including astronomy), chemistry, biology and Earth sciences (e.g., geology and atmospheric science). Social science consists of economics, sociology, psychology, linguistics, law, anthropology (which could include archaeology), etc., while the humanities are made up of "philosophy", religion, history,[19] arts (e.g., literature, visual arts and performing arts), languages, etc.

However, as shown elsewhere, this historical classification is unreasonable and unscientific [Lam, 2008a]. It is harmful to the healthy development of not just the humanities but also of social science and physical science. Since the humanities, social science and medical science are all about humans a logical and systematic approach would be to group them together in one (umbrella) discipline—*scimat*.

1.6 Scimat 2: Science, Scientist and the Science Room

With the historical developments in mind, here are the new definitions of science and scientist proposed by scimat, followed by an introduction to the Science Room.

1.6.1 *Science Defined*

According to scimat, *Science is humans' pursuit of knowledge about all things in Nature, which includes all (human and nonhuman) material systems, without bringing in God or any supernatural.* Some explanations are in order.

1. The pursuit of knowledge is the common denominator among philosophy, "philosophy", "science" and science (Fig. 1.4). "Pursuit" here means earnest and honest research aiming to find out what and why.

---

[19] Note that history is sometimes classified as social science. It depends on the scientific level achieved in the discipline as conceived by the classifier [Lam, 2008a].

1 About Science 1: Basics

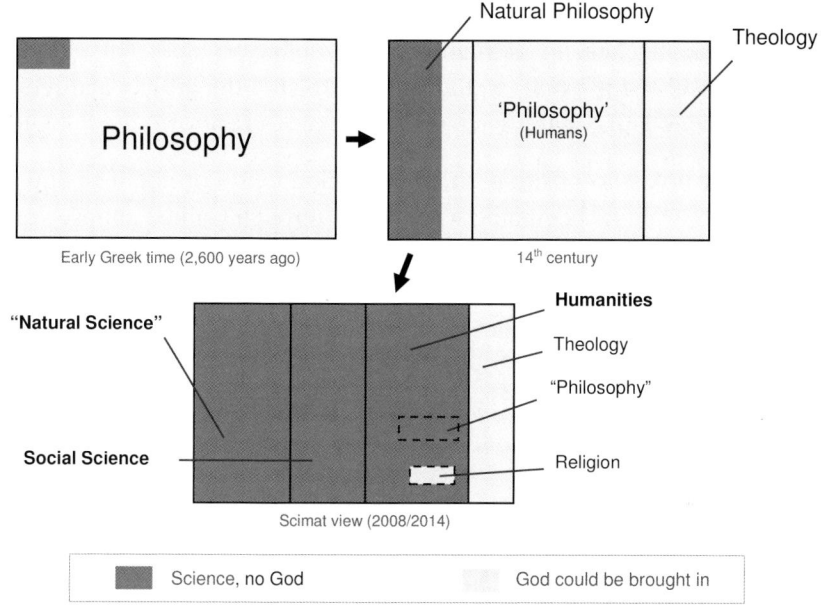

Fig. 1.4. Splitting of the discipline Philosophy in the last 2,600 years and the retreat of God in the disciplines over time (based on Fig. 1.2). The domain of Science, defined as humans' research in understanding Nature (human and nonhuman systems) without bringing in God/superstition, exists within the single discipline of Philosophy at Thales' time, the early Greek time. Science expands to about half of the discipline Natural Philosophy in the 14th century while Philosophy reduces to 'Philosophy', which further shrinks to "Philosophy" as we know it today (see text). The distinction between Religion studies in the Humanities from Theology is that God's existence is not assumed in the former but in the latter.

2. Nature here, after Darwin and Einstein as discussed in Sections 1.3 and 1.5.1, includes all the material systems—the human system and all (living and nonliving) nonhuman systems. It is identical to the Universe.

3. By excluding God in the scientific process we are in disagreement with philosophy but in agreement with modern "science".

4. That God is excluded is based on two considerations: (1) God, if exists, is beyond Nature since, e.g., according to the Bible, he creates everything. (2) God is a "game stopper." For example, if you

got a funding from the US National Science Foundation (NSF) to find the mechanism of high-T_c superconductors and you reported that "God did it", NSF would ask you to stop the project and return the money immediately.

5. Scimat holds no position on whether God exists or not. The aim of science, from its beginning with Thales through Galileo and Newton, was never to challenge the existence of God. Religion and science is in conflict with each other only when religion does not retreat fast enough as science advances, and when either side over claims [Lam, 2004].

1.6.2 Scientist Defined

Following the definition of science in Section 1.6.1, here is scimat's definition of scientists: *A scientist is a person who honestly seeks knowledge about Nature without bringing in God or any supernatural.* In other words, a scientist is simply a *researcher* (res for short). Here are some remarks.

1. By this definition, apart from the natural scientists and social scientists, the humanists are also scientists, a consequence of the nature of Nature which includes human and nonhuman (material) systems.

2. If the humanists do not look like scientists to some people, it is because most humanists are still carrying out their research at the empirical level, similar to what Aristotle did long time ago. (See Section 1.7 for the three levels/approaches of research.)

3. Honesty is a must in real research. It refers to the researcher's honesty in collecting and handling data, and in reporting results. It also refers to the res' readiness in admitting mistakes when the res knows a mistake has been made, even though we understand that researchers, being humans, could find it difficult to do so.[20]

---

[20] Honesty appears in two of the five items in the Scimat Standard [Lam, 2008a, p. 27].

4. If one wants to, scientists could be separated into two classes: professionals and amateurs. Both could contribute to the progress of science, like in astronomy. The distinction should be of concern to the administrators in universities and funding agencies since it is their job to bet on the success chance of a researcher, and to the general public who has to decide whom to trust more (which is not a simple matter [Lam, 2014]). For scientists, the works of professionals and amateurs are judged critically alike anyway.

5. When one is enjoying the beauty of a rainbow in the sky, one is not a scientist. But when the same person starts to wonder where those rainbow colors come from she is taking the first step in doing science. If she goes further and records the shape of the rainbow and the distribution of the rainbow colors, she is doing science at the empirical level and becomes a scientist. When she tries to figure out, by theory or experiment, the mechanism of rainbow formation she is doing science at a higher level. If she succeeds she is a good scientist. If she is the first one in history who discovers the mechanism and gets it published, she is a "successful" scientist. Publish or not, the real joy in doing science is to have fun in discovering or understanding something in Nature.

6. Similarly, an artist is not a scientist when she is going through the motion of creating an artwork. But before or during the process, if she tries to figure out seriously by herself how to make things work, e.g., what techniques applied will achieve the effect she wants or how the receiver's brain (through the senses) can be stimulated the way she intends it to be, she is doing science and could be called a scientist [Lam, 2011, p. 24]. That is why Leonardo da Vinci (1452-1519) is both an artist and a scientist even though he did not have formal training in science [Capra, 2007].

1.6.3 *The Science Room*

With scientists so defined in Section 1.6.2, we could also come around and say "science is what scientists do", borrowing the dictum from physics that "physics is what physicists do" [Lubkin, 1998, p. 24].

Science thus has two parts: (1) the results obtained by scientists, and (2) the process of doing science. (See [Warren, 2014] for an example.) Both parts evolve over time and are full of surprises (see Section 1.7).

To capture visually the essence of the above, we introduce the *Science Room* as a metaphor for science, which is represented by the right box in Fig. 1.5. For comparison, the conventional "natural science" room, which excludes the humanities and social science, is shown on the left. In the science room, over time, we see new scientists entering and old scientists leaving when they die or quit prematurely. Some, like Newton and Einstein, are bigger than others. Inside the room, results obtained by the scientists are kept, too. There are two kinds of results: those that are still in use (like the law of gravity, quantum mechanics and the special and general relativity theories) and those that are outdated (like Aristotle's mechanics). The latter are kept in a storage room (not shown). We do throw out junks (like alchemy[21]) from time to time.

The number of researchers inside the science room increases with time and also the size of the room. The former could be seen by counting the number of physicists: In ancient Greece the number could be in the tenths; 100 years ago, in the hundreds; today, 50,000 from the American Physical Society alone. The latter is evidenced by the rapid increase of research journals. Over all, the density of scientists in the room (i.e., number of scientists divided by room size) seems to become pretty high in the last 30 years, signaled by the difficulty of finding a worthwhile project and the "publish or perish" pressure felt by the professionals.

Observing the science room is like viewing a movie that never ends or not yet ends (we do not know which case it is), like the *Star War* series, with a new cast on the screen every 60 years or so (in historical time) if you happen to fall sleep intermittently. Like a good movie, science never repeats itself and is as intriguing as *War and Peace* while sometimes looks like the *Life of Pi*.

---

[21] Alchemy was a legitimate research topic in chemistry before the emergence of nuclear physics, only about 100 years ago. From nuclear physics we know that the identity of an element, like gold, is dictated by the number of neutrons and protons inside the nucleus of the atom. Thus, one can never obtain a gold atom from another metal by chemical method only, since chemistry only changes the electrons but not the nucleus.

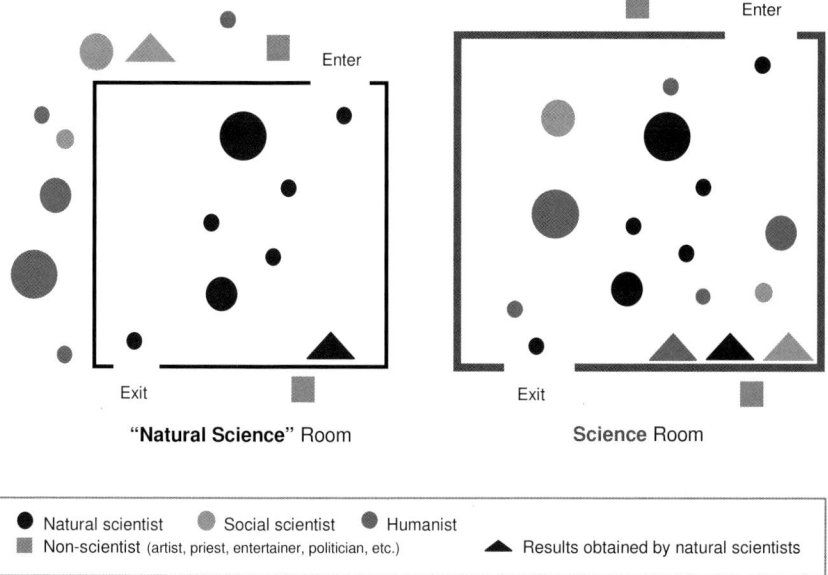

Fig. 1.5. The Science Room (right box) and the "Natural Science" Room (left box). The difference between the two is that humanists and social scientists (and their fruitful results) are included in the former but not in the latter. Note that "natural science" is mistakenly identified as science by many people. The walls of the room do not imply boundary of knowledge.

1.7 How Science Is Done

We here focus on the second part of science, i.e., how science was or is done.[22] A good way to explore this subject is to divide our discussion into two parts according to the subjects being studied, viz., simple systems and complex systems, since the methods of study though related, are sometimes quite different.

1.7.1 *Simple Systems*

Only a subset of nonliving systems and a small number of living systems could be considered simple systems (SS). They are the subjects studied

---

[22] Only basic science will be considered below. Applied science and technology are quite different and are excluded.

in physics (including astronomy), Earth science (excluding meteorology and climatology) and chemistry. Other living things such as vertebrate animals, humans included, belong to complex systems.[23]

The historical development of research in SS, particularly in physics and in later years, is rather well documented [Mason, 1962; Lindberg, 1992; Chen, 2009]. Essentially, it goes something like this.

1. Early period: from Thales to Galileo

With plenty of time and patience, ancient people *observed* and *recorded* what they saw in the sky about the movement of heavenly bodies and astronomy was born—science at the *empirical* level. Then with the aid of arithmetic and geometry, miraculously, they succeeded in predicting eclipses, for example, taking astronomy a big step forward.

After observation and data analysis, *theorizing* emerged, through guessing and logical thinking. Aristotle did it without experimentation about moving bodies, in the sky and on earth, and got it all wrong. Yet, the important point is that, in the absence of quantitative data, he was "right" at his time—an important criterion we use in judging scientists. Furthermore, through his classification works in biology, Aristotle did show us the need of empirical inquiry. His pioneering thinking and perceptions, the Aristotle tradition, was maintained for about 2,000 years after his death, helped by the ruling class. But this has nothing to do with Aristotle and is not part of the scientific process.

Archimedes was different. As the story goes, after the "eureka" moment triggered by water-level rising when he got into his bathtub, he run naked in the street and became the first streaker in history, which, luckily, did not become part of the scientific tradition. Importantly, after streaking, he did a few experiments in the lab and established the Archimedes' Principle about buoying bodies, which is still correct today [Hirshfeld, 2010]. Archimedes' Principle is an example of science at the *phenomenological* level, i.e., obtaining rules/laws about a phenomenon without knowing the mechanism.[24]

---

[23] Note that this demarcation is not sharp and is sometimes problematic. For example, a complex system, once understood, could become a simple system [Lam, 2008a].

[24] The mechanism behind the Archimedes' Principle is that water molecules keep on bombarding the body immersed in it as if the volume of water displaced by the body is

The Aristotle tradition was finally broken by Galileo.[25] It happened not because Galileo was much smarter than those before him but because he (1) picked simple systems (such as a small ball rolling down an inclined plane) to study, (2) did drastic and daring approximations in constructing theories (e.g., simplifying the body to a size-zero point particle), and (3) used detectors other than his bodily sensors to observe and record (e.g., using water fall as a timer in inclined plane experiments[26] and improving the telescope to look at the moon and beyond) [Lam, 2008a]. *Experimentation* became the hallmark of modern science; *mathematization*,[27] modern physics.

2. Modern period: four lessons

Everyone was happy, except for those living in the Vatican. And modern science flourished in every discipline (including biology and medical science in the complex-systems domain) except the humanities. Subsequently, for SS, two great theories, thermodynamics (1824) and Maxwell's electromagnetism (1873), were discovered. Near the end of the 19th century, it seemed to some that the "end of physics" has arrived. The sudden emergence of quantum physics in 1900, the black-body radiation experiments by Max Planck (1858-1947) [1900], was the party crasher. Newtonian physics no longer works and every able physicist was scrambling to find a new theory. Here is the *first* important lesson every modern scientist learns: *Nature is full of surprises and every dear theory may not be the final theory.*[28]

---

still there. Since the existence of molecules was not confirmed until 1905 there was no way that Archimedes could figure out the mechanism behind his Principle.

[25] Attempts to move away from Aristotelian physics since the 6th century and before Galileo are discussed by Peter Berg [2012].

[26] See: "Galileo's inclined plane experiment" (galileo.rice.edu/lib/student_work/experi ment95/inclined_plane.html, Dec. 1, 2013).

[27] Mathematization though desirable is sometimes over emphasized. Even within physics, Newton's third law of motion (action and reaction) and Thermodynamics' third law (it is impossible to reach absolute zero temperature in finite number of steps) are both written in words. Darwin's evolution theory, the most important result in biology, is expressed in words only, too.

[28] A similar case appeared at about the same time. Einstein in 1905 showed that Newtonian mechanics is a special case of his theory of special relativity when the body

As physicists are concerned, the next heart breaker showed up 57 years later: Madame Wu and her team discovered that parity is violated in weak interactions [Wu et al, 1957]. Parity is the mirror symmetry taken for granted by every physicist in the past, which says that if you set up an apparatus and put a mirror near it, construct a second apparatus according to what you see in the mirror, then the two apparatuses will give you identical results. Parity symmetry was regarded as basic as the time- and space-translational symmetries (which are still good, so far). Thus the *second* important lesson: *Never take any basic assumption for granted*.

Nature is kind to us. As if for compensation, in the same year of 1957, superconductivity discovered by Heike Onnes (1853-1926) 46 years ago was satisfactorily explained by the BCS theory proposed by John Bardeen (1908-1991),[29] Leon Cooper and Robert Schrieffer. Here is the *third* lesson: *If you wait long enough the answer may come*. However, it also shows that answer to your "prayer" does not come by keep on praying, but by more experiments and harder thinking. Also, Onnes did not live to see his discovery explained; the answer may not come within a lifespan. We kind of know this already since it took about 2,200 years for Democritus' atom to be confirmed, which is the case of theory preceding experiments while superconductivity is a case in reverse.

When the high-T_c superconductors were discovered in 1986 by Karl Müller and Johannes Bednorz and the Nobel Prize was awarded the next year—a lightning speed, quite a number of theoretical papers from famous authors were rushed to print in the prestigious physics journal *Physical Review Letters*. They turned out to be all wrong or unbelievable as more experimental results came in. In fact, instances like this happened before that a beautiful theory is not sustained by experiments and has to be rejected. The *fourth* lesson: Reliable *experimental results have the ultimate say in deciding the fate of theories in science*.

---

moves slowly in comparison with light. That is, what we see as a high mountain in the knowscape turns out to be part of a higher mountain.

[29] Bardeen is the only person in history who won two Nobel Prizes in *physics*, for co-inventing semiconductor and the BCS theory of superconductivity [Hoddeson & Daitch, 2002].

3. Three remarks

1. Disciplines (or subdisciplines) in SS could be divided into two types: non-historical disciplines (e.g., condensed matter physics and particle physics) and *historical disciplines* (e.g., astronomy and geology). Control experiments are possible in the former but not in the latter. Then how do we study the historical disciplines? It is done by comparing data collected in the historical disciplines with knowledge gathered through controllable, repeatable experiments and confirmed theories in the non-historical disciplines. For example, by comparing the color spectrum of light coming from a star with those from known elements obtained in the lab we could tell what kinds of element existing in the star.

2. We never check *all* consequences/predictions from a hypothesis before we accept it as a confirmed theory. For practical reasons, we just do enough number of checking to convince ourselves that it is correct. And that is why an established theory *could* be broken later when new contradictive findings (usually experiments), if any, show up. *This strategy is what makes the rapid progress of science possible* in the last few hundred years. It is like how countries are being conquered: The invading army occupies some strategic cities/places and then the capital, never the whole country, and declares the job done; then they move on to the next country—Napoleon did that; Hitler did that. In science, we do the same in the knowscape (see Section 1.2) instead of the landscape. Consequently and occasionally, disturbance might suddenly burst out from a not-yet-occupied place in a conquered country. The conqueror would be forced to look back and suppress it or tolerate it. And that was what happened to the conqueror, the physicists, in the knowscape, in the case of parity nonconservation, except that they could not suppress it because parity nonconservation sits in the human-independent part of knowscape (see Fig. 1.1). Instead, they update their map of the knowscape.

3. The existence of atoms and the discovery of quantum mechanics made it possible to study many-body systems with a new approach, the *bottom-up* approach, called the "microscopic picture" in physics. This is the third approach apart from the empirical and the phenomenological approaches. For closed systems (such as gas in a jar) the theory to

accomplish this is Statistical Mechanics, which links the microscopic world to the macroscopic world. Computer simulations could also do the job, for both open and closed systems, but are less enlightening [Tuckerman, 2010]. Each approach or level of study complements and reinforces the level above it. Thus, results from the phenomenological level, if correct, will have to reproduce that from the empirical level and give more; results from the bottom-up level will do the same for the phenomenological level [Lam, 2002]. The availability of the three approaches is like that of the army, navy and air force in a war situation; you want to use all of them, if necessary, to do a quick and thorough job.

1.7.2 Complex Systems

Complex systems (CS) consist of nonliving (e.g., the weather or climate system) and living systems. The latter consists of humans or human-related systems (e.g., the economy) and other nonhuman, biological systems (e.g., plant, insect, fish, bird and coyote). A central, time-honored method of tackling CS (and SS) is the *Socrates Method* due to Socrates (470/469-399 BC), i.e., "to solve a problem, it would be broken down into a series of questions, the answers to which gradually distill the answer a person would seek".[30] Later methods are described below.

1. Difficulties and successes

Complications and difficulties in studying CS arise from five sources.

1. *The potential of chaos.* Chaos is the phenomenon that the future of a system depends sensitively on the initial conditions [Lam, 1998]. It could happen in a system of three bodies (like the Earth, Moon and a rocket) or of many bodies (like the weather, even though, in this case, only a simplified model of three variables has been proved).

---

[30] Source: en.wikipedia.org/wiki/Socrates (Jan. 1, 2014). The Socrates tradition of "questioning and debate", so central and fruitful in advancing (scientific) knowledge, has been continued in the West but less so in the East. In China, a similar tradition starting in the Autumn and Spring period (770-476 BC) was broken early on after Confucianism became the official state ideology of the Han Dynasty (206 BC-220) and has not yet been restored completely today.

Given a CS, the potential of chaos is always there but is hard to prove. That is why chaos is so frustrating.

2. *Heterogeneous complexity.* Complex systems are mostly heterogeneous, in terms of its components and interactions among them. For instance, the human body consists of a large number of different organs, interacting directly or indirectly with each other; in a society, no two persons are the same. (In contrast, in the SS of electrons, all electrons are identical to each other and every two electrons interact the same with each other.)

3. *Ethical limitations.* There are ethical problems in experimenting with living systems [Rollin, 2006]. That has been always the case with humans as the experimental object (with well-known exceptions like dealing with war prisoners). It extends to experiments with nonhuman animals in recent years. In other words, not every informative experiment (like human cloning) that could be done can be done, a problem not associated with simple systems. But it does not mean that we cannot and do not experiment with humans. Psychologists do that all time, harmlessly, by passing out questionnaires; hospitals try new drugs or new treatments on volunteers everyday. (See [Venter, 2013] for an interesting discussion on this issue.)

4. *Historical irreproducibility.* Human-related happenings are all historical, like in astronomy, which are irreproducible, with some exceptions in medical research. Therefore, on-site and real-time collection of data is rare except in mass demonstrations these days; even so, the data are always incomplete. And being humans, recounting of events by the participants, the so-called first-hand data, is subject to memory deterioration and intentional personal considerations.[31]

---

[31] This is demonstrated vividly in the so-called "Lee-Yang Dispute". The two Nobelists Lee Tsung-Dao and Yang Chen-Ning, in their personal recounts of how the parity nonconservation work was done, cannot agree on which Chinese restaurant that the crucial idea was raised, not to mention who is the one who raised it [Yang, 1983; Lee, 1986; Chiang, 2002; Zi et al, 2004].

5. *Temporal-spatial localization.* Since humans are influenced by culture and environment (in addition to human nature [Wilson, 1978; Machery, 2008]) research on humans based on observation/data from a local place for a particular historical period may not be applicable to other period or to humans in other countries or continents. Ignoring or ignorance of this localization feature results in over claim by humanists and social scientists, West and East. The problem is lessened but still present in the globalization era.

How and to what extent these complications could be overcome will be discussed below.

While the Holy Grail in CS research is to find the universal law(s), similar to the Second Law of Thermodynamics, say [Waldrop, 1992], it is very difficult to do so and fails so far. Instead, much progress has been made in the study of individual systems or classes of systems. Two "universal" organizing principles applicable to a large number, but not all, of CS are found: fractals and active walks [Lam, 2008a]. In lieu of general laws, computer simulations are heavily used [Mitchell, 2009] and techniques developed in simple-system studies are borrowed [Castellano et al, 2009]. The former overcomes the heterogeneous complexity since heterogeneity can be programmed easily in computers.

2. Medical science and biology

Among CS studies, the transdisciplinary medical science (more a basic science than applied) is the most developed despite and because it is about humans. Medical science could easily be the most important among all the disciplines, for the obvious reason. Its development benefits from early start,[32] continuous attention, and heavy funding.[33] It also benefits from the rapid advances in physics, chemistry and biology

---

[32] In the West, it started with Shamans and apothecaries' "niche occupation" of healing and ancient Egyptians' system of medicine before Hippocrates (c. 460-c. 370 BC) became the "father of Western medicine" in early Greek time. In the East, medical knowledge dates equally early (en.wikipedia.org/wiki/History_of_medicine, Dec. 30, 2013).

[33] In the US, the Fiscal Year 2012 funding for the National Institute of Health is $30.860 billion, which is 4.3 times of the National Science Foundation's $7.105 billion [Sargent Jr., 2013].

since the Scientific Revolution. But more importantly, medical research in the West keeps a very open mind and employs all the three research approaches (i.e., empirical, phenomenological and bottom-up) as soon as it is feasible to do so. (Drug designs at the molecular level, and genetic and stem-cell treatments under test are examples in the bottom-up approach.) Unfortunately, the same cannot be said about Chinese medicine [Lam, 2008a, pp. 32-33].

In biology, the discovery of evolution theory (1859) by Darwin and of double helix (1953) by James Watson and Francis Crick (1916-2004) leading to the prospect of synthetic life [Venter, 2013] demonstrates the workings of the empirical, phenomenological and bottom-up research approaches in successive action. For many people, what we are witnessing in biology is comparable to what happened in physics in the early 20th century. Moreover, the fact that Crick is a physicist-turned-biologist exemplifies the early trend of an s-res morphing into a c-res. (Here, s-res means a "researcher in simple systems"; c-res, in complex systems.)

3. Social science

As humans are concerned, it is in fact easier to study a large number of humans than a single human because many approximations can be made and justified in the former but not in the latter.[34] And that is why the scientific level achieved is much higher in social science than in the humanities.

Economics is the most developed discipline in social science not merely because it was the first discipline invented in this field, but because (1) a lot of data are generated and kept (think stock index), and (2) the financial reward is huge. The economy being a CS, it is not

---

[34] For example, when a windowless room with one door containing a large group of males (or females) inside is on fire, everyone will rush to the door to escape. As an approximation, the different thoughts going on in their brains could be ignored. And a strategy to avoid jamming the door could be designed from computer simulations by treating the humans as point particles with simple interactions. This kind of research that concentrates on common human nature (escape from fire) avoids the localization problem. Pedestrian modeling has been developed successfully into a science, with applications ranging from pedestrian trail formation in a German campus [Helbing et al, 1997] to crowd control in Mecca [Helbing & Johansson, 2010].

surprising that no universal theory about macroeconomics has yet been found, not to mention that no one is able to predict the rise or fall of a stock market. This is in contrast to the success of the meteorologists who are able to predict pretty accurately the *local* weather, also a CS, of the next day. The difference is that the equations involved are known in the latter (solved by supercomputers) but not in the former. We therefore see our top economists revising their "prediction" of the national economy from time to time, if not every day, which in fact is a good sign showing that they are honest scientists [Lam, 2008a, p. 28]. But then the question: If the economists fail so miserably, why is there a Nobel Prize given out every year in economics? Well, the Nobel Prize rewards the solution of individual, significant problems, which is possible even in CS.

There is a long string of successful stories of physicists making contributions in economics and finance [Weatherall, 2013]. But Wall Street has been blamed for hiring physicists who helped to bring down the global economy in 2008 [Patterson, 2010]. This is an issue of s-res morphing into c-res, unsuccessfully in this case. But that is because Wall Street has hired the wrong kind of physicists.[35] There are different kinds of physicists, like there are different kinds of engineers. If one wanted to design a new bridge, one would not go out and hire electric engineers to do the job; right?

---

[35] The 2008 economic meltdown could be attributed to three causes: (1) the financiers themselves; (2) central bankers and regulators who failed to see it coming; and (3) the macroeconomic backdrop of low inflation, stable growth and plenty of cheap Asian money ("The origin of the financial crisis", *The Economist*, Sept. 7, 2013, pp. 74-75). The financiers part involves their hiring of mostly high-energy physicists, the "quants" [Derman, 2004], who helped to design financial "products" that were sold worldwide, making a lot of money for everyone before it collapsed. No one seemed to remember that for a "product" (like a toy for children) safety, called stability in physics, should be taken care of before you put the product in the market. In the expert's words, "the quants who devised the highly leveraged financial derivatives ignored systemic risk" [Stein, 2011]. And unfortunately, stability analysis is not part of a high-energy physicist's training. These days, the stability analysis is called "stress test" mandated by the government to the banks. Recommending Wall Street to hire more physicists [Weatherall, 2013] is not a bad idea, if only proper training/briefing is prescribed to these s-res before turning them into c-res and letting them play with real money. The case of physicists doing econophysics [Ball, 2006] is a different matter since they are just creating theory, not products.

Fortunately, there are also many successful examples of s-res doing fine in sociology.[36] And we see the emergency of a new field called Computational Social Science [Cioffi-Revilla, 2014; Epstein, 2014].

4. Humanities

The development in the humanities is lacking behind, due partly to the intrinsic complexity of humans as individuals but also to the inadequate scientific training of the researchers involved. Thus, it is not uncommon that we hear people saying that the humanities cannot be part of science. They are wrong, for two reasons: (1) Many humanists mistakenly identify science with Newtonian mechanics; (2) the humanists ask the wrong questions.

The mis-identification of science with Newtonian mechanics started early (since the Enlightenment) and is still with us today,[37] due to the failure of science education and science communication [Lam, 2014]. Since each human is an open system (that exchanges energy and materials with the environment) and the factors that could affect a human or a group of humans cannot be completely account for, the human system is a *stochastic* system (i.e., probability/chance is involved). If one identifies science with the deterministic Newtonian mechanics then, of course, science is inapplicable to handle the human system. But scientific theory (physics in particular) and techniques dealing with both deterministic and stochastic systems are available (see Fig. 1.3 and [Lam, 2011]).

Given a stochastic system the question one should ask is not what will surely happen in the future but with what probability something may happen in the future [Lam, 2002]. For example, in History, the question one can ask about the longevity of a dictatorship is not in which year it will end definitely, but what is the chance it will still be there five years later, say. Surprisingly, amid all the contingencies and historical

---

[36] For example, Duncan Watts has a BS in physics, PhD in engineering and is now a professor of sociology at Columbia University; Dirk Helbing, PhD in physics, is Chair of Sociology at Swiss Federal Institute of Technology Zürich.

[37] See, e.g., *The Counter-Revolution of Science* by the economics Nobelist Friedrich Hayek (1899-1992) [(1952) 1979] and the philosopher Peter Winch's *The Idea of a Social Science and Its Relation to Philosophy* [(1958) 2008].

irreproducibility a *law* about the lifetime of Chinese dynasties has been found [Lam, 2006; Lam et al, 2010]. And an e-journal dealing with the theoretical and mathematical aspects of history from the scimat perspective [Lam, 2008c], *Cliodynamics*, has been published by the University of California since 2010. What this demonstrates is that human matters can indeed be studied scientifically (by going beyond the narratives), irrespective of all the complex thinking and so-called "free will" going on in humans' brains.

Recently, a DNA study of 1,000 descendents of Cao Cao, an important Chinese general in the Three Kingdoms period (220-280), eliminates the possibility that Cao was the descendent of a famous aristocrat [Wang et al, 2013; Jiang, 2013]. This work showcases the bottom-up approach in history.

Studies at the empirical and phenomenological levels in the humanities have been going on for more than 2,400 years since Plato and Aristotle. What is new is that in the last decade or so we see the bottom-up approach being advocated by the humanists (including some in English Literature, e.g., [Hogan, 2003]) themselves. This includes the emergence of *Neurohumanities* and efforts to understand human matters from the cognitive and evolutionary perspectives (see [Lam, 2011] for details). As an example, the neurobiologist and Nobelist Eric Kandel's *The Age of Insight* [2012] shows how the Vienna portraiture from 1900 to present could be understood at the three research levels.

Even in "philosophy", the toughest discipline where metaphysics is studied, one finds serious attempts by its practitioners to do their trade with non-traditional methods. For instance, in addition to Neurophilosophy that tries to solve the mind/brain problem using cognitive science [Churchland, 1986; Churchland, 2007], Experimental Philosophy tries to solve philosophical questions by using empirical data (usually from surveys of ordinary people) [Knobe, 2011; Knobe & Nichols, 2008]. All these are very encouraging, from the scimat perspective.

5. Living with uncertainty

For inanimate CS, the existence or absence of chaos can be ascertained in rare cases while oversimplified models are used (e.g., the three-variable climate model of Edward Lorentz (1917-2008) [1993]). But the human system is different; we have no choice but to live with the potential of chaos. Similarly, the human system's intrinsic probabilistic nature cannot be circumvented. Worse, a small but finite probability for something to happen does not mean that it would not happen; on the contrary, it means it *could* happen. Some probabilistic events (like not winning a lotto) if happen, would be harmless; but others (like global warming or the danger of genetically modified foods) could have dire consequences.

It is true that consilience (i.e., the convergence of evidence) implies a conclusion is most likely to be correct [Wilson, 1998]. But since Nature is subtle and full of surprises, "most likely" means it is not safe to assume it is 100% valid. Uncertainties, big or small, are with us everyday.[38]

The research history of simple and complex systems shows that there exists no such thing called the *Scientific Method* [Bauer, 1994] even though this term is invoked frequently by scholars [Thurs, 2011; Gower, 1997] and laypeople alike, if the method means a recipe of steps that guarantee success when followed, like in the case of a cooking recipe. Instead, what we have are equally valuable, viz., "scientific experience" and "scientific tradition". The reason behind all this is that research is a very complex process that does not yield easily to a simple summary. Besides, there are emergent properties manifested on many levels in a given system which could be attacked through one or all of the three research approaches; there is no need for an oversimplified, universal scientific method. The absence or de-emphasis of a scientific method is exemplified by the criterion used in the Nobel Prizes in the sciences, which are awarded for confirmed, grand discoveries, irrespective of the method and reasoning used. In fact, if there was really such a powerful scientific method, research would be pretty boring and

---

[38] For example, whenever we set foot on the street (or even the sidewalk) we run the risk of being hit by a car. It does happen. In the first three months of 2014 in San Francisco, six pedestrians lost their lives this way (*World Journal*, Mar. 21, 2014, p. B1). The same goes for driving. Being careful yourself is not enough.

we would see many creative scientists walking away—doing art or go fishing.

1.8 The Essence of Science

About science the three features listed here, though incomplete, are the most important.

1. Science advances and lives with approximations

It is a myth that "exact science" ever exists, for the following three reasons. (1) Every theory constructed turns out to be an approximated theory. First, a theory is the model of the real thing; e.g., in Newton's laws of motion a body is approximated by a point mass. Second, the theory is the approximation of a bigger, later theory; e.g., Newtonian mechanics is an approximation of Einstein's theory of special relativity. The Standard Model in particle physics is the low-energy approximation of something not yet known [Weinberg, 2011]. (2) Even if a theory is exact, the solutions obtained usually involve approximations because exact solutions are rare.

(3) Even if the theory and the solution are both exact, it still has to be checked by experiments. And experiments involve instruments and measurements which always have finite resolution. For example, to check the equation $A = B$ we measure each quantity and try to show that the left-hand side equals the right-hand side. But let us say our measurements give $A = 2.5 \pm 0.1$ and $B = 2.5 \pm 0.1$, we can only conclude that $A = B$ within experimental uncertainty (due to that ± 0.1). Measurements' unavoidable finite resolution prevents a rigorous proof of any theory.

In other words, *science never proves anything, rigorously speaking* (in the mathematical sense of proof). There is nothing absolute and final in science, with profound implication for "philosophy" [Lam, 2014]. Thus, there *could* be room for improvement in any theory, which is known for sure when the new improvement appears. Approximation works because, like when lost in a forest, a rough map is all one needs to get out of the forest. More importantly, as mentioned above, it is precisely due to the use of approximations that science progressed so fast in the last few hundred years.

Science lives and thrives with approximations. Coupled with what we say in Section 1.7.2, we humans (i.e., everybody, scientists and non-scientists alike) simply have to go on living with uncertainty, more *wisely* and *humbly*.

2. Scientists are humans

Scientists are humans, like you and me.[39] Most if not all, including Einstein [Kennefick, 2005], do make mistakes [Youngson, 1998; Livio, 2013]. But some, like Newton, are not like you and I. They are not just brighter, but are tormented and darker—not entirely their fault.[40] Some genius not just contributed more but suffered more, too. And humanity owes them a lot.

3. There is always the Reality Check

If science is not and could not be exact, and scientists are just humans, why should anyone take science seriously? Everyone should, for two reasons: (1) *Science delivers.* We owe our air conditioning and cell phone plus many other goodies to science. Science is valued because it works. (2) *Science is the best game in town.* As Newton [(1730) 1952] puts it, "And although the arguing from Experiments and Observations by Induction be no demonstration of general Conclusions; yet it is the best way of arguing which the Nature of Things admits of".

Why is science able to deliver? It is because it has to pass the *Reality Check* (RC), like every legitimate driver has to prove herself by passing the road test. Reality check means "confirmed" by experiments or practices, or, at the minimum, is consistent with established data. Social theories, due to the imprecise and incomplete data collected, are hard to be 100% confirmed [Lam, 2008a]. Reality Check is the *necessary*, crucial step for a theory to be recognized as part of the knowscape. It is the RC that makes scientific knowledge unique among all forms of "knowledge".

---

[39] See [Kevles, 1987] and the many biographies of scientists.
[40] See PBS's "Newton's dark secrets" (www.pbs.org/wgbh/nova/physics/newton-dark-secrets.html, Sept. 1, 2013).

Among the RCs, the *Cell Phone Test* (CPT) stands out because the working of a cell phone depends on the validity of a large number of theories (Fig. 1.6). Any new theory, if it conflicts with what are behind the working of a cell phone, has to explain why and why it is better. A good answer would be the new theory contains the old ones as special cases. We strongly recommend the CPT to advocates of anything new.

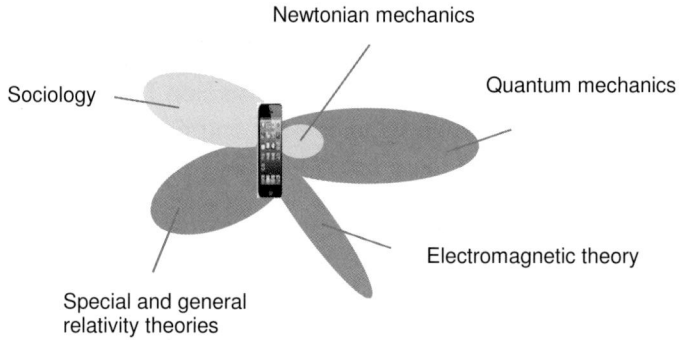

Fig. 1.6. Cell Phone Test: established, interrelated theories behind a cell phone. The working of a cell phone depends on Maxwell's equations, quantum mechanics, semiconductor theory, general relativity, and the sociological fact that there is enough number of humans who want to interact with others.

1.9 Scimat 3: Q & A and Ramifications

Scimat's spirit and essence as well as the action plan in the immediate future are summarized in Section 1.9.1. Some of the ramifications, including a new perspective on antiscience and pseudoscience, are discussed in Section 1.9.2.

1.9.1 Q & A

Q: What is scimat?
A: It is a new discipline that recognizes "everything in Nature is part of science".

Q: What do you mean by Nature?

A: Nature includes all living and nonliving material systems, humans in particular.

Q: What do you mean by Science?
A: Science is humans' pursuit of knowledge about all things in Nature, which includes all human and nonhuman systems, without bringing in God or any supernatural.

Q: What is scimat's position on God and religion?
A: Scimat holds no position on whether God exists or not. Personal choice of religion is respected.

Q: What are the topics that scimat covers?
A: All topics related to humans. That is, all the topics in the humanities and social science.

Q: Why does scimat put its emphasis on the humanities?
A: Because social science has been recognized as science but not yet the humanities.

Q: Why are the humanities so important?
A: All the world tragedies (poverty, war, race cleansing, injustice, corruption, etc.) are human-dependent matters and could be traced to the underdevelopment of the humanities in the last 2,400 years since Plato and Confucius.

Q: What new method or new tool is used by scimat in its research?
A: None. Scimat advocates the use of any method or tool that is available and applicable as long as honesty and ethics are respected. Reason: Scimat is about the search for knowledge which knows no boundaries and no pre-determined routes.

Q: Anything more?
A: Well, we do point out and want to emphasize that in any discipline there are three approaches—empirical, phenomenological (i.e., without knowing the mechanism) and bottom-up; they supplement and reinforce each other, like army, navy and air force in a war situation. This is well known in nonhuman studies but less so in the humanities.

Q: Then what is new about scimat?
A: The concept that "all human-dependent matters are part of science" is new.

Q: So what?
A: Humanity advances through new concepts. Examples: "All men are born free" brings down slavery; "all men are born equal", royalties and totalitarian regimes; "all women are born equal, too", restrictions on women's rights in education, employment and voting.

Q: Will all the world problems be solved if enough number of people become scimatists?
A: We don't know. But, in our judgment learned from history, scimat is the best, practical way to make the world better and more peaceful since people would be more enlightened and, hopefully, act more rationally, *humbly* and kindly toward others and the environment. The next step is to educate the decision makers.

Q: What is scimat's action plan for the future?
A: To set up 100 scimat centers worldwide, to ensure the development of and sustain scimat's ideals. The first step is to set up one such center. What the center can do is spelled out in "The *Science Matters* Program and a Proposal" (see: www.sjsu.edu/people/lui.lam/scimat).

Q: How can I help?
A: Buy this book. And tell others to do so.

Q: How can I help more?
A: Buy the other two books in the Scimat Series: *Science Matters* (2008) and *Arts* (2011). Or, ask your library to do that; better, do both.

Q: How can I help much more?
A: Be a sponsor or co-sponsor of our next scimat conference. It takes 20,000 USD to run a good conference, 10,000 USD to run a conference.

Q: What if I can't wait?
A: Gather, from your friends if necessary, 50,000 USD and contact the author: lui2002lam@yahoo.com. We will help to set up a Scimat

Center in the university or city of your choice (tax exempt in the US). The Center could be named after the person you prefer if more money is donated.

Q: How can I help without money involved?
A: Visit the scimat website and help spread the word.

Q: What is scimat's take home message?
A: Humanities are part of science.

1.9.2 Ramifications

Important implications of scimat including a new answer to the Needham Question have been given before [Lam, 2008a]; here are more.

1. Since science includes the humanities, "philosophy" in particular, the early Greek philosopher Socrates was a scientist, too. Socrates as a social and moral critic while practicing his philosophy became a "gadfly" of the state, in Plato's words. He was sentenced in court for "guilty of both corrupting the minds of the youth of Athens and of impiety (not believing in the gods of the state)" and persecuted.[41] Thus, the c-res *Socrates was the first "martyr of science"* while the s-res (in astronomy) Giordano Bruno (1548-1600) was the second one.[42] The former was ordered to swallow poison; the latter, burned at the stake.

2. Does antiscience really exist? It depends on how science is defined. Let us consider marriage and antimarriage first. The conventional definition of marriage is that (1) it is a legal piece of paper, (2) signed by a woman and a man (who promise to take care of each other). Antimarriage usually refers to the objection of item 2 but not item 1; i.e., an antimarriagist advocates that the legal paper could also be signed by a woman and a woman, or a man and a man. Similarly, it is hard to imagine anyone in her right mind would object to doing science per se unless science is defined in the narrow sense, i.e., science is about nonhuman systems only. *Antiscience* usually refers to the objection of certain things such as doing certain kinds of scientific research and the

---

[41] en.wikipedia.org/wiki/Socrates (Jan. 1, 2014).
[42] en.wikipedia.org/wiki/Giordano_Bruno (Jan. 1, 2014).

abuse of science; instead, putting the money in education, fighting poverty and human development are recommended [Holton, 1993]. All these debates are about over claims by some scientists and humans' choices, e.g., the priority and reallocation of resources. These are legitimate, human-dependent matters. Thus, if science is to include all human matters as it should, according to scimat, there is no such thing as antiscience. On the other hand, there does exist antiscientists—those who commit science frauds (in the humanities and other sciences) since they violate the first rule in doing science, i.e., being honest in their research.

3. The *science-pseudoscience demarcation* is a complex issue [Shermer, 2011] which is less about science per se [Pigliucci & Boudry, 2013] but more about the competition for attention, prestige and resources [Gordin, 2012]. In particular, the debate on "intelligent design" (ID, a form of creationism) being science or pseudoscience is due to the fuzzy definition of science used, either in the media or in court [Shermer, 2006]. If God is explicitly excluded from the definition of science, as in scimat, this debate would never happen, case closed. In fact, ID could be discussed in "philosophy" or theology classes. The problem for the ID advocates, of course, is that in America's public high schools, there is neither "philosophy" (unlike in France) nor theology classes. If religion-government separation works (in the US at least), why not religion-science separation?

4. The objectivity in and the reconstruction of past "reality" through historical narratives are considered impossible by some historians [Hayden, 1973] and deconstructionists [Derrida, 1976] because, they argue, the exact meaning of any writing is undecidable. These and other postmodern attacks led to a crisis in the history profession in the 1990s [Gilderhus, 2000; Evans, 1997]. That language, spoken or written, is an imprecise mode of communication is a well-known fact. The good news is that since science advances by approximations, history—a branch of science also does not need to be exact in its narratives, research at the empirical level. (See also [Lam, 2008c].)

1.10 Discussion and Conclusion

Here are some discussions and elaborations concerning what are written above.

1. The most negative effect derived from misconceptions about science is that the topics studied in the humanities are excluded from the domain of science. If that was the case, Aristotle would have two hats to wear. He would don on the "science hat" while he was studying the cosmos, plants and animals; and the "non-science hat" while he was talking about human affairs. Won't that drive Aristotle crazy? You bet!

2. A positive effect of recognizing humanities as part of science, apart from helping to reduce or eliminate human tragedies (such as ideological massacre or hunger), is to reverse the rapid decline of enrollment in humanities in the universities. According to a newspaper report (in San Jose Mercury News), from 1966 to 2009, the percentage of students receiving bachelor degrees in the humanities at Stanford University drops from 37% to 11%; for all institutes in the US, from 17% to 7%. By treating themselves as part of science, the humanities will attract science-inclined students who normally would opt for "natural science" and engineering for the wrong reasons; these students will also help to raise the scientific level of the disciplines and create more jobs.

3. We have come a long way to be able to declare freely that humans are animals made up of atoms. Only about 400 years ago, scientists in Naples were arrested and tried for maintaining "that there had been men before Adam composed of atoms equal to those of other animals" [Jacob, 1997, p. 28].

4. The existence and scale of the god-invoked part in natural philosophy and its influence on the no-god part is often overlooked by scholars in science. Consequently, some experts simply identify, wrongly, natural science with natural philosophy while only the no-god part of the latter should be identified as the former. (For more see [Cunningham & Williams, 1993, p. 421].)

5. Even some practicing scientists' characterization of today's science is wrong. When they say that controllable and reproducible experiments are a must in science they forget all those historical (scientific) disciplines like astronomy and archaeology (see Section 1.7.1).

An example is provided by the chemist Igor Novak [2011, p. 31] when he says, "Science uses reproducible, controlled experiments (as part of the 'scientific method') and draws conclusions on the basis of experimental results *and* logical reasoning *alone*" (emphasis in original). And, of course, he excludes social science, not to mention the humanities, in his use of the word science.

6. Some experts identify science with modern science. This choice causes unnecessary confusion and even some inner contradictions. If only modern science is science, what is the Archimedes' Principle which is still being used today? This problem could not be overcome by dividing the development of science into three levels, viz., prescience (e.g., Aristotle), parascience (Newton) and euscience (Galileo), as done in [Jaffe, 2010]. If so, Thales would be the "father of prescience". And it is odd to label Newtonian mechanics as parascience; it is good science, a special case of Einstein's relativity which in turn could be the special case of something else in the future. A better way is to view Aristotle's physics which was correct at his time but is obsolete today as one of the numerous outdated theories in the history of science. (For more see [Lam, 2014].)

7. Scientism has different definitions.[43] But essentially, *scientism* implies the universal applicability of the scientific method and the supremacy of scientific knowledge. Scimat is not scientism because we do *not* believe in the existence of a scientific method (see Section 1.7) and we do respect people's right to seek knowledge by any avenue they happen to choose as long as it is safe and ethical. We just say that judging from past history, science is the *most likely* way to succeed, from making a cell phone to curing cancer and to lessening humans' sufferings. In fact, science guarantees nothing and promises nothing except in those well-understood simple systems (like water boiling at 100 ºC). For unsettled problems in simple systems (like high-T_c superconductors) science thinks it can solve the problem given enough time; for complex problems related to humans (like love, ethics and soul) the promise of science remains a promise, until it delivers. Thus, scimat

---

[43] en.wikipedia.org/wiki/Scientism (Jan. 2, 2014).

does not belong to what Harry Collins calls the "first wave of science studies" [2014].

8. After this chapter is essentially completed, an interview of the Nobelist David Gross by Peter Byrne [2013] came to our attention. The non-finality of any scientific theory and scientists' willingness to abandon what they cheered and toiled with for a lifetime when convinced otherwise (described in Section 1.7.1) are concurred by Gross when he says, "We were all looking for the next overthrow, and we were willing to sacrifice existing theories at the drop of a hat". Other common or not-so-common believes presented above in this chapter are consistent with Gross' sayings, too:

> New discoveries tend to be intuitive, just on the borderline of believability. Later, they become obvious. ... A scientific "frontier" is defined as a state of confusion. ... Philosophers who contribute to making physics are, thereby, physicists! ... The public generally equates uncertainty with a wild guess. Whereas, for a scientist, a theory like the Standard Model is incredibly precise and probabilistic. *In science, it is essential never to be totally certain.* And that lesson is hammered into every scientist and reader of history. Scientists measure uncertainty using probability theory and statistics. And we have comfort zones when making predictions, error bars. *Living with uncertainty is an essential part of science*, and it is easily misunderstood..... [T]he human mind is a physical object. It's put together by real molecules and quarks. (My italics.)

Nevertheless, as stated in item 2 of the *Scimat Standard*, "We will not quote anyone's writing to support our own argument" [Lam, 2008a, p. 27]. The quotes by Gross should not be taken as evidence that the arguments in this chapter are correct, which can only be reached through the reader's own judgment.

9. Scimat provides the key in solving the two-culture problem [Lam, 2008a]. It is also the theoretical foundation of scientific culture studies [Liu & Zhang, 2014] because it provides a unified perspective and basis in describing human matters, and similarly for general-education courses which try to educate students on both the humanities and "science".

10. The Enlightenment's aim of making a science of humans was partially successful; it created social science but fails with the humanities (Section 1.5.2 and Fig. 1.3). Scimat could be viewed as *Enlightenment 2*, continuing with what the Enlightenment left unfinished but smarter.

11. In America, the way to raise the scientific level of the humanities is to ask the National Endowment for the Humanities (called the National Humanities Foundation, after NSF, in the planning stage) *and* NSF to fund more interdisciplinary research that merges the humanities with the "sciences".

Take home message: Science's characteristics are its secularity and the reality check. The necessary Reality Check is what makes science useful and distinctively different from humans' other types of inquiry.

References

Alexander, H. G. (ed.) [1956] *The Leibniz-Clarke Correspondence* (University of Manchester Press, Manchester).
Armstrong, K. [1993] *A History of God: The 4,000-Year Quest of Judaism, Christinity and Islam* (Ballantine, New York).
Ball, P. [2006] "Econophysics: Culture crash," *Nature* **441**, 686-688.
Bauer, H. H. [1994] *Scientific Literacy and the Myth of the Scientific Method* (University of Illinois Press, Urbana).
Berg, R. [2012] "Beyond the laws of nature," *Philosophy Now*, Issue 88, January/February, 24-26.
Bertman, S. [2010] *The Genesis of Science: The Story of Greek Imagination* (Prometheus Books, New York).
Buchwald, J. Z. & Feingold, M. [2013] *Newton and the Origin of Civilization* (Princeton University Press, Princeton).
Buckingham, W., Burnham, D., Hill, C., King, P. J., Marenbon, J. & Weeks, M. [2011] *The Philosophy Book* (DK Publishing, New York).
Burguete, M. & Lam, L. (eds.) [2014] *All About science: Philosophy, History, Sociology & Communication* (World Scientific, Singapore).
Buxton, R. [2004] *The Complete World of Greek Mythology* (Thames & Hudson, New York).
Byrne, P. [2013] "Waiting for the Revolution: An interview with the Nobel Prize-winning physicist David J. Gross," *Quanta Magazine*, May 24 (www.simonsfoundation.org/quanta/20130524-waiting-for-the-revolution/, Nov. 27, 2013).
Capra, F. [2007] *The Science of Leonardo* (Anchor, New York).
Castellano, C., Fortunato, S. & Loreto, V. [2009] "Statistical physics of social dynamics," *Review of Modern Physics* 81, 591-646.

Chen Fong-Ching [2009] *Heritage and Betrayal: A Treatise on the Emergence of Modern Science in Western Civilization* (SDX Joint Publishing, Beijing).
Chiang Tsai-Chien [2002] *Biography of Yang Chen-Ning: The Beauty of Gauge and Symmetry* (Bookzone, Taibei).
Churchland, P. S. [1986] *Neurophilosophy: Toward a Unified Science of the Mind/Brain* (MIT Press, Cambridge, MA).
Churchland, P. [2007] *Neurophilosophy at Work* (Cambridge University Press, Cambridge).
Cioffi-Revilla, C. [2014] *Introduction to Computational Social Science: Principles and Applications* (Springer, New York).
Collins, H. [2014] "Three waves of science studies," in [Burguete & Lam, 2014].
Collingwood, R. G. [1922] "Are history and science different kinds of knowledge," *Mind* XXXI, 443-451.
Collingwood, R. G. [(1945) 1960] *The Idea of Nature* (Oxford University Press, London).
Cornford, F. M. [(1912) 2004] *From Religion to Philosophy: A Study in the Origins of Western Speculation* (Dover, New York).
Cunningham, A. & Williams, P. [1993] "De-centring the 'big picture': The *Origins of Modern Science* and the modern origins of science," *The British Journal for the History of Science* **26**, 407-432.
Darwin, C. R. [1859] *On the Origin of Species by Means of Natural Selection, or The Preservation of Flavored Races in the Struggle for Life* (John Murray, London). 7-4
Derrida, J. [1976] *Of Grammatology* (Johns Hopkins University Press, Baltimore).
Derman, E. [2004] *My Life as a Quant: Reflections on Physics and Finance* (Wiley, Hoboken, NJ).
Donald, M. [2006] "Art and cognitive evolution," in *The Artful Mind: Cognitive Science and the Riddle of Human Creativity*, ed. Turner, M. (Oxford Univer-sity Press, Oxford).
Doren, C. van [1991] *A History of Knowledge: Past, Present, and Future* (Ballantine, New York).
Einstein, A. [1905] "On the movement of small particles suspended in stationary liquids required by the molecular-kinetic theory of heat," *Annalen der Physik* **17**, 549-560.
Epstein, J. M. [2014] *Agent_Zero: Toward Neurocognitive Generative Social Science* (Princeton University Press, Princeton).
Evans, R. J. [1997] *In Defence of History* (Granta Books, London).
Ferris, T. [2010] *The Science of Liberty: Democracy, Reason, and the Laws of Nature* (Harper Perennial, New York).
Gazzaniga, M. [2012] *Who's in Charge?: Free Will and the Science of the Brain* (HarperCollins, New York).

Gilderhus, M. T. [2000] *History and Historians: A Historiographical Introduction* (Prentice Hall, Upper Saddle River, N.J).
Godfrey-Smith, P. [2003] *Theory and Reality: An Introduction to the Philosophy of Science* (University of Chicago Press, Chicago).
Gordin, M. D. [2012] *The Pseudoscience Wars* (University of Chicago Press, Chicago).
Gower, B. [1997] *Scientific Method: An Historical and Philosophical Introduction* (Routledge, New York).
Harrison, P., Numbers, R. L. & Shank, M. H. (eds.) [2011] *Wrestling with Nature: From Omens to Science* (University of Chicago Press, Chicago).
Hayek, F. A. [(1952) 1979] *The Counter-Revolution of Science: Studies on the Abuse of Reason* (Liberty Fund, Indianapolis).
Helbing, D. & Johansson, A. [2010] "Pedestrian, crowd and evacuation dynamics," *Encyclopedia of Complexity and Systems Science* **16**, 6476-6495.
Helbing, D., Keltsch, J. & Molnár, P. [1997] "Modelling the evolution of human trail systems," *Nature* **388**, 47-50.
Henry, J. [1997] *The Scientific Revolution and the Origins of Modern Science* (St. Martin's Press, New York).
Hirshfeld, A. [2010] *Eureka Man: The Life and Legacy of Archimedes* (Walker & Company, New York).
Hoddeson, L. & Daitch, V. [2002] *True Genius: The Life and Science of John Bardeen* (Joseph Henry, Washington, DC).
Hogan, P. C. [2003] *Cognitive Science, Literature, and the Arts: A Guide for Humanists* (Routledge, New York).
Holton, G. [1993] *Science and Anti-Science* (Havard University Press, Cambridge, MA).
Jacob, M. [1997] *Scientific Culture and the Making of the Industrial West* (Oxford University Press, Oxford).
Jaffe, K. [2010] *What Is Science? An Evolutionary View* (www.dic.coord.usb.ve/WhatisScience.pdf, Jan. 2, 2014).
Jiang Jie [2013] "Research takes NDA path to historic figures," *Global Times*, Nov. 13 (www.globaltimes.cn/content/824454.shtml#.UqPOWdJDtIE, Dec. 7, 2013).
Kandel, E. R. [2012] *The Age of Insight: The Quest to Understand the Unconscious in Art, Mind, and Brain* (Random House, New York).
Kennefick, D. [2005] "Einstein versus the *Physical Review*," *Physics Today*, Sept., 43-48.
Kevles, D. J. [1987] *The Physicists: The History of a Scientific Community in Modern America* (Harvard University Press, Cambridge, MA).
Knobe, J. [2011] "Thought experiments," *Scientific American*, Nov., 57-59.
Knobe, J. & Nichols, S. (eds.) [2008] *Experimental Philosophy* (Oxford University Press, Oxford).

Läertius, D. [2011] *Lives of the Eminent Philosophers*, trans. Hicks, R. D. (Witch Books).
Lam, L. [1998] *Nonlinear Physics for Beginners: Fractals, Chaos, Solitons, Pattern Formation, Cellular Automata and Complex Systems* (World Scientific, Singapore).
Lam, L. [2002] "Histophysics: A new discipline," *Modern Physics Letters B* **16**, 1163-1176.
Lam, L. [2004] *This Pale Blue Dot: Science, History, God* (Tamkang University Press, Tamsui).
Lam, L. [2006] "Active walks: The first twelve years (Part II)," *Int. J. Bifurcation and Chaos* **16**, 239-268.
Lam, L. [2008a] "Science Matters: A unified perspective," in *Science Matters: Humanities as Complex Systems*, eds. Burguete, M. & Lam, L. (World Scientific, Singapore) pp. 1-38.
Lam, L. [2008b] "Science Matters: The newest and largest multidiscipline," in *China Interdisciplinary Science*, ed. Liu Zhong-Lin (Science Press, Beijing) pp. 1-7.
Lam, L. [2008c] "Human history: A Science Matter," in *Science Matters: Humanities as Complex Systems*, eds. Burguete, M. & Lam, L. (World Scientific, Singapore) pp. 234-254.
Lam, L. [2011] "Arts: A Science Matter," in *Arts: A Science Matter*, eds. Burguete, M. & Lam, L. (World Scientific, Singapore) pp. 1-32.
Lam, L. [2014] "About science 2: Philosophy, history, sociology and communication," in [Burguete & Lam, 2014].
Lam, L., Bellavia, D. C., Han, X.-P., Liu, C.-H. A., Shu, C.-Q., Wen, Z., Zhou, T. & Zhu, J. [2010] "Bilinear Effect in complex systems," *Europhysics Letters* **91**, 68004.
Lee, T. D. [1986] *Selected Papers*, Vol. 3 (Birhauser, Boston).
Lestienne, R. [1998] *The Creative Power of Chance*, trans. Neher, E. C. (University of Illinois Press, Urbana).
Lindberg, D. C. [1992] *The Beginnings of Western Science: The European Scientific Tradition in Philosophical, Religious, and Institutional Context, 600 B.C. to A.D. 1450* (University of Chicago Press, Chicago).
Linder, D. [2004] "The Vatican's view on Evolution: The story of two popes" (law2.umkc.edu/faculty/projects/ftrials/conlaw/vaticanview.html, Nov. 30, 2013).
Liu Bing & Zhang Mei-Fang [2014] "Scientific culture in contemporary China," in [Burguete & Lam, 2014].
Livio, M. [2013] *Brilliant Blunders: From Darwin to Einstein—Colossal Mistakes by Great Scientists that Changed our Understanding of Life and the Universe* (Simon and Schuster, New York).
Lloyd, G. E. R. [1970] *Early Greek Science: Thales to Aristotle* (Norton, New York).

Lorentz, E. [1993] *The Essence of Chaos* (University of Washington Press, Seattle).
Lubkin, G. B. [1998] "A personal look back at Physics Today," *Physics Today*, May, 24-29.
Machery, E. [2008] "A plea for human nature," *Philosophical Psychology* **21**(3), 321-329.
Mason, S. F. [1962] *A History of the Sciences* (Macmillan, New York).
Mitchell, M. [2009] *Complexity: A Guided Tour* (Oxford University Press, Oxford).
Natterson-Horowitz, B. & Bowers, K. [2013] *Zoobiquity: The Astonishing Connection between Humans and Animal Health* (Vintage Books, New York).
Newton, I. [(1730) 1952] *Opticks*, eds. Cohen, O. B., Roller, D. H. D. & Whittaker, E. (Dover, New York) p. 404.
Novak, I. [2011] *Science: A Many-Splendored Thing* (World Scientific, Singapore).
Oldroyd, D. [1986] *The Arch of Knowledge: An Introductory Study of the History of the Philosophy and Methodology of Science* (Methuen, New York).
Patterson, S. [2010] *The Quants: How a New Breed of Math Whizzes Conquered Wall Street and Nearly Destroyed It* (Crown Business, New York).
Pigliucci, M. & Boudry, M. [2013] *Philosophy of Pseudoscience: Reconsidering the Demarcation Problem* (University of Chicago Press, Chicago).
Planck, M. [1900] "Entropy and temperature of radiant heat," *Annalen der Physik* **1**(4) 719-737.
Pollard, K. S. [2009] "What makes us human?," *Scientific American*, May, 44-49.
Pope John Paul II [1996] "Message to the Pontifical Academy of Sciences: On evolution" (www.ewtn.com/library/papaldoc/jp961022.htm, Nov. 29, 2013).
Porter, R. [2001] *The Enlightenment* (Palgrave, New York).
Rollin, B. E. [2006] *Science and Ethics* (Cambridge University Press, New York).
Ross, S. [1962] "'Scientist': The story of a word," *Annals of Science* **18**, 65-86.
Sargent Jr., J. F. [2013] "Federal research and development funding: FY2013," (www.fas.org/sgp/crs/misc/R42410.pdf, Dec. 30, 2013).
Shermer, M. [2006] *Why Darwin Matters: The Case Against Intelligent Design* (Owl Books, New York).
Shermer, M. [2011] "What is pseudoscience?" *Scientific American*, Sept., 92.
Smith, A. [(1776) 1977] *An Inquiry into the Nature and Causes of the Wealth of Nations* (University of Chicago Press, Chicago).
Stein, J. L. [2011] "The crisis, Fed, quants and stochastic optimal control," *Economic Modelling* **28**(1-2), 272-280.
Suddendorf, T. [2013] *The Gap: The Science of What Separates Us from Other Animals* (Basic Books, New York).

Thurs, D. P. [2011] "Scientific methods," in [Harrison et al, 2011].
Tse, P. U. [2013] *The Neural Basis of Free Will: Criterial Causation* (MIT Press, Cambridge, MA).
Tuckerman, M. E. [2010] *Statistical Mechanics: Theory and Molecular Simulation* (Oxford University Press, Oxford).
Venter, C. [2013] *Life at the Speed of Light: From the Double Helix to the Dawn of Digital Life* (Viking, New York).
Waldrop, M. M. [1992] *Complexity: The Emerging Science at the Edge of Order and Chaos* (Simon & Schuster, New York).
Wang, C.-C., Yan, S., Yao, C., Huang, X.-Y., Ao, X., Wang, Z.-F., Han, S., Jin, L. & Li, H. [2013] "Ancient DNA of Emperor CAO Cao's granduncle matches those of his present descendants: a commentary on present Y chromosomes reveal the ancestry of Emperor CAO Cao of 1800 years ago," *Journal of Human Genetics* **58**, 238-239.
Warren, R. [2014] "*Helicobacter*: The ease and difficulty of a new discovery," in [Burguete & Lam, 2014].
Weatherall, J. O. [2013] *The Physics of Wall Street: A Brief History of Predicting the Unpredictable* (Houghton Mifflin Harcourt, New York).
Weinberg, S. [2011] "Particle physics, from Rutherford to the LHC," *Physics Today*, Aug., 29-33.
White, H. [1973] *Metaphysics: The Historical Imagination in Nineteenth-Century Europe* (Johns Hopkins University Press, Baltimore).
Williams, R. [1983] *Keywords: A Vocabulary of Culture and Society* (Oxford University Press, New York).
Wilson, E. O. [1978] *On Human Nature* (Harvard University Press, Cambridge, MA).
Wilson, E. O. [1998] *Consilience: The Unity of Knowledge* (Alfred A. Knopf, New York).
Winch, P. [(1958) 2008] *The Idea of a Social Science and Its Relation to Philosophy* (Routledge, New York).
Wu, C. S. , Ambler, E., Hayward, R. W., Hoppes, D. D. & Hudson, R. P. [1957] "Experimental test of parity conservation in beta decay," *Physical Review* **105**, 1413-1415.
Yang, C. N. [1983] *Selected Papers 1945-1980 with Commentary* (Freeman, New York).
Youngson, R. [1998] *Scientific Blunders: A Brief History of How Wrong Scientists Can Sometimes Be* (Carroll & Graf, New York).
Zi Cheng, Liu Huai-Ju & Teng Li (eds.) [2004] *Solving the Puzzle of Competing Claims Regarding the Discovery of Parity Nonconservation: Lee Tsung-Dao Answering Questions from ScienceTimes Reporter Yang Xu-Jie and Related Materials* (Gansu Science and Technology Press, Lanzhou).
Ziegler, J. [2013] *Betting on Famine: Why the World Still Goes Hungry* (New Press, New York).

2

About Science 2: Philosophy, History, Sociology and Communication

Lui Lam

Within the last century, four new (sub)disciplines related to science were added to the humanities. They are Philosophy of Science, History of Science, Sociology of Science, and Science Communication. While these disciplines did contribute positively, they had also caused all sorts of problems towards people's understanding of science. What happened and why it happened? This chapter tries to answer this question with new insights gleaned from our historical and cultural heritage of thousands of years. The aim here is not to give a full review of the four disciplines but to analyze them from the perspective of scimat, coming from a humanist and physicist with experience in simple and complex systems. In particular, the mistakes of Ernest Mach, Karl Popper, Thomas Kuhn, Paul Feyerabend and David Bloor and why they occurred are analyzed from a new angle. It has to do with the time-evolving nature of the scientific process, obliviousness of the differences between simple and complex systems, failure of the educational system, and the underdevelopment of the humanities. Suggestions for the near future are provided.

2.1 Introduction

Within the last century, three new subdisciplines and a new discipline related to science were added to the humanities. The three interrelated subdisciplines that analyze the nature and development of science are Philosophy of Science (PS), History of Science (HS) and Sociology of Science (SS). The new discipline is Science Communication (Scicomm) which depends heavily on the other three. In China these four disciplines are loosely lumped under the umbrella "scientific culture" [Liu & Zhang, 2014].

While these disciplines did contribute positively and valuably, they had also caused all sorts of problems towards people's understanding of science. What happened and why it happened? This chapter tries to answer this question with new insights gleaned from our historical and cultural heritage of thousands of years. The early part of this heritage of 195,000 years is recorded in our genes (and fossils) and the last 5,000 years or so in written words. It turns out that the problem arose from intertwined issues involving both simple and complex systems. The distinction between these two kinds of systems is not always appreciated by the practitioners, partly due to the shortcomings in our educational system. The aim here is not to give a full review of the four disciplines but to analyze them from the perspective of scimat (see Section 2.2), coming from a humanist and physicist with experience in simple and complex systems.

To ease the discussion below we classify the practitioners into three groups according to the amount of training and experience the person has in "natural science": insider, outsider and marginer.[1] An *insider* is a "scientist"-turned-philosopher; examples are Ernst Mach (1838-1916), Arthur Eddington (1882-1944), and John Ziman (1925-2005) [2000]. An *outsider* is someone who does not have an academic degree in "natural science", such as Karl Popper (1902-1994), Paul Feyerabend (1924-1994) and most of the philosophers in science. A *marginer* is a person with professional training in "natural science" but does not become a practitioner in this field for a long period of time. The training could include a PhD in physics—like in the case of Thomas Kuhn (1922-1996), but, more often, a BS or MS in "natural science". It does not matter whether this person has published a few professional papers in "natural science" or not.

---

[1] The quotation marks indicate the conventional use of the term, which is not exactly correct. In our terminology, science = natural science = scimat + "natural science", with scimat = humanities + social science + medical science; "science" = "natural science" + social science. (See Section 2.2.)

One may think that an insider should have the best chance of getting things right, but that was not the case in history. However, the nature of the mistakes they made did differ considerably according to which group they belong. And it was the marginers that got the most following even when they were wrong because people thought, mistakenly, that the marginers should know how science actually works. The reasons will become clear later after representing cases are analyzed.

In the following, the issues in these four disciplines are elaborated in turn (Sections 2.3-2.6) after a brief summary of science is given in Section 2.2. An outline of what happened and suggested actions for the near future are provided in Section 2.7. Section 2.8 concludes the chapter. To our knowledge, this chapter is the first time that PS, HS, SS and Scicomm are examined together and from a unified perspective.

2.2 Science in a Nutshell

Science is humans' pursuit of knowledge about all things in Nature, which includes all (human and nonhuman) material systems, without bringing in God or any supernatural.[2] Science thus has two parts: (1) the results obtained by scientists, and (2) the process of doing science. Both parts evolve over time and are full of surprises. The scientific process could involve induction, deduction or intuition, but not necessarily all three of them. Scientific results that remain on the scene are those that passed the *Reality Check* (RC). Reality check means "confirmed" by experiments or practices, or, at the minimum, is consistent with established data [Lam, 2014]. *Science thus has* two *anchors (or signatures): its secularity and the reality check.*

---

[2] It is by this definition that we retroactively call the Godless part of works by Thales (c. 624-c. 546 BC) and Isaac Newton (1642-1727) science and these two scientists, irrespective of their belief in God; we further identify Thales as the father of science. For this reason, Andrew Cunningham and Perry Williams [1993] are wrong to locate the beginning of science at the Age of Revolution (1775-1848) because the *secular* study of the natural world (the signature of science) has been going on since Thales, first as part of philosophy and later as part of natural philosophy [Lam, 2014]. How the secularity guideline is made explicit and the scientist's personal motivation (to glorify God or not) are immaterial as long as God is left out in the research's reasoning.

Nature, same as the universe, includes all material systems (plus the invisibles such as the spacetime fabric and energy fluctuations). Humans, a kind of animal called *Homo sapiens*, are material systems made up of atoms. Science thus is the study of the human system and all nonhuman systems. Science of human, or human science, has three parts: the humanities, social science and medical science. Human science is also called *scimat* (Science Matters) [Lam, 2008a]. Science of nonhuman systems is commonly called "natural science"; "science", the sum of "natural science" and social science. Accordingly, an appropriate metaphor for Science is two linked animals—a *Homo sapiens* and her/his dog (Fig. 2.1).[3]

Fig. 2.1. A metaphor for Science: Two linked animals—a *Homo sapiens* and her/his dog. The human being controls the dog and can direct it to do good things (pick up a newspaper, say) or bad things (bite the neighbor). The dog represents "natural science" while the human being, scimat (humanities, social science and medical science). [A dog leash is added to a rescaled picture of Abujoy's *Size Comparison between a Beagle and a Man* (2008), Wikimedia Commons, May 11, 2014.]

Note that the aim of science is to understand Nature as a whole. The boundary between different disciplines is not fundamental and should not be taken seriously. Different disciplines share common techniques and tools and can learn from each other. For any discipline there are three research levels/approaches: empirical, phenomenological and bottom-up [Lam, 2011, p. 20]. The three levels are routinely used in "natural

---

[3] Harry Collins and Trevor Pinch's metaphor for science is one (imagined) animal, the golem [1998]. It is wrong on two counts: they confuse science with "natural science", and the *application* of "natural science" (more in the domain of the humanities) with "natural science" itself.

science" studies while the bottom-up level in the form of computer simulations is getting more common in social science research. Most of the humanities research is at the empirical and phenomenological levels but we start seeing works at the bottom-up level in the last decade or so [Lam, 2014]. In principle, the "level of scientific development" of a discipline can be measured by its *scientificity*.[4] Generally speaking, scientificity is high for disciplines in "natural science", medium in social science and low in the humanities.

2.3 Philosophy of Science

Philosophy was the one and only one academic discipline in ancient Greece, starting with Thales about 2,600 years ago. It represents humans' attempt to understand the world around them, from the stars in sky to their fellow citizens in the same state, by reasoning (with or without bringing in God) and by abandoning the gods/heroes [Lam, 2014]. It was the most successful discipline in history; every discipline we know of in the universities today branched out from it, directly or indirectly [Morris, 2002].

Philosophy of Science established itself as a branch of "philosophy" from 1925 to 1965 [Nickles, 2013]. The latter is a discipline within the humanities after the early Greeks' philosophy was split into humanities, social science, "natural science" and theology in the 18th century (see Fig. 1.2 in [Lam, 2014]). The aim of PS is to understand everything about science, sometimes called "metascience" [Oldroyd, 1986], e.g., what scientists actually do in their investigations, meaning of the scientific concepts and theories, nature of scientific laws and principles as well as explanations, etc.[5] Books with the name PS are usually about philosophy

---

[4] Scientificity is a quantity that could be but not yet defined. For example, we could define scientificity as a number S ranging from 0 to 10, with 10 the highest in scientific development. And so, for every discipline, we have a S. For instance, $S = 8.5$ for physics, $S = 6.2$ for economics, $S = 0.5$ for philosophy, say. Like the Dow Jones index for stocks, S depends on a number of factors (such as consistency with available data, predictability, etc.) with different weights. Scientificity could be updated from time to time. It would be a useful index for funding agencies and researchers to gauge the direction of their works.

[5] For beginners in PS we recommend David Oldroyd's *The Arch of Knowledge* [1986] which adequately and critically covers the subject from the ancient Greeks to the present

of physics (and of simple systems) while biology (complex systems) [Johnstone, 1914] and chemistry [van Brakel, 2000] have their own extra philosophical questions since they are about emergent phenomena.

We do have prominent philosophers of science (physics actually) who are insiders, marginers or outsiders. To understand how their ideas came about and why they were (almost) all wrong, it is imperative to know about their background, personally and historically. Here are some representative cases.

2.3.1 Ernst Mach (1838–1916)

Ernst Mach (Fig. 2.2) studied physics and for one semester medical physiology at University of Vienna and received his doctorate in physics at age 22.[6] He was a well-established physicist who also contributed to physiology and psychology. He was looking for a unified perspective on the science of inanimate systems (physics) and animate systems (psychology), and proposed a phenomenalist/positivist philosophy. A central idea is that in science, anything that cannot be observed should be excluded from physical laws which he considered as descriptions of sensations [Oldroyd, 1986].

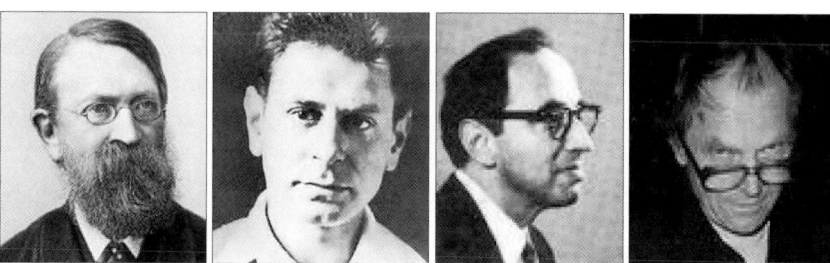

Fig. 2.2. Four philosophers of science: Ernst Mach, Karl Popper, Thomas Kuhn and Paul Feyerabend (left to right).

---

sociologists; for more in the last 100 years, see Peter Godfrey-Smith's *Theory and Reality* [2003]; for a modern, accessible overview of the nature of science from the perspective of a scientist, see Moti Ben-Ari's *Just a Theory* [2005].

[6] The biography of Mach and others below, when not specified otherwise, come from en.wikipedia.org.

Mach wielded tremendous influence, counting the young Albert Einstein (1879-1955) among his many admirers. He is credited with inspiring the formation of the Vienna Circle, a science movement based on logical positivism-empiricism that eventually failed [Holton, 1993]. What brought Mach down as a philosopher is the "trivial" mistake he made about the existence of atoms. Forgetting that the invention of a new instrument like the optical microscope can turn something un-seeable before (such as a single-cell amoeba) into seeable, he *betted* that atoms can never be observed and thus not be allowed into any theory. And he lost. These days we can show directly the image of a single atom through a scanning tunneling microscope which magnifies the atom 100 million times [Greenemeier, 2013]. But in 1905 and the few years afterward, the existence of atoms was demonstrated indirectly but convincingly through Einstein's theory of Brownian motion and Jean Perrin's confirming experiments [Kennedy, 2012]. The mistake committed by Mach was a little bit curious. He did not exclude atoms from theoretical grounds but on technical grounds (that they can never be observed technically) [Holton, 1993, p. 58]. The lesson here is that *betting on the future of anything is precarious and should proceed with caution.*

2.3.2 Karl Popper (1902–1994)

Karl Popper (Fig. 2.2) was born in Vienna, in 1902. At 17, he was attracted by Marxism and joined the workers' party but abandoned the ideology the same year; he remained a supporter of social liberalism throughout his life. At 26, he earned a doctorate in psychology. In 1937, he immigrated to New Zealand upon Nazism's rise in Europe. Nine years later after World War II (WWII) he moved to England and became professor at London School of Economics. He died at 92 in UK, in 1994.

Popper is famous for his idea of what constitutes a scientific theory: The criterion of the scientific status of a theory is its falsificability, not its confirmability. The idea was formulated in "the winter of 1919-1920", the years he messed around with Marxism as a teenager and well before he had a professional training in psychology [Popper, 1963]. He was motivated to find, in a rush, the difference between the established scientific theories from Marxism which he just abandoned and viewed as pseudoscience.[7] His "theory" looked nice on paper (ironically, like Marxism did) but it is easy to see that it could not be right and is impossible to be upheld in practice (see below). Instead of abandoning it when he grew older Popper spent all his life in defending it.

Rigorously speaking, a scientific hypothesis cannot be falsified empirically in the manner Popper wants it. As already pointed out by Pierre Duhem (1861-1916) [1906 (1954)] when Popper was four years old, it is because auxiliary assumptions (like energy conservation) always come with a hypothesis' explicit assumptions. Therefore, when a (reliable) experimental result is found to disagree with the hypothesis' prediction, it is impossible to decide by deduction alone whether it is the auxiliary assumptions or the hypothesis' assumptions that are at fault. Unfortunately, Duhem's analysis was not "fully absorbed" by Popper even though he was aware of it [Worrall, 2003, p. 72]. Here are more remarks on falsificationism.

1. Demarcation between science and pseudoscience is a messy business which could be a non-issue if science is defined properly and broadly [Lam, 2014]. Instead, how to identify the "best available theory" should be a topic in PS [Berg, 2012].

2. Falsification can only be applied to mature theories; it excludes many early theories if taken seriously. That is like defining a human being to be a person who can walk; that would make all crawling babies to be nonhumans. In fact, it may take several or hundreds generation of scientists to modify and develop a theory from its inception to its mature form of being predictable. It is a time consuming process and more often than not, a collective process. We have to be patient, very patient.

3. Popper had mainly physics in mind when he formulated his falsification hypothesis. It is more difficult for non-physics "theories" to make predictions, even when they are good. For

---

[7] Sigmund Freud's psychology, Mach's other pseudoscience target, was a different story. Freud (1856-1939) falsified his patient's record and covered it up by burning all his private papers in 1907 [Dewdney, 1997]. There is not much point in examining a researcher's proposition if that person is less than honest.

example, Darwin's evolutionary theory hardly predicted anything precisely when it was first proposed [Zimmer, 2001]. It did predict that the Earth has to be very old (so biological systems have enough time to evolve) but could not tell how old. For the same matter, Marxism did predict that given enough time, every society on Earth will become communism.

4. The problem is that Popper ignored or was unaware of the basic differences between the theories of deterministic, simple systems (like Newtonian mechanics) and that of a stochastic, complex system (humans in this case, to which Marxism and psychology belong). For example, for a deterministic system like a piece of stone falling from sky, you can ask how long it will take to hit the ground. But for the simplest stochastic system, a random walk, you cannot ask where it will be after its next step, not to mention after ten or 100 steps; you have to ask different questions [Lam, 1982].

5. Falsification is not the way theories are developed in practice. (An example is given in [Ben-Ari, 2005, p. 68].) It is routine get a theoretical paper published which explains a new experimental finding without showing any predictions.

6. There are many legitimate laws of Nature that cannot be falsified. Here is an example. Equation (2.1) is a *new* law with the H term added to Newton's law of gravity, where F is the gravitational force between two point masses m_1 and m_2 separated by a distance r. The point is that the added H term is perfectly consistent with everything known. Experimental tests, with high-enough resolutions in the future, may show the additional $1/r^4$ behavior (confirming but not falsifying this new term); or, failing to find any evidence of it, can only give an upper limit to the constant H but can never conclude that $H = 0$ (failing to falsify it). Why scientists do not keep this H term? Because there is no need to do so, *so far*, as experimental evidence is concerned—the Occam's Razor at work.

$$F = G\frac{m_1 m_2}{r^2} + H\frac{m_1 m_2}{r^4} \qquad (2.1)$$

2.3.3 Thomas Kuhn (1922–1996)

Thomas Kuhn (Fig. 2.2), born in 1922, earned his BS (1943), MS (1946) and PhD (1949) in (solid-state theoretical) physics, all from Harvard University. He has published a few physics papers around 1949. But he changed field to history of science before that. In 1948-1956, he taught HS at Harvard, first as a general-education instructor and later as Assistant Professor of General Education and the History of Science at Harvard.[8] In 1956, after promotion failed, Kuhn moved to University of California, Berkeley, as an assistant professor of HS in history and philosophy departments. In 1961 he was discontinued by the philosophy department but promoted to full professor in the history department [Hufbauer, 2012].

Kuhn's break came in 1962 when he published *The Structure of Scientific Revolutions* (*SSR* for short; 2nd ed., 1970; 3rd ed., 1990). Two years later, he joined Princeton University as chair professor of Philosophy and History of Science. In 1979 he joined the Massachusetts Institute of Technology (MIT), not Harvard, as chair professor of Philosophy, remaining there until 1991. Kuhn was diagnosed with lung cancer in 1994 and died in 1996. Throughout his life, he published three monographs in HS and one in PS. What made him famous are not his HS writings but his PS book, the *SSR*. And he spent his life since its publication to revise it and confuse his critics and friends alike, ending, apparently, as an unhappy professor.[9] To understand why Kuhn, someone with a rigorous training in physics, failed to clarify himself in a span of 34 years (1962-1996) will be a major topic here and in

---

[8] George Sarton (1884-1956) and others' one-dimensional historiography of science (see Section 2.4.2) was criticized by Kuhn [1962] without mentioning Sarton's name. (See [Kuhn, 1984] for Kuhn on Sarton and you will understand why; see also [Pinto de Oliveira & Oliveira, 2013].) It is at this stage Kuhn was disappointed by Sarton's brand of historiography and set out to do differently. Sarton worked intermittently at Harvard University from 1916 to 1951, appointed tenured HS professor in 1940. He established the HS discipline with the support of Carnegie Institution and Harvard (see Section 2.4).

[9] A succinct description of Kuhn's life and career is given by Thomas Nickles [2003, pp. 8-12]; his early years (1940-1962) are detailed by Karl Hufbauer [2012], Kuhn's Berkeley advisee. According to Wes Sharrock and Rupert Read [2002, p. 1]: "Thomas Kuhn died in 1996, convinced that his lifework had been misunderstood, and failing to complete a categorical restatement of his position before his death".

Section 2.7.4. But first, let us look at the simple mistakes that Kuhn committed in *SSR*.

For one, Kuhn soon found out that the "paradigm shift" he proposed in 1962 was so problematic that, instead of abandoning the concept as a good physicist would do, he abandoned the term and replaced it by "disciplinary matrix" and "exemplar" he coined in the second edition of the book in 1970 (see [Oldroyd, 1986, p. 324]). In other words, he started playing with words. Another, the "incommensurability" he also proposed in 1962 was equally in trouble; the mistake of this one is easier to see. For example, contrary to what Kuhn claimed, the mass in Newtonian mechanics is recognized as the "rest mass" in Einstein's special-relativity theory, which is Einstein's mass when the velocity of the body approaches zero (Fig. 2.3).

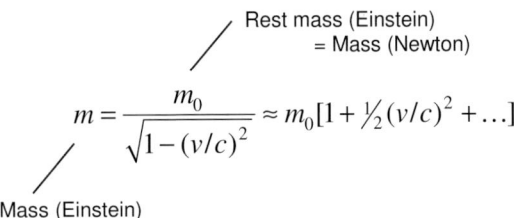

Fig. 2.3. Relationship between Newton's mass and Einstein's mass. (See text.)

What happened? According to Kuhn as told to Steven Weinberg [2001, pp. 203-204], a physics Nobelist and Kuhn's colleague at MIT, in 1947 he was studying Aristotle's work in physics as a young physics instructor at Harvard, and it suddenly dawned on him that Aristotle's mechanics was not "bad physics" when judged in Arisotle's time, even though it is wrong by today's knowledge. And, according to Weinberg [p. 204], the (obvious) paradigm shift from Aristotle's mechanics to Newton's mechanics could be what led Kuhn to his later ideas. If so, here lies the key in understanding Kuhn's fallacy in his incommensurability. The concepts, terms and results of Aristotle's mechanics are nowhere to be found in Newton's mechanics because the former does not pass the RC—i.e., agreeing with the real world, while that of Newton's mechanics are retained or recognizable in Einstein's relativity because they do. In other words, anything not passing the RC will disappear eventually and incommensurability happens. In the opposite case, anything passing the check will somehow remain and be recognized, and no incommensurability exists.[10] In fact, RC is what distinguishes science from other forms of humans' inquiry [Lam, 2014], but strangely, it is never mentioned in Weinberg's discussion of Kuhn nor by Kuhn and others in PS, HS or SS.

Two questions remain: (1) If Kuhn's premises in *SSR* are so obviously wrong why the book "has sold over a million copies in two dozen languages" [Nickles, 2003, p. 1]. (2) If Kuhn is so wrong why he "was perhaps the most influential philosopher writing in English since 1950, even the most influential academic" [Sharrock & Read, 2002, p.1]? The answer to question 2 is easy: The status of a philosopher is judged not by the correctness of his ideas (which mostly are wrong by later judgment; think Plato) but by how much a stir he created during and after his lifetime. And Kuhn created quite a stir, as evidenced by the fact mentioned in question 1. The answer to question 1 is less trivial, but nothing subtle. This will be discussed in Section 2.7.4.

2.3.4 Paul Feyerabend (1924–1994)

Paul Feyerabend (Fig. 2.2) was born in Vienna, in 1924.[11] He developed an interest in theater and started singing lessons in his teens. In 1943, he joined the army; his spine was hit by a bullet. This made him impotent; severe pain accompanied him daily and he had to walk with a stick for the rest of his life [Feyerabend, 1995]. At age 24, he met Karl Popper. The next year he became a founding member of the Kraft Circle, a post-WWII extension of the Vienna Circle. He studied with Popper at the

---

[10] A theory that passes the RC is like a boat got anchored; that does not is like a boat that will drift away and disappear given enough time. Another way to visualize this is that before Galileo, Aristotle's mechanics was thought to be a component of the human-independent part of the *knowscape*—the landscape of knowledge [Lam, 2014]. But after Galileo, Aristotle's mechanics is shifted to the human-dependent part—the HS part.

[11] A collection of informative essays from Feyerabend's friends and students is given in [Preston et al, 2000].

London School of Economics in 1952, but he was critical of Popper's falsificationism. He started working at University of Bristol in 1955.

Since 1958, he was professor at University of California, Berkeley; he and Kuhn overlapped with each other there. He liked to travel. He has been visiting professor at University College London, in Berlin and at Yale University. He taught at University of Auckland, New Zealand (1972 and 1974). In the 1980s, he alternated between ETH, Zürich and Berkeley. He left Berkeley for good in 1989, first to Italy, then Zürich, and retired in 1991. He died of brain cancer in 1994. He has written two major books, *Against Method: Outline of an Anarchistic Theory of Knowledge* (1975) and *Killing Time* (1995), an autobiography just before he passed away.

The essence of *Against Method*, written like a long manifesto, is that there are innumerable different methods in scientific inquiry and each is worth trying. Thus, science cannot be regarded as a strictly rational enterprise; it may depend upon people thinking counter-intuitively. So far so good, even though this is common knowledge among the good scientists. But he went on to say, "[T]he time is overdue for adding separation of state and science to the...separation of state and church" [Feyerabend, 2010, p. 164]. This is bad and wrong because the high-school science he talked about, like mathematics and English, is simply a skill the citizens need to know to survive in today's technological society (to understand the electric bills and read the nutrition labels on food products, say). Since Feyerabend has written "I hope...the reader will remember me as a flippant Dadaist..." [p. xiv] and begged us not to take him seriously, we should honor his wish and forget his sayings except for the central statement he made against method in doing science.

2.4 History of Science

History of science is a branch of history. The latter is about all things happened to humans in the past [Lam, 2002] and the former is about what and when scientific results are obtained and how and when scientists go about in obtaining them. Obviously, HS would and did suffer from the lack of a clear definition of science, a failing of the PS.

The beginning of History is credited to the ancient Greek, Herodotus (c. 484-425 BC). Called the "Father of History", he was "the first historian known to collect his materials systematically, test their accuracy to a certain extent, and arrange them in a well-constructed and vivid narrative".[12] Similarly, HS has an equally long history associated with the names of Eudemus (c. 370-316 BC), Plato (427-347 BC) and Aristotle (384-322 BC).[13] But the Belgian George Sarton, the "Father of History of Science", was the one who made HS into an autonomous academic discipline in the early part of the 20th century [Liu, 2008; Kragh, 1987] while professionalization of HS speeded up in the 1950s [Kuhn, 1984]. Sarton studied chemistry, crystallography and mathematics; he obtained his PhD at University of Ghent (1911) with the thesis "The principles of Newton's mechanics" [Garfield, 1985]. The next year, he founded *Isis*, a review devoted to HS *and* PS aimed at philosophers, historians, sociologists and "scientists"; he was its editor for 40 years. He immigrated to the United States in 1915 and taught at Harvard University. He founded the History of Science) and its Society (1924two official journals, *Osiris* (1936) apart from *Isis* (1912). Today, HS is a well established discipline, with the International Union of History and Philosophy of Science's Division of History of Science and Technology (IUHPS/DHST) founded in 1947. As we see it, there are three central issues facing HS, described next below.

2.4.1 Scope of History of Science

The history of human-related complex systems (in social science and the humanities) is not yet included in HS. The fact that Sarton failed to recognize the humanities and social science as part of science is unfortunate since he was born long after Darwin's evolutionary theory first appeared (see Section 2.2). Consequently, he ended up advocating a "New Humanism" which tried to connect the sciences to the humanities using HS [Sarton, 1962]—Sarton's way of solving the (not-yet-named)

---

[12] en.wikipedia.org/wiki/Herodotus (Mar. 21, 2014).
[13] A concise introduction to the historical development of science (including "natural science" *and* social science) is given in en.wikipedia.org/wiki/History_of_science (Nov. 1, 2013).

"two-cultures" problem, a viable but inefficient and insufficient approach –a mission not endorsed by Kuhn [1984]. A better approach is to increase the humanities' scientificity through collaborations between humanists and "scientists" [Lam, 2008a]. Presently, partly due to Sarton's misconception of science, HS institutes cover only "natural science", medical science and mathematics,[14] nothing on social science and humanities.

2.4.2 *How Much Detail?*

How much detail the historian of science wants or needs in her historical description depends on the circumstances and personal taste. Generally speaking, more details are provided as time progresses and HS has experienced four levels of details so far. This issue is not always appreciated by HS researchers.

Since science covers everything in Nature, HS is about the historical exploration of the knowscape—the "landscape" of knowledge [Lam, 2014], which includes two major mountain ranges corresponding, respectively, to human-independent knowledge ("natural science") and human-dependent knowledge (scimat). The description of the discovery of an extended part of the landscape depends on how much details one wants. As a *first-order approximation*, Sarton's kind of HS gives a very rough description—like who had done what, in a chronological manner. It is similar to reporting who are the firsts to reach the high mountain peaks on earth, from the Carstensz Pyramid, Papua to Kilimanjaro, Africa and to Mount Everest, say. For a more local description, take Mount Everest for example. A rough history would say the British "discovered" it in 1856, and Tenzing Norgay and Edmund Hillary did the first official ascent in 1953.[15] A more detailed description would include all the previous attempts that failed. A much more detailed version would have to include the names of all the members of that expedition, and what equipments they brought with them, etc., etc.

---

[14] Even though in the common view, mathematics is *not* part of science. See [Livio, 2011] for a discussion of whether mathematics is invented or discovered by humans.
[15] en.wikipedia.org/wiki/Mount_Everest (Mar. 21, 2014).

Kuhn's HS amounts to a *second-order* approximation within which personal factors such as psychology are considered. The *third-order* approximation is provided by sociologists who emphasize interpersonal interactions and the social forces at play (see Section 2.5). The feminist approach in HS, by noticing that there are two sexes, male and female, among knowscape's explorers, takes us to the *fourth-order* approximation [Schiebinger, 1999].[16] A higher-order approximation always brings in new insights, apart from more details. Yet *there is no need to discredit or disown the previous approximation(s) while taking it one order higher in the approximation ladder, and drop the RC along the way.* History of science, like any other discipline, is accumulative.

Here is the *fifth-order* approximation, not yet done by anybody: the role of supporting actors/actresses in HS. The characters who provide a scientist's personal needs (food, sex, etc.) have no effect on the scientific results obtained but do affect heavily the scientist's choice of place to live, selection of topics and his career. For example, a detailed history of Einstein's discovery of general relativity could include who cooked for him during those ten years (1905-1915) when the work was done, since without a good cook and good food Einstein might never made it. But, of course, the importance of a historical figure is judged by whether this person is replaceable. Einstein was not replaceable; the cook was (or could be—only Einstein would know).

2.4.3 Beyond Narrative

History laws do exist and history research has gone beyond the narrative with studies carried out at three levels [Lam, 2008c]. At the empirical level, statistical analysis and Zipf plot (borrowed from complex-system studies) have been used; at the phenomenological level, we have

---

[16] These approximations kind of parallel those used in building physics models. If HS is compared to a gas system, Sarton's approximation is like the independent-particle approximation used in the ideal gas model; Kuhn's is to include the internal states of the particles; sociologists' is like adding in the interparticle interactions and external fields; feminists' is to take note of the two kinds of particle existing in the gas. What kinds of factors to be included depend on the aim of the model; e.g., whether it is for illustrating the mechanism or explaining a trend, or for detailed comparison with existing data.

computer modeling; at the bottom-up level, computer simulations, differential equations solving and DNA tracing (see [Lam, 2014]). There is no reason that these approaches cannot be applied in HS studies.

2.4.4 *An Open Problem*

The most interesting open problem in HS, in our opinion, is this one. Aristotle has proposed that heavenly bodies move in circles while territorial bodies move in straight lines [Henry, 1997, p. 16] (Fig. 2.4). We know that Aristotle, unlike his teacher Plato, did get his hands dirty in empirical studies; he pioneered biology by classifying 540 and dissecting at least 50 animal species [Mason, 1962, p. 44]. Why then Aristotle never hurled a piece of stone in air and observed that the path is curved, and not straight lines? More strangely, why there is no record to show that people after him in the next 2,000 years have done this "test" and put down in writing that Aristotle is wrong? It is such an easy test that can be done by a child, unlike the case of disproving the circular motion of the heavenly bodies which has to wait for Copernicus.

Fig. 2.4. Projectile path according to Aristotle (left) and in reality (right), respectively.

2.5 Sociology of Science

Sociology of science is the study of the scientific process and the content of scientific knowledge from the sociological viewpoint [Sismondo, 2004]; it has a short history of a few decades. It is part of the "sociology of knowledge" trend associated with the Germans, Karl Marx (1818-1883), Friedrich Nietzsche (1844-1900), Karl Mannheim (1893-1947), etc. [Oldroyd, 1986]. It is perfectly alright if the sociologists decide to give a fair and "symmetrical" treatment to competing theories and competing scientists within the historical period in time they choose to study, as advocated in the "strong program in the sociology of knowledge" [Bloor, (1976) 1991]—an important part of the so-called Sociology of Scientific Knowledge (SSK) [Collins, 2014]. But there is an *intrinsic limitation* to this kind of approach which is ignored by the practitioners: *From the sociological study of a particular period* alone, *one can* never *generalize the finding to cover a long period of time beyond the period under study.* Here are the explanation and related issues.

2.5.1 *The Scientific Process*

Science is done by humans. The scientific *process* involves these steps: (1) Find a problem; (2) ask questions. Then decide (3) on a project for investigation; (4) to get funding or not; (5) who to collaborate with, if any; (6) who to discuss with, if any; (7) where to do it; (8) how to do it; (9) who's previous results to take seriously; (10) to call a news conference or not, if the result obtained looks important; (11) to publish or not; (12) if yes, the manuscript will be reviewed by referee(s) and decided by the journal's editor; (13) if not, apply for a patent or not. All these 13 steps are *human-dependent*, and thus involve sociology. In this regard, SS is a legitimate branch of science studies.

2.5.2 *The Scientific Results and the Book Drop Test*

For simplicity, let us assume that the scientific *results* obtained are about nonliving systems, such as the "free" fall of a piece of rock. (Free fall means the only force acting on the rock is gravitational; it is the very opposite of the word free—physicists' humor.) The scientific result is that it takes 0.45 s (s for second) for the rock to hit the ground if it is let go from a height of 1 meter. Now, because this result was obtained after the 13 steps above which happened to involve repeated fierce competition, bitter bickering and ugly rivalry, as observed or learned by a sociologist, this sociologist concludes that the 0.45-s result is not a "true belief" but is socially constructed. In other words, this sociologist does not believe in the existence of *human-independent* knowledge [Lam, 2008a; 2014].

For such sociologists—called "Class F" sociologists (F or fanciful or fantastic, your pick),[17] here is the *Book Drop Test*: Pick a heavy book around you; remove your right shoe; stand up and drop the book from your waist above your right foot; do not move for 1 s. A true Class-F sociologist will not try to move his right foot before the 1 s ends. (What happens to the right foot in that late 1 s involves Newton's second law of motion and medical science and is too complicated to be explained here.)

2.5.3 The Generalization Trap

The scientific process spans a period of time and is time evolving, and is thus a historical process [Lam, 2014]. It is like a growing human being, *a history-dependent and time-evolving system*, unlike the case of an electron which is the same all the time. Let us say that the life of a child at age three is observed closely (24 hours per day, every day) for one year. Whatever one learns from this one-year study, it is obvious that one cannot make generalization from it alone to say much useful about the child's other years, in the past or in the future. Unfortunately, that is what some of the SS researchers did about scientific studies. What usually happens is that if one examines their writings, they look alright and conventional in what they observed but it becomes troublesome when they draw conclusions by generalizing the observations beyond their validity. This is the intrinsic limitation that forbids one to generalize a slice of history to the whole history of a long time period.

Take the acceptance of Einstein's general relativity as an example. In *The Golem*, Collins and Pinch [1998] recount the happenings

---

[17] A class-F person is one who believes all knowledge is socially constructed; e.g., the gravitational law could not be discovered by smart aliens. Class-F people include self-proclaimed relativists as well as some sociologists and postmodernists, but not any practicing "scientists"—deceased or alive. It is not clear that there are really any truly class-F persons in practice because no one has claimed that he has passed the book drop test. Adolf Hitler (1889-1945) is identified as a class-F person [Ferris, 2010, p. 245] when he supposed to have said, "Science is a social phenomenon...The slogan of objective science has been coined by the professorate..." [Rauschning, 1939, p. 221]. But the credibility of Hermann Rauschning's quotes is not established.

surrounding the 1919 eclipse experiments[18] that "prove" the validity of the general theory of relativity. They show correctly that the scientific process does involve human-dependent, professional judgments[19] and is not as neat as the so-called "Scientific Method" leads one to believe. But they go on to conclude that the "culture of science" is formed by the establishment's consensus, and is comparable to what happened in the Soviet Union. Here are several problems: (1) The simple-minded, recipe-like "scientific method" [Thurs, 2011] does not really exist [Lam, 2014]. (2) It is true that *sometimes* the mainstream consensus on a scientific topic is heavily influenced by those in powerful positions, either in the West (such as Aristotle's mechanics, for 2,000 years) or elsewhere. But (3) the difference is that, eventually, only those results that pass the RC will remain in the scene. More importantly, (4) what *The Golem* fails to point out is that the (conclusive) acceptance of a scientific theory is not based on one experiment, even though it may seem like that to outsiders like the reporters.[20]

2.5.4 The Outsider Problem

It seems that all sociologists of science are "outsiders" (defined in Section 2.1), reflected in the kind of mistakes they committed. David Bloor, one of the pioneers of SSK, is such an example. Bloor, born 1942 in UK, was director of Science Studies at University of Edingburg. His book *Knowledge and Social Imagery* in 1976 introduces the strong

---

[18] *The Golem* mislabels 1919 as 1918 (p. 48).
[19] The discarding of discrepant data points in the eclipse experiments was less arbitrary than *The Golem* conveys; it involves careful, technical considerations [Kennefick, 2009].
[20] The 1919 eclipse experiments did result in newspaper headlines like "Einstein's theory confirmed", and maybe so even for some scientists at that time. But as science goes on, more people learned to be more careful. One experimental proof is just a tentative confirmation while a "conclusive" confirmation needs a number of independent positive experiments. Unfortunately, in the case of eclipse experiments, one has to wait many years before the same experiments can be repeated. In fact, in the physics community, the validity of general-relativity theory (or *any* theory) is not built on that one result, but on a long string of confirming experiments (on different aspects of the theory) that have poured in over the years [Bederson, 1999, pp. 74-83] and are still coming in. For the laypeople, the working of the Global Positioning System (GPS) could be the most important confirmation. (See also [Mermin, 1996].)

program of SSK, with the conclusion that knowledge is not "true belief" but whatever men take to be "knowledge" [Bloor, (1976) 1991]. This may be true for some "knowledge" about human affairs that the RC is difficult or impossible to perform due to insufficient or unavailable data—a fact that sociologists were familiar with in their own discipline before they ventured into science studies. As explained above, it is absurd to call those scientific results (about inanimate systems) that pass the RC to be socially constructed, but *the RC is not mentioned by the sociologists*.

In his book Bloor discusses Pythagoras' abandoning of the study of irrational numbers (after he found them objectionable philosophically) and concludes that mathematics is an empirical enterprise. Here, it seems that Bloor does not understand that mathematics, like science, has two parts. The abandonment belongs to the choosing/terminating of a research topic and, of course, is human-dependent. But the Pythagoras' theorem (also discovered by the Chinese) and the irrational numbers are mathematical results that exist independently of Pythagoras (and could be discovered by smart aliens). In fact, the particular research topics picked by the scientists are indeed dictated by personal considerations (e.g., she wants to work with this beautiful person) and professional judgments that the time is ripe to tackle them—both are human decisions and subject to sociological and cultural influences.

As outsiders, some sociologists, like many others, have misconceptions about science. For example, they mixed up scientific proof with mathematical proof, without realizing that science is built on approximations and thrives on approximations [Lam, 2014]. [21] They criticized scientists for fault or over claim when a rigorous, mathematical-like proof is absent.[22]

---

[21] For example, we know from Newton's universal gravity that a piece of rock falling from the sky is attracted by *all* other masses in the universe. But when we calculate how fast the rock will hit the ground, we ignore all other masses except that of the nearby Earth—an approximation.

[22] For these sociologists, Steve Mirksy's article, "Physics uncowed" [2012] is recommended. It describes how physicists "prove" that the moon is not made of cheese without going to the moon to sample it, similarly for the impossibility of bending spoons with the mind. For spoon bending, see [Randi, 1982; Collins, 1992].

2.5.5 The Laboratory Visits

Scientific works are classified into two types: theory and experiment. Theory is further divided into three types according to how much computer is used: (1) pure theory (no computer calculation), (2) theory with computer calculation, and (3) computer simulation.[23] Experiment has three types, too, according to the size of the team: (4) big-lab experiment (hundreds or more collaborators, like those in Fermi Lab or CERN), (5) medium-size experiment (with one to a few collaborators in the same room, like those in condensed matter research in universities or the old Bell Labs), and (6) table-top experiment (one or two persons are usually enough). From my own experience,[24] I know that these six forms of research are very different. And that is why graduate students in science have to decide early on which type they want to do and are trained accordingly. But it seems this is not known to the outsider sociologists.

At a certain point, some researchers in SS decided to visit the labs, type 5 above, and observed what was going on first hand [Sismondo, 2004, pp. 86-96]. And that is good even though what they learned is common knowledge among graduate students in experiments at good universities.[25] For example, tinkering, skills and tacit knowledge are involved in experiments. That is exactly why the graduate training is in the form of apprenticeship, and why it is extremely difficult for science in developing countries to play catch up without sending their students abroad. The students learn this when they cannot get the apparatus working and the professor comes in and makes it work by touching here and there.

Tacit knowledge and required tinkering in a lab could arise in two circumstances: (1) The measurements, such as the separation between

---

[23] Some people call computer simulations as computer experiments. That is confusing; it is actually wrong since experiment means you have to get your hands dirty [Lam, 2006].

[24] I have worked at the Nevis Labs of Columbia University and Bell Labs, Murray Hill (where semiconductor was invented); I have done table-top experiments on pattern formation at San Jose State University. And I have published in all three kinds of theory.

[25] This kind of "common" knowledge is available in scientists' memoirs and popular-science (popsci) books. What one needs is to read enough number of them, or just talk to the graduate students

two components or the size of a component in an apparatus, can only be specified with finite resolution (usually up to two decimal points in centimeters) but the working of the real setup requires higher precision. (2) When three control parameters, from one or a few components, are involved in an apparatus there is the possibility of chaos; i.e., the outcome depends sensitively on the initial conditions [Lam, 1998]. Both cases come from the problem of finite resolution which is unavoidable in the real world. And that is a distinctive feature of delicate experiments (like replicating a laser [Collins, 1992])—unlike those in undergraduate labs—which require the presence of an experienced experimentalist; not even a good theorist can handle that.[26]

This problem of course will present limitations on the development of expert system and artificial intelligence [Collins, 1992; Sismondo, 2004] but needs not be so if the computer is capable of learning (using genetic algorithms, say). Moreover, the real world can tolerate and work with approximations. For example, you make an appointment to meet a friend at 2:30 pm; you are right on time if you arrive between 2:30:00 pm and 2:30:59 pm. Or, when we say this electronic appliance works with a current of 1 Ampere, it actually works with small fluctuations around this number. Contrary to some people's belief, the real world does not rely on mathematically exact precisions; another way of saying this is that the basin of attraction of the attractor of a real event is not fractal [Lam, 1998].

Collins investigated several early gravitational wave[27] experiments of the Weber type and came up with what he called "experimenter's regress" [Collins, 1992; 2014]. It argues that "one knows an experimental system is working when it gives the right answer, and one

---

[26] An example: The physics Nobelist Yang Chen-Ning was advised by Edward Teller to switch from experiment to theory in his PhD training after he had shown himself to be obviously inept in the former [Chiang, 2002].

[27] Einstein first submitted a manuscript to *Physical Review* titled "Do gravitational waves exist?" with a negative answer, as inferred, erroneously, from the general-relativity theory he invented. After mistakes were pointed out by an anonymous referee, Einstein reversed his answer, reworked his manuscript, changed the title to "On gravitational waves" and got it published in another journal, without thanking the original referee [Kennefick, 2005].

knows what the answer is only after becoming confident in the experimental system" [Sismondo, 2004, p. 92] and so there is a circle. That may be the case with the few experiments that Collins studied (and is short term). But a good experimentalist like Wu Chien-Shiung would point out why and where the previous experimenters are wrong when presenting her own results that are different [Chiang, 2014]. This practice is especially essential if similar experimental setups are used by the challengers. Collins falls into the generalization trap here. Yet, history is full of twists and turns. The latest news (March 17, 2014) is that gravitational waves have been detected via a new route, completely different from the Weber type. They were detected indirectly in the wrinkles of spacetime from the earliest moment of the universe. This time, while champagne was indeed opened at a house in Stanford, the rest of the world is waiting for independent verifications [Chodos, 2014].[28]

In fact, the generalization trap applies to most lab-visit studies which are unavoidably limited in scope and time. Moreover, as in physics experiments, taking data is only half the story; *the more difficult part is in interpreting the data.* Sometimes, even a two-year field trip [Latour & Woolgar, 1986] is insufficient [Oldroyd, 1986, pp. 353-356]. In the case of Aritotle's mechanics, it takes 2,000 years for the issue to be settled. And that is how science works.

2.5.6 *The Role of Theory*

The role of theory in commonly misunderstood by outsiders. As described above, the scientific process is like the exploration of an unknown landscape, except that it is done on the knowscape. In the very

---

[28] In Alan Cholos' report on the gravitational wave discovery, he writes: "The *good* news is that many experiments are poised *not* to repeat the measurements exactly but to add potentially confirmatory or contradictory evidence" (my italics). Except for simple table-top experiments (like Faraday's and those in high-temperature superconductors), discovery experiments were rarely repeated. Sometimes it is because of the difficulty (like Wu Chien-Shiung's parity-nonconservation experiment) but the deeper reason is subtle. To confirm that a person is coming out of a hotel, the same picture from two different cameras is not as strong and informative as more pictures taking from different angles. This is a point that some SS sociologists failed to understand.

beginning, people try to understand things visible to them (like the heavenly bodies). But after the obvious objects or phenomena are exhausted or do not yield to analyses (e.g., why the sky is blue became known only after quantum mechanics was invented), where in the knowscape do scientists set their foot on? They cannot wait idly for the rare accidental discoveries to show up. Instead, they pick their research topics guided by the theories. Theories' predicting and retrodicting ability (e.g., Einstein's cosmology) is like a telescope that allows the explorer to look around at long distances.

If science is compared to a vertebrate (or two vertebrates; see Section 2.2) then (interrelated) theories form the backbone and control experiments the other bones of this animal.[29] Without the bones, the animal will not be full of vitality and strength. Theories enable scientists to explore swiftly the knowscape without visiting every spot on it [Lam, 2014].

Theory could appear after experiments (to explain them) or ahead of experiments (to predict them). Theory and experiments are like two brothers conversing, helping and competing with each other while running together. They correct each other's mistakes so that both will be better runners. But experiments (a form of RC) are the big brother who has the final say.

2.6 Science Communication

Science communication, called "science popularization" in China [Lam, 2008b], is a discipline in its infancy. As international English journals are concerned, it has only two printed journals, *Science Communication* (quarterly, since 1979, USA) and *Public Understanding of Science* (now eight issues per year, since 1992, UK), and an e-journal, *JCOM* (quarterly, since 2002, Italy). Only a few books exist (e.g., [Gregory & Miller, 1998; Sanitt, 2005; Broks, 2006]).[30]

---

[29] Apart from confirming theories and bringing in unexpected discoveries, control experiments enable the building of reliable equipments and comparison with data collected from uncontrollable sources (such as the optical spectra originating from the stars).

[30] A useful, brief introduction to Scicomm is given in en.wikipedia.org/wiki/Science_comunicaton (April 17, 2014).

Scicomm does not aim to analyze science but to communicate science to the public, and vice versa. It depends heavily on the success of PS, HS and SS. Despite its obvious importance in a modern society that the citizens have to be informed and consulted in governmental decision making, scicomm is a discipline gravely underdeveloped. Scicomm, like any other discipline, has two aspects: pure and applied. Pure scicomm is scicomm studies; applied scicomm is the public understanding (or engagement) of science. The difficulty of Scicomm, the most complex discipline on earth, derives from its very nature and comes from several sources:

1. Scicomm involves every other discipline, from physics to PS and HS, and from cognitive and learning sciences to mass communication. It is hard for anyone to master such a wide range of knowledge.

2. Consequently, we end up with many research papers of substandard quality. They are written by people with knowledge in any other field but little training in scicomm.

3. The practical (or application) side of scicomm is most difficult. Unlike a researcher who only has to convince one or two referees of a journal to call her work done, a scicomm worker has to convince a large number of people, the public, and may never know when her work is done.

But the advantage of point 1 is that everyone can join in and, hopefully, most of them would improve given enough time. As for point 3, there are a lot of practical experiences but, it seems, there are no working theories; after all, we are dealing with the complex system of communication here.

The reason of doing scicomm studies is the same as that for any other discipline, i.e., for the sake of knowledge. But the reasons of communicating science to the public are less obvious; there are several:

1. Promote the scientific spirit. (This is easier said than done, especially in societies where the scientific spirit is not generally observed by those in power.)

2. Promote the scientific way of thinking. (This is better done by telling scientific-discovery stories. The term "scientific method" is misleading and should be avoided.)

3. Popularize scientific knowledge. (Something practical like why people should floss their teeth daily and something more abstract like no one could mend the Earth after it exploded—as claimed by the founder of some new "religion". That is, the aim is to promote citizens' ability to deal with practical problems and avoid being preyed upon by people.)

4. Awaken citizens' inert curiosity about the world/universe. (This should be done in school, but scicomm can supplement that since not every student has a good science teacher.)

5. Stimulate children's interest to be future humanists/"scientists". (For example, many eminent scientists decide to have a science career after reading a good popsci book as a youth.)

6. Promote citizens' ability to participate in public-affair discussions and decisions. (This is most tricky; see Section 2.6.3.)

Not listed here are: to convince the public that the government should support scientific research; to share with the public the splendid of doing science (suggested by Carl Safina [2012]). The former is a non-issue since the public is already convinced; e.g., 5 in 10 Americans say the government's spending on science and technology research is about right [Lucibella, 2014]. The latter is not restricted to science; artists and religious believers do that, too.

2.6.1 Two Brief Histories: United Kingdom and China

The importance of scicomm was recognized only in the last few decades (as attested by the short history of its journals), even though scicomm activities can be traced to earlier times [Broks, 2014; Yin, 2014]. Scicomm now exists around the world [Schiele et al, 2012]. However, the scicomm histories in UK and in China are quite different from each other.

According to Peter Broks [2014], in UK, in the early 19th century, there was no separation between science and the public; all could join— the "Republic of Science" period. In the late 19th century, rise of the experts resulted in the separation of scientific experts and the lay public; scicomm was needed to bridge the gap. Then, in the 1950s, Soviet Union's Sputnik satellite triggered fear about the low level of scientific literary in the West. In the 1980s, the program of Public Understanding of Science (PUS) appeared, and scicomm was viewed as the new duty of scientists. In 2000, the emphasis shifted from PUS to PEST (Public Engagement with Science and Technology); a whole new set of questions emerged:

1. What counts as being a scientist?

2. Where do we draw the boundaries between science and nonscience, between scientists and nonscientists?

3. What counts as expertise? Who are the experts? What about lay experts?

4. Why scientists should listen to what the public has to say?

5. Engagement presupposes particular social and political relationships which in turn raises questions about authority and democracy.

But the situation is more complex than these questions indicate; see below.

China's story is different. Science was first introduced piecemeally to China by Western missionaries in the late Ming Dynasty (1368-1644) while scicomm in modern China started in the 1840s [Yin, 2014]. The turning point happened in the late Qing Dynasty (1644-1912). The importance of technology dawned suddenly on the Chinese after the country suffered humiliating defeats in the hands of the Western powers and Japan, culminating in the invasion by the Eight-Power Allied Forces in 1900. After Qing was replaced by the Republic of China, the May Fourth Movement (1919, the same year general relativity's eclipse experiments were performed) intellectuals concluded that technology was insufficient; China needed science (and democracy). And so scicomm was pursued earnestly amid the Republic's rough years that followed. After the establishment of the People's Republic of China in 1949, scicomm in mainland China was transformed from grassroots efforts to government-controlled activities, resulting in a restrictive mode of operation that differs from that in the West [Li, 2008].

2.6.2 Two Modes of Operation: United States and China

In the United States, scicomm is a grassroots effort, coming from individuals (including students [Sanders, 2011] and private citizens) and learned societies (e.g., the American Physical Society has an Outreach Program and *Focus* on its website—the latest physics research written by science writers for the public); it is a *bottom-up* approach. Scicomm degrees, with one exception, are not offered by the universities but a few scicomm programs exist (e.g., at Cornell University). The exception is the Department of Life Sciences Communication at the University of Wisconsin-Madison which offers BS to PhD degrees. The grassroots approach seems to work in such a vast country.

On the other hand, the *top-down* approach is adopted in China [Li, 2008; Shi & Zhang, 2012]. This is due partly to the fact that most people, professionals and citizens alike, are still too busy with their daily jobs since the "big bang"—the country's "reform-and-opening up" policy change of 1978—and have not much time and energy to spare, and partly because no large non-government organizations are able to flourish.[31] No scicomm degrees are offered in China. The major duty of the one and only one research institute, China Research Institute on Science Popularization (CRISP) under the China Association for Science and Technology, is to carry out government scicomm projects. CRISP publishes *Study on Science Popularization* (formerly *Science Popularization*), the only scicomm journal in China. Apart from other things, a major problem with China's scicomm is that there is not enough number of scientists got involved, even though China is the only country with a scicomm *law* [Zhang & Ren, 2012].

---

[31] A top non-government scicomm organization is the Science Squirrels Club (songshuhui.net), which favors biological, health and medical topics [Chen et al, 2014], all complex systems. The Club's microblog attracts one million fans vs. government counterparts' thousands or tens of thousand [Wang & Tang, 2014].

2.6.3 Why Communicating Science Is Difficult: A Few Examples

There are several reasons, some technical and some fundamental, that communicating science to anybody is difficult, which are illustrated here by a few examples.

1. Newton's three laws of motion: even the physics textbooks are wrong

In all physics textbooks, Newton's three laws of motion are described as about physical "bodies", a word adopted from Newton's *Principia*. However, Newton's "body" actually means "point mass" (a mass of zero size which does not exist physically)—a point ignored by textbook writers and even some physicists, causing confusion among the students. The reason behind this is that many physicists are not aware that physics is built on approximations [Lam, 2014], a fact that they should be proud of instead of sweeping it under the carpet. If even scientists cannot communicate physical laws accurately and clearly to their own students you can imagine how difficult it would be for scicomm people to deliver them to the public.

2. The essence of Copernicus' heliocentric theory

Our Universe is more like the surface of a balloon instead the balloon itself. In other words, every point in the universe can be considered the center. Thus, we can call our Earth the center of the solar system, meaning that the Earth, like any other point in space, can be used as the vantage point in observing the sky. What Copernicus' heliocentric theory actually says is that the paths of the planets are pretty complicated if the observer is located on Earth, but the planets' orbits will be simpler to be described if the observer sits at the Sun: the planets move in concentric circles [actually ellipses as shown later by Johannes Kepler (1571-1630)].[32] That is, the essence of Copernicus' heliocentric theory is about the *convenient* choice of an observational point. It is not about the "correct" choice, a point missed in scicomm.

---

[32] See, e.g., "Copernicus to Kepler" in Johann Sommerville's "History 351: Seventeenth century Europe" (faculty.history.wisc.edu/sommerville/351/351-182.htm, May 25, 2014).

3. Try to convince someone astrology is wrong

Let us say you try to explain to someone that when two particular stars line up in the sky, they will not have particular effect on a person located on earth. For that, you have to say the two stars' only force on that person is gravitational which, according to Isaac Newton (1642-1727), is very weak and decreases inversely proportional with the square of the distance; thus, the forces can be ignored. This someone may reply: It is not zero, right? You have to admit that he is right. A more sophisticated someone may add: Don't you know the "butterfly effect" that says an extremely small force can cause big effect? You have to admit that no one has ever proved the butterfly effect, from chaos theory, does not apply to this three-body (two starts plus the person on earth) problem. We in science have great faith in Newton's laws after we spent many years learning them and using them, but not this someone. Ultimately, a non-expert— "scientist" and layperson alike—will decide to accept the conclusion on a certain topic not because he understands thoroughly the reasoning behind it (which is always technical and only the experts can ascertain it; that is why they are called experts) but because he trusts this person who is explaining it to him. And trust is not built on one day or through one conversation.

4. Lessons of the cold fusion saga

The year 1989 was eventful. China's Tiananmen "incident" occurred on June 4; California's Loma Prieta earthquake of magnitude 6.9 happened on October 17; Berlin Wall fell on November 9. Before that, on March 23, two chemists gave a press conference in Salt Lake City (SLC), Utah, and announced that they had *discovered* cold fusion, without mentioning their competitor in Provo of the same state. The media and the scientific community around the world went crazy. After a serious of experiments that failed to confirm the claim, the saga ended within one year, resulting in two of the professors resigning from their universities (the third one was already in retirement when the saga began) [Dewdney, 1997]. Different lessons from this saga were drawn by people from various quarters, including those from scicomm. Some say the story illustrates the sociological factors in the doing of science, which is obviously right since the scientific process is a human endeavor. Some say the press conference should not be called. This is wrong because press conferences on scientific breakthroughs have been held before, e.g., parity nonconservation in 1957 announced by Columbia University featuring Wu Chien-Shiung's experiment [Chiang, 2014]. And some say the story confirms that the scientific community can seek out the bad science from good science; this is right, too, as usual.

But two fundamental lessons are missed by these commentators: (1) When the SLC professor(s) applied for funding from Department of Energy to further his research, he was opting for academic glory such as a Nobel Prize which will require him to reveal enough details of his experiments for others to confirm the discovery down the road. Submitting a proposal with referees to review it was his first act of revealing information, and he knew that. But at the press conference, his aim has changed since he refused to give out any information; he was now opting for wealth (apart from his 15 minutes of fame). In modern time, getting a Nobel Prize and getting rich could not be had by the same person at the same time because the former requires openness while the latter, secrecy (since commercial applications will be involved). Changing his aim in mid-course was fatal as the SLC professor was concerned. (2) At the press conference, he knowingly over claimed (and misled the press by not mentioning the Provo competitor). Without detecting neutrons, what he could claim most is that there is a 99.9 % chance he had discovered cold fusion. He could remain a viable scientist in the face of a 0.1% error if he turned out to be wrong, which was indeed the case. Instead, he claimed 100% and there was no way out. In short, the cold fusion saga demonstrates bad human choices apart from bad science (sloppy science in this case). Now how could a scicomm worker "educate" the public properly if the lessons of the story are completely misrepresented?

5. Interpretation of the Sokal hoax

The fact that the Sokal hoax [Sokal & Bricmont, 1998], amid the "science wars" of the 1990s, stirred up such a strong reaction is due to people's misinterpretation of the event itself. There is no guarantee that the editor would reject Sokal's manuscript if he added in a physics referee, because in the face of two opposing referee reports, assuming that was the case, the editor could exercise his judgment and power and choose to publish the paper. In fact, even prestigious science journals such as *Science* and *Nature* failed to filter out papers with fraudulent claims, as demonstrated in the Bell Labs incident [Agin, 2006]. What Sokal did is to show that he can punch a hole in this postmodern journal, but that is the case with every other journal. No big deal! Once again, in scicomm, as in any other matter, getting the facts right is not enough; getting the interpretation right is equally or more important.

6. Climate change: decision making

Climate change is about a complex system, the long-term weather, which comes with inevitable uncertainties [Ken, 2001] and decision making. It is perhaps the most complex and interesting issue facing the public and policymakers worldwide [Somerville & Hassol, 2011]. The debate centers on how reliable computer modeling of the climate is and, if the doom prediction of global warming is correct, what decisions we have to make to prevent it from happening or getting worse.[33] *This is the first time in human history that we are asked to make big decisions based on the computer prediction of a* messy *system.* The system addressed in climate change is mostly physical—the circulation of Earth's atmosphere and ocean—but includes uncertain chemical and biological factors. The computer simulations involve solving a huge number of coupled differential equations with probabilistic considerations. It belongs to the class of what we call "very messy physics".[34]

---

[33] According to "Climate change impacts in the United States", the government's National Climate Assessment released in May 2014, global warming is already here (nca2014, globalchange.gov, May 28, 2014).

[34] Classical physics, relativity theories and quantum physics are "neat physics" which are well tested. We have worked with them for over hundred(s) of years and have confidence on them. The equations involved are deterministic (even though the predictions in quantum mechanics are probabilistic). In contrast, weather forecasting is relatively new and is "messy physics". The weather equations are well known but chaotic, and can only be solved by powerful computers. Because of the "butterfly effect", weather forecasting is accurate up to two or three days only. In comparison, the coupled equations in climate change are much larger in number and are less certain; it is "very messy" in content and in practice. Arguing that ten computer models converge to the same result is not good enough. A computer model is as good as what one puts into the model; garbage in, garbage out (see Freeman Dyson's interview in [Lemonick, 2009]). The climate-change

Decision making, also a complex system, is a branch of the humanities [Bird & Ladyman, 2013] which is part of (human) science. Here, the public has to choose who to trust and the policymakers have to make risky decisions knowing the uncertainties, with potentially huge consequences. The good news is that if we act now to reduce carbon dioxide emission but the computer prediction turns out to be wrong,[35] we still have healthier air to breathe. If we do nothing and the prediction turns out to be true, there will be a lot of human hardships. Either way, there will be hardship but the human race will survive. Of course, there is the third scenario that we do nothing and the computer's prediction is wrong; we will save ourselves a lot of trouble. In climate change, we are asked to *bet* on a messy complex system with probabilistic predictions. It is unlike the making of an atomic bomb, a simple system where $E = mc^2$ is an exact result from well established, deterministic equations.

Unfortunately, there is not much scicomm can do here. Not too many scicomm people are equipped to deal with such a situation that involves *two* complex systems—climate change and decision making. Basically, it is because scicomm's focus in the past is on simple systems, and most scicomm workers are not educated in the science of complex system. But is it really nothing scicomm can do? See Section 2.7.2.

2.7 What Happened and What to Do

Here is a historical, cultural and social outline (or first-order approximation) of what happened since the Big Bang, with emphasis on

---

predictors owe the public the explanation of why their system is not chaotic, why they can predict for 20 or 100 years while the much simpler weather forecasting is good for only three days, and how do they ascertain the unimportance of the many factors left out in their models.

[35] There is such a possibility. No matter how good a computer model is the calculations are always done approximately since the numbers in a computer are expressed with a finite number of digits. Or, probability is involved in the model building. In either case, predictions can only be given in terms of probabilities. Contrary to many people's understanding, an event with a small probability, no matter how small, could still happen [Hand, 2014]. Therefore, if global warming is predicted to be 99% certain, there is still a 1% chance that it will not happen if nothing is done. And if that 1% prediction is actually realized people would say the prediction is "wrong". That, of course, is not true; people just misunderstand probabilities.

the last 100 years. It is followed, accordingly, by recommendations for PS, HS, SS and Scicomm in the near future.

2.7.1 *The Beginning*

It all began with the Big Bang, or earlier, about 13.7 billion years ago (bya). At 10^{-5} seconds later, "long" after the cosmic inflation (which happened about 10^{-37} seconds after the bang), protons and neutrons appeared. They combined to form nucleus at 0.01-300 seconds after the bang, with *atoms* formed 380,000 years later. Stars appeared 13.4 bya; our solar system, of which Earth is a member, 4.7 bya; life on earth, 3.7 bya [Turner, 2009]. The important point is that *all living and nonliving material systems on earth are made up of atoms.*

While the origin of life on earth is an unsolved problem, we do know from Darwin's evolutionary theory (1859) that humans were evolved from simpler lives. However, we are not descendants of the monkey, but of fish.[36] In fact, six million years ago, the chimp and human lineages split. We, *Homo sapiens*, appeared 195,000 years ago in Africa, moved out from there 60,000 years ago and spread all over the world [Lam, 2011]. That is, all living human beings are relatives from the same family tree. We share the same genes, more or less.

A turning point appeared 10,000 years ago while villages were created and agriculture began. Written history is known for only 8,000 years or so. Long before that, with plenty of free time available, our ancestors began wondering what happened around them—everything, down on earth and up in the sky, including, very likely, the crucial question of why we are here. Without much objective information, they came up with a reasonable answer for every question they asked: God did it. That was perfectly alright except that it was not very enlightening.

2.7.2 *Ancient Time*

Then about 2,600 years ago, Thales and others like him showed that they could do better by reasoning. And that was the beginning of Philosophy.

---

[36] See PBS's Nova program, "Your inner fish".

With slaves doing all the mundane works in ancient Greece, the citizens not just were financially secure once they were born but were able to live a lifestyle with plenty of leisure. They exercised their bodies and gave us the Olympic Games. But others—like Socrates and Plato but especially Aristotle—did more and gave us all kinds of knowledge about Nature, including the human system (ethics, political science, logic, etc.) and nonhuman systems (physics, biology, etc.); i.e., they studied both complex and simple systems. As pioneers, they were really ambitious and fearless. They asked all sorts of basic question such as what is truth, reality, etc. and demanded *absolute* knowledge. Soon after, Archimedes made the breakthrough in a simple system by discovering the principle on buoyancy force [Lam, 2008a]. The works by Thales, Aristotle and Archimedes on inanimate systems are later recognized as part of "natural science".

Meanwhile, at about the same time as Plato and Aristotle, in the Warring States period in ancient China, Confucius and Lao-tse chose to work on the complex system of humans, essentially on ethics and political science at the empirical level. Since then the Chinese paid dearly for this "unlucky" choice of ignoring the simple systems; they are still playing catch up in science. The reason is that modern science since Galileo got its breakthrough by focusing on simple (inanimate) systems, while complex systems like human matters are so difficult that it has not advanced much since Aristotle. In other words, modern science did not arise in China because the Chinese picked the wrong topic in their research [Lam, 2008a].[37, 38]

---

[37] The other ingredient essential to the development of science in the West is "debating with each other" as advocated by Socrates [Lyold, 1970]. Debating means "getting to the bottom of things" or "finding out what happens". Unfortunately, debate is not encouraged by Confucianism. Ironically, more than 2,000 years ago, the Chinese already realized, correctly, that "heaven and man are oneness" [Lin, 2010]. What it means is that "heaven" (i.e., Nature) and humans can influence each other and are organized similarly. The former is trivial now and the latter is borne out by science in the last few decades; i.e., there exist three "universal" organizing principles that cut through living and nonliving systems—chaos, fractals and active walks [Lam, 2008a]. By this "oneness" doctrine, the Chinese could equally choose to study "heaven" (such as buoyancy or ants) in order to understand humans and they would invent "science" but they did not. The above is our answer to the so-called Needham Question [Lam, 2008a].

2.7.3 Modern Time

Knowledge acquired by the ancient Greeks was preserved by the Arabs and passed to the Europeans in Renaissance time, forming the basis of Western civilization and, in fact, a major part of all civilizations today. In particular, Aristotle's *theoretical* cosmic system was enforced by the Christians until it was overthrown by Galileo's *experimental* findings. But it was the publication of Newton's *Principia* in 1687 that was the dividing line that ushered in the rapid development of the "natural sciences" which are about (mostly simple) nonhuman systems and the gradual decline of the humanities which are about the (complex) human system, as we have witnessed in the last 300 years. Why was that? How did this happen?

It all has to do with a misunderstanding of what Newtonian mechanics is about and the misconception of the human system. The former is less so today but the latter persists to a large extent. Largely encouraged by Newton's success, the Enlightenment started the next year and lasted 101 years, the goal of which was to make a "Science of Man", presumably in the fashion of Newtonian mechanics. The intention was good but the timing was wrong. It was wrong on two accounts [Lam, 2014]: (1) Newtonian mechanics is about deterministic systems while the human system is probabilistic. The "clockwork" worldview derived from the former does not apply to the latter. (2) The tool of probability required to handle the human system was not yet ready [Lestienne, 1998]. Fortunately, everything about humans is an emerging property and can be handled without knowing the mechanisms involved; they can be studied at the empirical or phenomenological level, say. As it turned out, Adam Smith (1723-1790) succeeded in creating Economics (1776), the first discipline in social science, by staying away from Newton. On the other hand, after the Enlightenment, Auguste Comte created Scociology (1844), which he initially called "Social Physics", along the line of Newton—a mistake corrected later by others. Amazingly, as in the case of psychology, a totally new discipline can be created with

---

[38] Similar experience occurs all the time in physics research. If you pick a difficult research topic you may get nowhere for a long, long time, if not forever.

everything wrong in the beginning; what is needed is a good concept, persistence and enough number of followers who make it right later.

While all this was going on in the social sciences, not much progress had been made in the humanities. The reason is that a single human being is much more complicated than a large number of human beings, in the sense that many approximations that could be made in the latter could not be made in the former (see [Lam, 2014] for examples). And as breakthroughs in the "natural sciences" appeared one after the other, the humanities' scientificity remained kind of flat. It appeared that the humanities had been relegated to the backyard. In the midst of this, in 1859, Charles Darwin published *On the Origin of Species*—a publication as relevant to the humanities as Newton's *Principia* is irrelevant, and as controversial as the latter was uncontroversial when it first came out. Out of the many implications coming out of Darwin's evolutionary theory, the one most important to the humanities has been overlooked by almost everybody; i.e., since humans are a kind of animal evolved from the fish and beyond, then *all studies on humans (like in the case of bees)—the humanities in particular, are part of science.*

2.7.4 Last Century

The first three decades of the 20th century was the peak years in physics. Einstein's theories of special and general relativity (1905 and 1915, respectively) and quantum mechanics (1925/1926) appeared, which, however, are irrelevant to the humanities [Lam, 2002]. What is relevant is Einstein's proof (indirectly via Brownian motion) that atoms actually exist, published in 1905. The reason is that since humans are made up of atoms which do exist, then, like any other material system in Nature, all human-related matters are part of science.

The Vienna Circle, originated in Vienna (1922) and inspired by Mach, tried to pick up where the Enlightenment failed. They tried to constrict science into a jar (see Section 2.3.1) and failed, too, done in by Einstein and Kurt Gödel [Byers, 2011]. Out of Vienna came Popper and Feyerabend, teacher and student. The mistakes made by these two humanists are opposite to each other, at two extremes. Popper, by restricting scientific theories to those that can be falsified, kept science's door open by a slit. Feyerabend, favoring no restrictions, flinged the door wide open. Moreover, Feyerabend, by saying science is not strictly rational without mentioning the RC, opened the door for misunderstanding and misinterpretation about science. What he should say carefully is that the "scientific *process*" may not be "rational" all the time,[39] but the "scientific *results*" which pass the RC are rational results". Well, maybe he was not aware of the difference or maybe he did not care. Telling people that science is not strictly rational would get him more public attention in the 1960s era of antiwar and antiestablishment, especially in Berkeley where he was teaching [Gitlin, 1993]. And he loved attention and theatrical presentations; remember that he was interested in theater and singing as a teenager?

Kuhn's case was different. With his PhD physics training that emphasizes clarity, it would not be too difficult for him to come out and say something like "Newton's gravitational law or its equivalent could be discovered by smart aliens, too", meaning that the law has validity beyond human constructs.[40] Why he did not do so? My *conjecture* is that Kuhn enjoyed so much his sudden fame coming with his 1962 book of ambiguity that he decided to prolong it by keeping his positions ambiguous.[41] It is like someone "hiding in the corset", with a secret to keep. And that explains why he ended up feeling frustrated, being misunderstood by others—a condition he could remove easily but "couldn't".

An academic book could become a best seller if (1) it is not too technical (no equations, say), (2) not too thick, (3) happens to satisfy the

---

[39] If "rational" means induction and deduction only then science is definitely not always rational since good science involves intuition or (educated) guessing [Lam, 2004].

[40] And it would not be too difficult for him to realize, after 1962, say, that the difference between Aristotle's mechanics and Newton's mechanics is that the latter passes the RC (within its applicability domain) while the former does not. But if so and if he admitted it openly, he would have to give up his incommensurability that helped to make him famous. Was Kuhn ever aware of the RC difference? This will be an interesting project for historians of science.

[41] Feyerabend, Kuhn's Berkeley colleague, has written: "I venture to guess that the ambiguity is *intended* and that Kuhn wants to fully exploit its propagandistic potentialities" (my italics) [Hoyningen-Huene, 2000, pp. 109-110]. A detailed and thorough biography of Kuhn could help to settle this conjecture.

need of a large number of readers, and it helps if (4) the book is full of ambiguities.[42] Kuhn's 1962 book could be the only one that satisfies all these four conditions. Before we talk about who were among the readers and what are their needs let us recall the two "revolt against (natural) science" in the 20th century America [Kevles, 1987]. The first revolt occurred in the 1930s after the great depression of 1929. The complains were twofold: (1) "Natural science", technology in fact, was blamed for dehumanizing the society—machines displacing workers, etc., and (2) the simple-minded opinions offered by some "natural scientists" on social matters and the over respect paid to their opinions by the press. The second revolt happened in the 1960s, the anti-Vietnam War and student movement years [Gitlin, 1993; Farber, 1994], for similar reasons except that this time it was also against the over-blown influence of the physicists who helped to end WWII with atomic bombs. On top of that, postmodernism had been raging in the 1950s, going strong in the 1960s and spreading from France to American campuses, Berkeley included.[43] In short, the 1960s is the agitative era of antiwar, antiauthority and antiestablishment. Here, a lot of misunderstandings are involved, on the part of the "natural scientists" and also the critics:

1. The training of a "natural scientist" is very narrow. One who excels in superconductors has not much useful to say about black holes, say, not to mention climate change or human matters. Moreover, contrary to the claim by the physics Nobelist Robert Millikan (1868-1953) [Kevles, 1987, p. 183] and others, the critical thinking these

---

[42] That is why no physics book ever became a best seller. Stephen Hawking's *A Brief History of Time* (1988) is a different story. It is a popsci book, found incomprehensible by most people; many bought it out of sympathy for the author. The book is incomprehensible not because the author tried to write ambiguously but because he was a novice popsci writer. The publisher should bear the blame because as professionals, they knew that they should find Hawking a coauthor but they did not. Of course, it is easier to come up with a best seller by avoiding scientific details. Examples are Eric Segal's *Love Story* (1970) and Robert Waller's *The Bridges of Madison County* (1992), which are about human relationships, based on the science of love at the observational/empirical level; no dopamine mentioned. (See [Fisher, 2004].)

[43] Postmodernism is a late-20th century movement in the humanities. It is anticonvention and antiauthority by playing on words and "nonwords" in the name of semantics. See [Sokal & Bricmont, 1998] and en.wikipedia.org/wiki/Postmodernism (May 1, 2014).

"scientists" gained through their experiences in (mathematical and physical) simple systems helps but does not automatically make them experts in dealing with human matters. Human problems are complex systems: Each problem is nonlinear and has multiple solutions; each outcome involves multiple (correlated) factors and cannot be precisely predicted. More importantly, every human problem is history dependent while the only history-dependent system most physicists have learned in their trade is the magnet, nothing comparable to a human being in terms of complexity.

2. Many "natural scientists" are not even aware of the basic differences between the (deterministic) simple systems they are familiar with and the (probabilistic) complex systems to which human matters belong, and cross the line inadvertently when making public comments on the latter.[44] Some comments bordered on scientism and enraged the public.

3. The over attention paid by the press to the "natural science" stars, after the triumphs of physics in the early part of the 20th century and again after its demonstrated contributions in ending WWII, was not the fault of the "scientists". It had to do with the nature of the press, reflecting the public's keen interest on stars of any kind. They, the press and the public, also misunderstood the differences between simple and complex systems. Fortunately, things have improved. They no longer asked a "science" Nobelist for his opinion on fashion

---

[44] Yet some *very* good "natural scientists" were aware of the problem. For instance, Einstein opinioned a lot about matters beyond physics but he was extremely careful in his uttering [Einstein, 1982]. When he turned down the offer to be nominated for Israel's presidency, upon the death of Chaim Weizmann (1874-1952), the first president and a chemist, his reason was that since he had devoted his life to objective matters, he lacked "both the natural aptitude and the experience to deal properly with people and to exercise official functions" [Burko, 2013]. Enrico Fermi (1901-1954), another physics Nobelist, after the completion of the Manhattan Project, turned down an offer to serve on a government committee on the ground that human matters involve multiple solutions while his training and experience equipped him to deal only with problems with unique solutions like those in physics [Fermi, 1954]. The same cannot be said about the other two Nobelists, Robert Millikan and Niels Bohr (185-1962), and the lesser ones (see [Kevles, 1987]).

trends, say, nor did they ask Lady Gaga for her opinion on gravitational waves.

4. The adverse effects of "natural science" came from human decisions on when and how to use the scientific results, if at all. And *human decisions are in the domain of the humanities*. Even though advances from the nonhuman sciences may help, wiser decisions are made by humans and thus could only come from improving the humanities. So *the problem lies in the underdevelopment of the humanities*. The "revolt against science" actually is a "revolt against humanities" prompted by the aggressive growth of physical science.

And so there was a huge number of potential readers in waiting before Kuhn's book came out in 1962. Kuhn, an insider in the eyes of the public but actually a marginer, with the more than proper credentials of a physics PhD from Harvard and professorship from Berkeley,[45] confirmed to them what they suspected all along: That scientists are just humans, working sometimes with selfish motivations, unlike Superman who lives to save the world; the scientific process is like other human endeavors that could be influenced by social forces. All these are in fact trivially true. They now came out with a deeper research, just like a second-order approximation shows more details than a first-order approximation of anything. The book's true impact outside of PS lies not on whether "paradigm shift" or "incommensurability" are correct or not because if wrong, that will affect only one person, the author's academic reputation. The impact lies on what the book hints at but refuses to make clear,[46] that some scientific results (like the gravitational law) that passed the RC do have objective validity and are not human constructs.

---

[45] Kuhn's motivation in finishing his PhD in physics while already determined to shift his career plan to HS was that he thought, correctly, the degree will enhance his credentials as a historian in science. Besides, he took some courses in philosophy and found that the gap was too large for him to cross if he wanted a degree in Philosophy, and so a physics degree was his only option since there was no degree in HS yet [Hufbauer, 2012].

[46] The game is played somewhat like this. The author writes that when someone's worldview changes, the world changes. The reader asks: When you say "the world changes" do you mean it in the literary sense or do you mean the physical world changes? And the author refuses to give a straight answer; instead, he replies: It depends on what "physical" means. Etc,, etc.

The sociologists of science picked up the hint and elaborated on the human aspects by doing lab visits, etc. Opposite to the anthropologist Margaret Mead's working style of trusting too much what the natives told her [Hellman, 1998], the sociologists did the opposite by not trusting anything the "scientists" told them. This is not good since the scientific process is history dependent. Why a scientist makes a certain judgment depends heavily on what he picked up all the years before doing that particular experiment. And this information can be provided by the scientist if properly asked; otherwise, it has to be dug out through tedious research of the scientist's past. Working without accumulating enough number of case studies (directly by observation or indirectly from HS) and rushing to draw conclusions from insufficient data characterize the research works of many of these sociologists. And it appears that many were not aware of the intrinsic limitation of sociology's fieldworks and fell into the generalization trap (see Section 2.5.3). As outsiders, some of these researchers had no idea how science actually works, and insisted that a scientific proof to be as rigorous as a mathematical proof. Such a demand bears the footprint of the absolutism of knowledge dating back to Plato which, in fact, was found to be unsupportable and unnecessary by modern science which is anchored in observations and experiments [Lam, 2014; Shapiro, 1983].

Furthermore, some sociologists (see, e.g., [Collins, 2014]) and others cling to the outdated misconception that social science which studies the collective behavior of humans, a kind of animal, is fundamentally different from "natural science" which they implicitly and mistakenly take it to mean the study of only inanimate systems. In fact, (1) for example, the bees or chimps studied in "natural science" has an "actor's categories" which are distinct from the scientist's "technical categories" but that does not prevent humans from understanding them, and, similarly, for humans to understand other human groups. (2) Common quantitative laws (e.g., power laws) are found in both inanimate systems and human history [Lam, 2008c] which are independent of any researcher's technical categories. (3) The same modeling and analyzing methodologies are shared successfully between "natural science" and social science [Lam, 2014]. Therefore, these sociologists are wrong.

Sarton had the chance of getting the contents of HS right since HS as a discipline was established by him after 1905, the year the existence of atom was proved by Einstein. But he failed to pick up the message implied by atoms' existence and, before that, Darwin's evolutionary theory, that social science and the humanities are also part of science. Consequently, he led HS into the narrow alley of "natural science" (and medical science) as we see it today. Moreover, Sarton's characterization of "scientists" is so naïve and wrong [Sarton, 1962]. It seems that he was not aware of Newton's giving up physics soon after he assumed the Lucasian professorship at Cambridge and his bitter dispute with Leibniz. Of course, this could be far from the case. We have no idea what happened on this account, a possible project for historians of science.

Scicomm is not in the business of finding out what science is or was. It depends on the other disciplines, PS, HS and SS, to provide the answer. And so when those disciplines fall short, Scicomm became their victim. Yet there are a few mistakes that are Scicomm's own doing. One is the narrow definition of "scientific literacy" which may not be definable or useful [Bauer, 1994; Broks, 2014], even though it is a convenient tool for scicomm administrators.

In *summary*, the shortcomings of PS, HS and SS are due to three reasons: (1) The content and practice of science change with time. Old writings that pay no attention to the changes are bound to be wrong or soon become obsolete and thus mislead the unknowing readers. (2) Some early aims of PS (e.g., insistence on absolute knowledge) are misguided; SS suffers from intrinsic limitations in their methodology. (3) Overconfidence of and over claims by some SS researchers who are not or insufficiently trained in science are aplenty.

On the other hand, Scicomm's central problem is (1) the fuzzy/incorrect concept of science held by and (2) the mixed background of its practitioners. As a discipline that deals with the public, a good scicomm researcher needs knowledge not just in "natural science" but in human-dependent matters provided by the humanities. Yet, researchers trained in both "natural science" and the humanities are rare.

On top of these difficulties, all four disciplines suffer from the common misperception that science is about "natural science" (and medical science) only, with social science and the humanities excluded. In other words, these disciplines presently are dealing with science in its very narrow, outdated sense. This fundamental misconception hurts not just the development of the disciplines but also humanity, jeopardizing the future of humankind [Lam, 2014].

2.7.5 Near Future

Confucius said something like this: To get things straight, the first step is to get the names straight. Thus, it would help if people are more careful in their use of words. The word science should be reserved to mean what it is: the sum of nonhuman science and human science. It will be misleading if we continue to call nonhuman science by natural science, even thought it is not straightly wrong if quotation marks are included in the latter (though it would be clumsy to pronounce the quotation marks every time you utter the term). But *it is wrong to call nonhuman science by science*, with or without quotation marks (since "science" includes social science, part of human science). Similarly, the humanities and social science (and medical science) should be recognized as components of human science. This kind of clarification should start with the curricula in the educational system, from grade 1 on, but this will take time.

Meanwhile, what we can do is to do what we professionals can do immediately. For researchers in PS, HS and SS, it is time and in fact fruitful to turn their attention to the human sciences, going beyond biology. For example, the history of the humanities from the vantage point of HS would be surely different from those written by the humanists. This change of direction not merely will enlarge the scope of the disciplines but will make them more relevant to the pressing problems faced by our modern societies.

As for scicomm practitioners, time is long overdue for them to shift their focus from nonhuman science to human science. The public need and want information on matters close to their lives, loud and clear. Another thing they can do: Apart from communicating the importance of critical thinking, they could change their emphasis on explaining the mechanisms, the why in science, to the appreciation of science. Start something called "Science Appreciation". Learn from the art people! Finally, it is not too late for scicomm to initiate some crash courses on climate change for themselves and the public, and enable a dialogue between the climate-change scientists and the public.

2.8 Conclusion: An Old but New Frontier

In the last 400 years or so since Galileo, the study of nonhuman systems under the name of "natural science" or modern science did enlighten deeply our understanding of Nature (e.g., Big Bang), make our living easier (cell phone) and help to prolong our lives (for good or bad). But that is not enough as the future of humanity is concerned, as the so-called "revolt against science" tried very hard to remind everybody. *It is the humanities that determine our quality of life* (e.g., to pollute or not to pollute) *and bring us genuine happiness* (human relationships, arts). While the study in "natural science" should be continued, it is time for us to return to the Aristotle tradition of treating the human system and nonhuman systems as equally important in our search for knowledge. This tradition was interrupted by the phenomenal success of modern science.

Deepening humanities' research and taking it to the next level do not require large increase of the research budget. No smashing machines needed to be built. *What is needed is a change of our concept of science and our perception of priority.* (See scimat in [Lam, 2014].) For example, for the four science-related humanities disciplines covered in this chapter, shifting the focus from simple systems to complex systems, from nonhuman systems to the human system, can be started immediately.

The humanities were the frontier for the ancient Greeks and Chinese but are the new frontier for the rest of us. Go complex! Go humanities!

References

Agin, D. [2006] *Junk Science: An Overdue Indictment of Government, Industry, and Faith Groups that Twist Science for Their Own Gain* (Thomas Dunne Books, New York).

Bauer, H. H. [1994] *Scientific Literacy and the Myth of the Scientific Method* (University of Illinois Press, Urbana).

Bederson, B. (ed.) [1999] *More Things in Heaven and Earth: A Celebration of Physics in the Millennium* (American Physics Society & Springer, College Park & New York)
Ben-Ari, M. [2005] *Just a Theory: Exploring the Nature of Science* (Prometheus Books, Amherst, NY).
Berg, R. [2012] "Beyond the laws of nature," *Philosophy Now*, Issue 88, January/February, 24-26.
Bird, A. & Ladyman, J. [2013] *Arguing about Science* (Routledge, New York).
Bloor, D. [(1976) 1991] *Knowledge and Social Imagery* (University of Chicago Press, Chicago).
Brakel, J. van [2000] *Philosophy of Chemistry: Between the Manifest and the Scientific Image* (Leuven University Press, Leuven, Belgium).
Broks, P. [2006] *Understanding Popular Science* (Open University Press, New York).
Broks, P. [2014] "Science Communication: A history and review," in [Burguete & Lam, 2014].
Burguete, M. & Lam, L. (eds.) [2008] *Science Matters: Humanities as Complex Systems* (World Scientific, Singapore).
Burguete, M. & Lam, L. (eds.) [2014] *All About science: Philosophy, History, Sociology & Communication* (World Scientific, Singapore).
Burko, L. [2013] "Einstein and the presidency of Israel," *APS News*, January, 4.
Byers, W. [2011] *The Blind Spot: Science and the Crisis of Uncertainty* (Princeton University Press, Princeton).
Chiang Tsai-Chien [2002] *Biography of Yang Chen-Ning: The Beauty of Gauge and Symmetry* (Bookzone, Taibei).
Chiang Tsai-Chien [2014] *Madame Wu Chien-Shiung: The First Lady of Physics Research*, trans. Wong Tang-Fong (World Scientific, Singapore).
Chen Mei-Ting, Chen Lu-Yao & Chong Lu [2014] "The bias of disseminating content and audience attention about popular science on network: Taking Science Squirrels Club's microblog, blog, website for example," *Study on Science Popularization* **9**(1), 39-45.
Chodos, A. [2014] "Polarization measurement detects primordial gravitational waves," *Physics Today*, May, 11-12.
Collins, H. [1992] *Changing Order: Replication and Induction in Scientific Practice* (University of Chicago Press, Chicago).
Collins, H. [2014] "Three waves of science studies," in [Burguete & Lam, 2014].
Collins, H. & Pinch, T. [1998] *The Golem: What You Should Know about Science* (Cambridge University Press, Cambridge).
Cunningham, A. & Williams, P. [1993] "De-centring the 'big picture': The *Origins of Modern Science* and the modern origins of science," *The British Journal for the History of Science* **26**, 407-432.

Dewdney, A. K. [1997] *Yes, We Have No Neutrons: An Eye-Opening Tour through the Twists and Turns of Bad Science* (Wiley, New York).
Duhem, P. [1906 (1954)] *The Aim and Structure of Physical Theory* (Princeton University Press, Princeton).
Einstein, A. [1982] *Ideas and Opinions* (Three Rivers Press, New York).
Farber, D. (ed.) [1994] *The Sixties: From Memory to History* (University of North Carolina Press, Chapel Hill).
Fermi, L. [1954] *Atoms in the Family: My Life with Enrico Fermi* (University of Chicago Press, Chicago).
Ferris, T. [2010] *The Science of Liberty: Democracy, Reason, and the Laws of Nature* (Harper Perennial, New York).
Feyerabend, P. [1995] *Killing Time* (University of Chicago Press, Chicago).
Feyerabend, P. [2010] *Against Method*, 4th ed. (Verso, New York).
Fisher, H. [2004] *Why We Love: The Nature and Chemistry of Romantic Love* (Owl Books, New York).
Garfield, E. [1985] "The life and career of Geroge Sarton: The Father of the History of Science," *Journal of the History of Behavioral Sciences* **21**, 107-117.
Gitlin, T. [1993] *The Sixties: Years of Hope, Days of Rage* (Bantam Books, New York).
Godfrey-Smith, P. [2003] *Theory and Reality: An Introduction to the Philosophy of Science* (University of Chicago Press, Chicago).
Greenemeier, L. [2013] "Lights, camera, atoms: IBM creates the world's tiniest movie," *Scientific American*, July, 26.
Gregory, J. & Miller, S. [1998] *Science in Public: Communication, Culture, and Credibility* (Perseus, Cambridge, MA).
Hand, D. J. [2014] *The Improbability Principle: Why Coincidence, Miracles, and Rare Evens Happen Every Day* (Scientific American/Farrar, Strass and Giroux, New York).
Hellman, H. [1998] *Great Feuds in Science: Ten of the Liveliest Disputes Ever* (Wiley, New York).
Henry, J. [1997] *The Scientific Revolution and the Origins of Modern Science* (St. Martin's Press, New York).
Holton, G. [1993] *Science and Anti-Science* (Havard University Press, Cambridge, MA).
Hoyningen-Huene, P. [2002] "Paul Feyerabend and Thomas Kuhn," in [Preston et al, 2000].
Hufbauer, K. [2012] "From student of physics to historian of science: T. S. Kuhn's education and early career," *Physics in Perspective* **14**, 421-470.
Johnstone, J. [1914] *The Philosophy of Biology* (Cambridge University Press, Cambridge).
Ken, R A. [2001] "Global warming: Rising global temperature, rising uncertainty," *Science* **292**, 192-194.

Kennedy, R. E. [2012] *A Student's Guide to Einstein's Major Papers* (Oxford University Press, Oxford).
Kennefick, D. [2005] "Einstein versus the *Physical Review*," *Physics Today*, Sept., 43-48.
Kennefick, D. [2009] "Testing relativity from the 1919 eclipse—a question of bias," *Physics Today*, Mar., 37-42.
Kevles, D. J. [1987] *The Physicists: The History of a Scientific Community in a Modern America* (Harvard University Press, Cambridge, MA).
Kragh, H. [1987] *An Introduction to the Historiography of Science* (Cambridge University Press, Cambridge).
Kuhn, T. S. [1962] *The Structure of Scientific Revolution* (University of Chicago Press, Chicago).
Kuhn, T. S. [1984] "Professionalization recollected in tranquility," *Isis* **75**(1), 29-32.
Lam, L. [1998] *Nonlinear Physics for Beginners: Fractals, Chaos, Solitons, Pattern Formation, Cellular Automata and Complex Systems* (World Scientific, Singapore).
Lam, L. [2002] "Histophysics: A new discipline," *Modern Physics Letters B* **16**, 1163-1176.
Lam, L. [2004] *This Pale Blue Dot: Science, History, God* (Tamkang University Press, Tamsui).
Lam, L. [2006] "Active walks: The first twelve years (Part II)," *International Journal of Bifurcation and Chaos* **16**, 239-268.
Lam, L. [2008a] "Science Matters: A unified perspective," in [Burguete & Lam, 2008] pp. 1-38.
Lam, L. [2008b] "SciComm, PopSci and The Real World," in [Burguete & Lam, 2008] pp. 89-118.
Lam, L. [2008c] "Human history: A Science Matter," in [Burguete & Lam, 2008] pp. 234-254.
Lam, L. [2011] "Arts: A Science Matter," in *Arts: A Science Matter*, eds. Burguete, M. & Lam, L. (World Scientific, Singapore) pp. 1-32.
Lam, L. [2014] "About science 1: Basics—knowledge, Nature, science and scimat," in [Burguete & Lam, 2014].
Latour, B. & Woolgar, S. [1986] *Laboratory Life: The Construction of Scientific Facts* (Princeton University Press, Princeton).
Lemonick, M. D. [2009] "Freeman Dyson takes on the climate establishment," *Yale environment 360* (e360.yale.edu/feature/, May 12, 2014).
Lestienne, R. [1998] *The Creative Power of Chance*, trans. Neher, E. C. (University of Illinois Press, Urbana).
Li Da-Guang [2008] "Evolution of the concept of science communication in China," in [Burguete & Lam, 2008] pp. 165-176.

Lin Ke-Ji [2010] *Oneness of Heaven and Man and the Subject-Object Dichotomy: An Important Vantage Point in Comparing Chinese and Western Philosophies* (Social Sciences Academic Press, Beijing).

Liu Bing & Zhang Mei-Fang [2014] "Scientific culture in contemporary China," in [Burguete & Lam, 2014].

Liu Dun [2008] "History of science in globalizing time," in [Burguete & Lam, 2008] pp. 177-190.

Livio, M. [2011] "Why math works," *Scientific American*, August, 80-83.

Lloyd, G. E. R. [1970] *Early Greek Science: Thales to Aristotle* (Norton, New York).

Lucibella, M. [2014] "Gaps widen in attitudes toward science," *APS News*, Mar., p. 3. (The source of this article is the biennial *Science and Engineering Indicators* report issued by the National Science Foundation, which brings together numerous surveys.)

Mason, S. F. [1962] *A History of the Sciences* (Macmillan, New York).

Mermin, N. D. [1996] "The golemization of relativity," *Physics Today*, April, 11-13.

Mirsky, S. [2012] "Physics uncowed: You don't have to say cheese to get the picture," *Scientific American*, Jan., 86.

Morris, R. [2002] *The Big Questions: Probing the Promise and Limits of Science* (Times Books, New York).

Nickles, T. (ed.) [2003] *Thomas Kuhn* (Cambridge University Press, Cambridge).

Nickles, T. [2013] "The problem of demarcation: History and future," in *Philosophy of Pseudoscience: Reconsidering the Demarcation Problem*, eds. Pigliucci, M. & Boudry, M. (University of Chicago Press, Chicago).

Oldroyd, D. [1986] *The Arch of Knowledge: An Introductory Study of the History of the Philosophy and Methodology of Science* (Methuen, New York).

Pinto de Oliveira, J. C. & Oliveria, A. J. [2013] "Kuhn, Sarton, and the history of science," preprint (phisi-archive.pitt.edu/id/eprint/10078, May 12, 2014).

Popper, K. [1963] *Conjectures and Refutations: The Growth of Scientific Knowledge* (Routledge & Kegan Paul, London).

Preston, J., Munévar, G. & Lamb, D. (eds.) [2000] *The Worst Enemy of Science? Essays in Memory of Paul Feyerabend* (Oxford University Press, Oxford).

Randi, J. [1982] *The Truth About Uri Geller* (Prometheus Books, Buffalo, NY).

Rauschning, H. [1939] *Hilter Speaks: A Series of Political Conversations with Adolf Hitler on His Real Aims* (Thornton Butterworth Ltd., London).

Sanders, N. [2011] "How to succeed at engaging the public's interest in science," *Physics Today*, May.

Sanitt, N. (ed.) [2005] *Motivating Science: Science Communication from a Philosophical, Educational and Cultural Perspective* (The Pantaneto Forum, Luton, UK).
Sarton, G. [1962] *The History of Science and the New Humanism* (Indiana University Presss, Bloomington).
Schiebinger, L. [1999] *Has Feminism Changed Science?* (Harvard University Press, Cambridge, MA).
Schiele, B., Claessens, M. & Shi Shun-Ke [2012] *Science Communication in the World: Practices, Theories and Trends* (Springer, New York).
Shapiro, B. J. [1983] *Probability and Certainty in Seventeenth-Century England: A Study of the Relationships between Natural Science, Religion, History, Law, and Literature* (Princeton University Press, Princeton).
Sharrock, W. & Read, R. [2002] *Kuhn: Philosopher of Scientific Revolution* (Blackwell, Malden, MA).
Shi Shun-Ke & Zhang Hui-Liang [2012] "Policy perspective on science popularization in China," in [Shiele et al, 2012].
Sismondo, S. [2004] *An Introduction to Science and Technology Studies* (Blackwell, Malden, MA).
Sokal, A. & Bricmont, J. [1998] *Fashionable Nonsense: Postmodern Intellectuals' Abuse of Power* (Picador, New York).
Sommerville, R. C. J. & Hassol, S. J. [2011] "Communicating the science of climate change," *Physics Today*, October, 48-53.
Thurs, D. P. [2011] "Scientific methods," in *Wrestling with Nature: From Omens to Science*, eds. Harrison, P., Numbers, R. L. & Shank, M. H. (University of Chicago Press, Chicago).
Turner, M.S. [2009] "The universe," *Scientific American*, Sept., 36-43.
Wang Yu-Hua & Tang Shu-Kun [2014] "A comparative study on microblog science communication between government and civil science popularization organization," *Study on Science Popularization* **9**(1), 32-38.
Weinberg, S. [2001] *Facing Up: Science and Its Cultural Adversaries* (Harvard University Press, Cambridge, MA).
Worrall, J. [2003] "Normal science and dogmatism, paradigms and progress: Kuhn 'versus' Popper and Lakatos," in [Nickles, 2003, pp. 65-100].
Yin Lin [2014] "Popular-science writings in early modern China," in [Burguete & Lam, 2014].
Zhang Yi-Zhong & Ren Fu-Jun [2012] "The development and prospects of legal system construction for science popularization in China since the science and technology popularization law was issued," *Science Popularization* **7**(3), 5-13.
Ziman, J. [2000] *Real Science: What It Is, and What It Means* (Cambridge University Press, Cambridge).
Zimmer, C. [2001] *Evolution: The Triumph of an Idea* (HarperCollins, New York).

Lin Ke-Ji [2010] *Oneness of Heaven and Man and the Subject-Object Dichotomy: An Important Vantage Point in Comparing Chinese and Western Philosophies* (Social Sciences Academic Press, Beijing).
Liu Bing & Zhang Mei-Fang [2014] "Scientific culture in contemporary China," in [Burguete & Lam, 2014].
Liu Dun [2008] "History of science in globalizing time," in [Burguete & Lam, 2008] pp. 177-190.
Livio, M. [2011] "Why math works," *Scientific American*, August, 80-83.
Lloyd, G. E. R. [1970] *Early Greek Science: Thales to Aristotle* (Norton, New York).
Lucibella, M. [2014] "Gaps widen in attitudes toward science," *APS News*, Mar., p. 3. (The source of this article is the biennial *Science and Engineering Indicators* report issued by the National Science Foundation, which brings together numerous surveys.)
Mason, S. F. [1962] *A History of the Sciences* (Macmillan, New York).
Mermin, N. D. [1996] "The golemization of relativity," *Physics Today*, April, 11-13.
Mirsky, S. [2012] "Physics uncowed: You don't have to say cheese to get the picture," *Scientific American*, Jan., 86.
Morris, R. [2002] *The Big Questions: Probing the Promise and Limits of Science* (Times Books, New York).
Nickles, T. (ed.) [2003] *Thomas Kuhn* (Cambridge University Press, Cambrid-ge).
Nickles, T. [2013] "The problem of demarcation: History and future," in *Philosophy of Pseudoscience: Reconsidering the Demarcation Problem*, eds Pigliucci, M & Boudry, M. (University of Chicago Press, Chicago).
Oldroyd, D. [1986] *The Arch of Knowledge: An Introductory Study of the History of the Philosophy and Methodology of Science* (Methuen, New York).
Pinto de Oliveira, J. C. & Oliveria, A. J. [2013] "Kuhn, Sarton, and the history of science," preprint (phisi-archive.pitt.edu/id/eprint/10078, May 12, 2014).
Popper, K. [1963] *Conjectures and Refutations: The Growth of Scientific Knowledge* (Routledge & Kegan Paul, London).
Preston, J., Munévar, G. & Lamb, D. (eds.) [2000] *The Worst Enemy of Science? Essays in Memory of Paul Feyerabend* (Oxford University Press, Oxford).
Randi, J. [1982] *The Truth About Uri Geller* (Prometheus Books, Buffalo, NY).
Rauschning, H. [1939] *Hilter Speaks: A Series of Political Conversations with Adolf Hitler on His Real Aims* (Thornton Butterworth Ltd., London).
Sanders, N. [2011] "How to succeed at engaging the public's interest in science," *Physics Today*, May.

Sanitt, N. (ed.) [2005] *Motivating Science: Science Communication from a Philosophical, Educational and Cultural Perspective* (The Pantaneto Forum, Luton, UK).
Sarton, G. [1962] *The History of Science and the New Humanism* (Indiana University Presss, Bloomington).
Schiebinger, L. [1999] *Has Feminism Changed Science?* (Harvard University Press, Cambridge, MA).
Schiele, B., Claessens, M. & Shi Shun-Ke [2012] *Science Communication in the World: Practices, Theories and Trends* (Springer, New York).
Shapiro, B. J. [1983] *Probability and Certainty in Seventeenth-Century England: A Study of the Relationships between Natural Science, Religion, History, Law, and Literature* (Princeton University Press, Princeton).
Sharrock, W. & Read, R. [2002] *Kuhn: Philosopher of Scientific Revolution* (Blackwell, Malden, MA).
Shi Shun-Ke & Zhang Hui-Liang [2012] "Policy perspective on science popularization in China," in [Shiele et al, 2012].
Sismondo, S. [2004] *An Introduction to Science and Technology Studies* (Blackwell, Malden, MA).
Sokal, A. & Bricmont, J. [1998] *Fashionable Nonsense: Postmodern Intellectuals' Abuse of Power* (Picador, New York).
Sommerville, R. C. J. & Hassol, S. J. [2011] "Communicating the science of climate change," *Physics Today*, October, 48-53.
Thurs, D. P. [2011] "Scientific methods," in *Wrestling with Nature: From Omens to Science*, eds. Harrison, P., Numbers, R. L. & Shank, M. H. (University of Chicago Press, Chicago).
Turner, M.S. [2009] "The universe," *Scientific American*, Sept., 36-43.
Wang Yu-Hua & Tang Shu-Kun [2014] "A comparative study on microblog science communication between government and civil science popularization organization," *Study on Science Popularization* **9**(1), 32-38.
Weinberg, S. [2001] *Facing Up: Science and Its Cultural Adversaries* (Harvard University Press, Cambridge, MA).
Worrall, J. [2003] "Normal science and dogmatism, paradigms and progress: Kuhn 'versus' Popper and Lakatos," in [Nickles, 2003, pp. 65-100].
Yin Lin [2014] "Popular-science writings in early modern China," in [Burguete & Lam, 2014].
Zhang Yi-Zhong & Ren Fu-Jun [2012] "The development and prospects of legal system construction for science popularization in China since the science and technology popularization law was issued," *Science Popularization* **7**(3), 5-13.
Ziman, J. [2000] *Real Science: What It Is, and What It Means* (Cambridge University Press, Cambridge).
Zimmer, C. [2001] *Evolution: The Triumph of an Idea* (HarperCollins, New York).

Part I
Philosophy of Science

3

Towards a Phenomenological Philosophy of Science

Guo-Sheng Wu

Opposite to the objectivity and the readiness-made of scientific thinking, phenomenology in philosophical thinking manifests the intentionality and the constructivity. And the analysis of the constructivity of intentionality is exactly the reflection. The traditional philosophy of science is a "positive thinking." A phenomenological philosophy of science is a reflective philosophy of science which makes it possible to understand science in the broad sense, to investigate the metaphysical foundations of modern science, and to integrate questions concerning foundation of science with the broader mode of life. It also allows us to understand the trend of modern society and the reflection of modernity, and change the problem of demarcation between science and non-science into the problem of investigating the conditions of possibility that enables "non-science" to be science. With the phenomenology approach, philosophy of science transits from methodology and epistemology to ontology.

3.1 Introduction

In the Anglo-American world, philosophy of science is a mature subject belonging to traditional, analytic philosophy. The community of philosophy of science in contemporary China has largely followed the Anglo-American tradition characterized by logical analysis and empirical justification. For such a mature subject, *phenomenology* means a new "turn." By turning towards phenomenological philosophy of science, traditional philosophy of science would broaden its scope in research.

Phenomenology is often linked with the name of Edmund Husserl (1859-1938), but not always and necessarily associated with him.

Phenomenology is diverse [Spiegelberg, 1965]. In addition to Husserl's stress of the analysis of the intentionality of consciousness, finally approaching a transcendental phenomenology concerning transcendental subject, there are Max Scheler's anthropological phenomenology, Martin Heidegger's existential phenomenology and hermeneutic phenomenology, and Maurice Merleau-Ponty's phenomenology of perception. By combining the masters' phenomenological works with individual interests and backgrounds, different scholars could approach phenomenology from various points of views. For example, starting from Husserl could lead to more emphasis on logic; departing from Heidegger, more emphasis on history.

Has phenomenology, held by different phenomenologists, a universal essence (such as the so-called phenomenological method, or eidetic reduction and intentional description of backing to things per se) or just a familiar resemblance? The prevailing view is that all the phenomenological methods are one, but their applications end with various conclusions. This standpoint, however, is more like that of the Husserlian. Scheler and Heidegger, who perhaps would even deny there is a united phenomenological method, only recognize there is a same "attitude," i.e., "an unconditional seeing concerning things per se," although the "seeing" and "see what" could be different.

What is phenomenology? Obviously, different phenomenological approaches would give different answers. For those of us working in philosophy of science and technology and having trained in natural science, the most important issue, compared with scientific thinking and philosophical thinking familiar to us, is: What is the unique characteristic of phenomenology?

Combined with my own learning experience I would like to propose an essential characteristic of phenomenological thinking: opposite to the objectivity and the readiness-made of scientific thinking, phenomenology manifests the intentionality and the constructivity, and the analysis of the constructivity of intentionality is exactly the reflection.

3.2 Phenomenology as Reverse Thinking (Reflection)

For phenomenologists, the analyzed object is not only the thing intuited by the eyes, but also something more associated with it. If your theme is consciousness, then the intentionality of consciousness will bring out intentional act, intentional object, and intentional content all together. If your theme is a thing, then the thing as thing per se depends on its mode of presence, its horizon, and its world; i.e., it depends on those absent things. In the appearance of the thing, the presence and the absence are always in a dynamic correlation; moreover, the "world," in which the thing appears, is pre-given. If your theme is the man's existence, then its "Being-in-the-world" is always the first phenomenon. Because of its "in-the-world," the man has "solicitude" and "concern"; therefore, in the investigation of man it is impossible to treat "man" as a thing that is present-at-hand and isolated. Otherwise, the possibility of the investigation of man will surely be missed. The saying of José Ortega y Gasset (1883-1955), a famous Spanish philosopher, "I am me plus my environment," is the expression of such meaning. In phenomenologists' eyes there are full of potential "mutually affected" fields. Those objects, seemingly indepen-dent, in fact, are in a covert potential field. Only by grasping the potential field (the "pre-givenness") our understanding of the Being of the object (the given) is possible.

Exactly because all the given always carry a "pre-givenness," the mission of phenomenology is to seek the pre-givenness, so as to explore the truth of things or the things themselves. Considering that the approach of phenomenological thinking is to seek the "pre," it is a kind of reverse thinking. It is thus called "reverse *thinking* (reflection)." In any case, eidetic intuition, the archaeology of knowledge, tracing pre-structure and the construction of meaning, or the investigation of "the condition of possibility," are all of such reflection. As a result of the omnipresence of intentional construction the pre-structure (such as background, motivation, context, etc.) is also ubiquitous; therefore the task of reflection is endless. In the reflection, phenomenology, as logical analysis and empirical generalization, can provide a new knowledge. Furthermore, it is clear that, in the strict sense, the new knowledge provided by phenomenology is not "new," but "old." However, such an inherent "old" knowledge always remains covert, which is hard to lighten. Long time ago Plato had recognized that all true knowledge can only be obtained by virtue of "reminiscence." The so-called "reminiscence" is nothing but in some way a knowledge in us potentially; however, we are not aware of it (i.e., forgotten), so that the reminiscence is also the emergency of the pre-givenness from its hidden state. As a reflection, therefore, at the outset, phenomenology is the authentic task of philosophy. The authentic philosophy is phenomenology.

At first, phenomenology needs to break through naturalistic thinking and ready-made thinking, both of which are ideal stereotypes unconsciously held by the community of philosophy of science and technology. Naturalistic thinking and ready-made thinking treat a lot of things as "natural" and put their "Being" aside. By "Epoché," phenomenology strives to break the "natural attitude," to take eidetic reduction of things treated as a matter of course, and to clarify the pre-givenness among them, the structure of meaning, the construction of the world, the affected correlation of the absent, and so on. If our experience constructs a ground plane, natural science is just like a giant tree growing up from such soil. The tree absorbs nutrients from the soil, growing upward further, stretching around and branching out. People often forget that the full-grown tree can be so luxuriant is due to its roots grasping the soil deeply. The process of rooting is concealed but actually is a "pre-given" and is prerequisite. Phenomenology is just like a kind of mission of seeking roots; it is also a process starting from the ground plane of experience, digging into the underground, searching for the process of rooting of the giant tree. Partially because of this, phenomenology has often been compared to an archaeological work.

3.3 Philosophy of Science from "Positive Thinking" to "Reverse Thinking"

The introduction of phenomenology will inevitably bring about a kind of "turning" in traditional philosophy of science, i.e., from "positive thinking" to "reverse thinking."

The traditional philosophy of science is a "positive thinking." The tree analogy above produces a growing model of the tree of natural science; the traditional philosophy of science is a modeling work. The basic feature of the modeling work is "imitation," so that it is a kind of "positive thinking." It basically adopts the ideal mode of natural science, viz., investigating the science itself with scientific methods; let the light of science enlightens itself. The fact that many fathers of traditional philosophy of science are scientists themselves also shows that "positive thinking" is the characteristic of traditional philosophy of science.

The traditional philosophy of science has its own particular tradition; hence the so-called "philosophy of science" is not "philosophical investigation of science" in the general sense, but a philosophical investigation of science according to a research program of specific philosophical school and tradition. In short, traditional philosophy of science started with the scientific program of logical positivism-empiricism: the program of rejecting metaphysics and limiting the work of philosophy of science to analyze scientific concepts, propositions and judgments at the logical and meaning level. On this basis, the work of philosophy of science has been gradually established to reconstruct logically the process of scientific discoveries and the growing process of the scientific knowledge, to determine logical and positive relationships between empirical observation and theoretical statements, so as to distinguish between the science and the non-science and to confirm the primary reason by which the scientific knowledge is to be the only true knowledge. In the entry of philosophy of science in *Encyclopedia Britannica*, Stephen Toulmin writes that "the discussion of philosophy of science is the methodology and epistemology" [1973]. John Losee proposes that philosophy of science should concern itself with the following four questions:

> 1. What characteristics distinguish scientific inquiry from other types of investigation? 2. What procedures should scientists follow in investigating nature? 3. What conditions must be satisfied for a scientific explanation to be correct? 4. What is the cognitive status of scientific laws and principles? [Losee, 1980, p. 2]

In general, in the first half of the 20th century, philosophy of science is basically equivalent to logic and methodology of science, which is also why the International Union of History and Philosophy of Science, a Division of Philosophy of Science, was named "Logic, Methodology and Philosophy of Science." This reflects manifestly the specific meaning of "philosophy of science" before the 1960s: It is a subject closely related to scientific logic and scientific methodology.

Beginning in the 1960s, traditional philosophy of science entered the era of post-empiricism, post-positivism or historicism. Some inherent flaws of positivism were increasingly exposed. The historicism school, represented by Thomas Kuhn, accused the positivism of acting blindly and devoting in rational reconstruction without facing the historical facts in the actual development of science. The traditional scientific logic and the methodological procedure of science have experienced a process of being deconstructed. "Philosophy of science" in the sense of scientific logic and scientific methodology almost become unsustainable. Paul Feyerabend, ironically, calls it "a subject with a great past" [1970]. Philosophy of science was in a dangerous situation with the possibility of being replaced by the history of science.

After the 1980s, a group of sociologists of scientific knowledge who come from sociology, moving on along Kuhn's path, swarmed into a more radical scientific reflection. It was dangerous that scientific "philosophy" would be replaced by scientific "sociology." In the face of the challenges of the history of science and the sociology of scientific knowledge (SSK), philosophy of science constantly adjusted itself. According to the interior terms of the tradition of analytic philosophy and philosophy of language, philosophy of science completed a paradigmatic shift from syntax to semantics. Traditional philosophy of science, dominated by the tradition of analytic philosophy in Anglo-American circles, was defined in both the broad and the narrow senses. In the narrow sense it is referred to as syntactic philosophy of science and in the broad sense as semantic philosophy of science. Except for these two, in their horizon, all others cannot be defined as "philosophy of science."

For philosophy of science, phenomenology has at least two significances. Firstly, as one of intellectual resources, with the SSK, feminism, and post-modernism, etc., it can contribute to the self-rebuilding of semantic philosophy of science. In this regard, hermeneutics, as the descendant and a branch of phenomenology, plays a significant role. People increasingly realize Kuhn's work has a strong hermeneutic style. There is a similarity between the two propositions—the proposition of "theory-ladenness of data" and the hermeneutics proposition, viz., the "necessary prejudice." All in all, in the research style and terminological framework of traditional philosophy of science, phenomenology, especially hermeneutics, can be accepted as a new intellectual resource to promote the solution of this traditional problem.

Secondly, in my opinion, in addition to the traditional philosophy of science of sequent thinking or positive thinking, the philosophy of science with reflection should be established. I once referred to it as the "second philosophy of science." The second philosophy of science is beyond the horizon of traditional philosophy of science in the following five aspects, showing again that it is indeed a new field of study.

1. It enables the understanding of science in the broad sense. "What is this thing called science?" is an exciting problem.[1] It is a very serious and very important issue, especially for the modern Chinese. However, the problem was not positively responded in traditional philosophy of science. Although traditional philosophy of science has also raised such a question and has regarded finding the answer to it as the fundamental task, for traditional philosophy of science, "science" in this issue has distinct referent, viz., modern science. The doubts aroused by this question were strictly limited to the discussion of modern science, as the knowledge in the strictest sense and as the paradigm of knowledge. How does it have the truth and the validity (i.e., it is no doubting that it is true and valid)? In consequence, it calls up a logical "justification"; the philosophy needs to justify for the science. In other words, in the question of "What is science?" the task of traditional philosophy of science is to elucidate the "meaning" of science in possession of distinctive "referent."

For phenomenology: What does "Science" mean? Does modern science have the undoubted truth? Is it the knowledge in the strictest sense? Is logical justification the highest, the most effective and

---

[1] A book on philosophy of science with the same title has been written by Chalmers, an Australian philosopher of science [1978].

legitimate among justifications? All these questions need to be analyzed reflectively. "Science" in Husserlian sense does not firstly refer to modern science at all. On the contrary, for Husserl, modern science is certainly *not* the science in the strictest sense. Furthermore, the crisis of European science firstly has originated from the modern science started by Galileo. For Husserl, "science" is a self-founded knowledge system, which has inherence, certainty, and evidence. "Self-foundation," "inherence," and "evidence" are the signs of science as science. Actually these are the unique characteristics in Western culture, which are also covert "pre-givenness" in modern science. If one does not understand this point, the understanding of the differences between Chinese and Western culture is impossible. Such a question "Why did modern science not born in China?"—a question usually implied in the question of "What is science?"—would not even be able to have a preliminary answer.

2. It enables the investigation of the metaphysical foundations of modern science. Traditional philosophy of science thinks of itself as an essentially closed system. The textbooks of traditional philosophy of science favor to begin with "philosophy of science is not such things"; e.g., it is not the history of science, not the sociology of science, not the psychology of science, etc. For phenomenology, if its focus of study was confined to modern science the harvest would still be pathetic. Alexandre Koyré, a phenomenologist to inherit authentic Husserlian thought and a historian of intellectual history of science, once said,

> At the beginning of my study, I have been inspired by the faith of the unity of human thought, especially in its supreme form. It seems impossible for me to separate the intellectual history of philosophy from the intellectual history of religion to be independent sections. The former has always been permeated by the latter, for the sake of reference, or for rivalry.
>
> The evolution of science idea, as least during that period I focus on, has not been self-contained, but on the contrary it has associated closely with those thoughts beyond science, philosophy, metaphysics and religion. [Koyré, 1973]

In fact, as found by Koyré, research programs regarding the investigation of the metaphysical foundations of modern science as its main characteristic are the ones that had enlightened Kuhn greatly and, soon after, opened the door to historical philosophy of science.

3. It enables the integration of questions concerning foundation of science with the broader mode of life, the trend of modern society and the reflection of modernity. Traditional philosophy of science had consciously given up answering such type of questions as "What does science mean to the life of human beings?" At the beginning of *The Philosophies of Science*, Rome Harry from Oxford writes,

> Most people suppose that philosophers think about very general and very deep questions, at the heart of which is the problem of the relation of Man to the Universe. Philosophers are popularly thought to offer ideas about the general purpose of living, and even the more particular aims one should set oneself in one's ordinary life. Philosophy of science in this sense would be a discussion of the place of the scientific enterprise in the whole pattern of life. It would probably be concerned with providing an ultimate justification for doing science, that is, with whether science was worth doing at all. It might be argued, for instance, that the accumulation of scientific knowledge is destructive of the conditions for living the best possible human life. It might be thought that the effort expended in the pursuit of scientific knowledge might be better employed in the cultivation of artistic sensibility, in the refinement of manners, and in the embellishment of the environment. I am not going to pursue that kind of discussion, though I am very far from thinking that discussions of such general questions have no value. In this book I shall be discussing a great many detailed question which arise in the actual practice of science itself. ...I shall be trying to make clear what principles are assumed in the use of time-honoured methods of acquiring knowledge. We shall find that certain principles are operative in scientific work. It is the aim of this book to make these principles manifest. [Harry, 1989, p. 1]

Obviously, modern science has a decisive influence on modern life. It is incredible that the orthodox philosophy of science lays these influences aside without reviews and evaluations in depth. The analysis of phenomenological reverse thinking can break through the self-limitation in traditional philosophy of science and enable the question of "science" beyond the science so as to associate with the spirit of times and the mode of existence.

4. It changes the problem of demarcation between science and non-science to the problem of investigating the conditions of possibility of enabling "non-science" to be science. The concern of traditional philosophy of science is the problem of demarcation. Even today, after suffering the impact of historicism, this problem is still the one philosophy of science strives to deal with. The motive hidden behind the demarcation problem is "to justify the truth of science." Science tells us the truth but non-science tells us non-truth or fallacy. For anti-pseudo-science warriors in China the demarcation problem is particularly important and cannot be treated carelessly. For phenomenology, both the demarcation of science and the anti-pseudoscience have their own non-scientific motives. It is just these non-scientific motives that dominate respective works. Additionally, non-science as the possible condition is also prerequisite to scientific research. Therefore, our study of the rise of modern science has to consider also the significance of the Christian religion, the meaning of alchemy, and the idea of natural philosophy held by all sorts of mysticisms in Renaissance. In the research of the 20th-century science, we need to consider the significance of scientific worldview and cosmology as ideologies, the meaning of isomorphism between the setting of time-space and operation procedures in the laboratory, and the social operation of various ranks in contemporary societies. Generally speaking, the problem of the conditions of possibility that make "non-science" to be "science" is the distinctive problem of the second philosophy of science. Apparently, the problem is the same as the SSK's, whereas the latter claims clearly that their work is not philosophy but sociology.

5. Philosophy of science transits from methodology and epistemology to ontology. The focus of traditional philosophy of science is basically on epistemology of science and methodology of science while ignoring ontology of science. The so-called ontology of science is not an ontological commitment in the epistemology and methodology of

science—an ontological commitment on nature, the object of natural science. The ontological commitment is still subordinate to the category of epistemology and methodology while appearing as disregard and non-criticalness of the ontological status of science per se. Such a non-criticalness did not begin with philosophy of science, but the latter had inherited it from the modern epistemological tradition. In modern epistemology, natural science and the objective world explored by it are non-reflective and predetermined. The two questions of "Why is science effective?" and "Why is the outside world real?" are already laid out in advance. Both are a priori principles of epistemology. Kant had profoundly revealed this fact. As a perspective of science, natural ontology is not so much to refer to the ontology of science per se but to cancel the ontology.

The ontology of science in question aims to discuss the status of ontological-existential status of science per se. Ontology is a philosophical enquiry concerning Being. However, in modern philosophy, as Heidegger puts it incisively, people have mixed up Being with beings, treating Being per se as a special beings, object, substance to pursue with the hope of building a metaphysical system [1962]. The point is that because Being is concealed in the dark, the existential (ontology) is fallen to the knowledge system of entities. Entities present themselves in different horizons, but their situations of appearance (viz., existential situations) escape from the ontological-existential reflection. The ontology is reduced to the statement of things, but the authentic ontological problem, which is the existential situation that makes a thing to be thing per se, is being ignored. In the philosophical texts in Chinese, the word "ontology" (translated as "real objects") often reminds people of the natural ontology and of an old ontology referring to an entity.

Science always belongs to human beings. It is created and understood by us, so the ontology of science firstly understand science as a kind of existential mode of human beings, which is a certain mode of their Being-in-the-world. According to Heidegger's analysis of the existential structure of Dasein, science as a theoretical attitude concerning the world is not the fundamental and the a priori mode of Being-in-the-world.

As ways in which man behaves, sciences have the manner of Being which this entity—man himself—possesses. This entity we denote by the term 'Dasein'. Scientific research is not only manner of Being which this entity can have, nor is it the one which lies closest. [Heidegger, 1962, p. 32]

The basic state of Dasein is Being-in-the-world, is "Being-in", and "Being-in" is concern (Besorgen). In our dealings of entities within-the-world, the first encountered is something ready-at-hand, viz., some equipment or its manipulability. Only if it is unusable the equipment becomes conspicuous, which is prerequisite to theoretical knowing. Hence, man's "Being-in-the-world" is primarily a pragmatic attitude. As a mode of projection concerning pure presence-at-hand, scientific knowledge is based on direct pragmatic experience and the pre-scientific ordinary life.

Science as a projection is to theorize a thing to an object, to demarcate a specific domain of objects, and to develop a horizon—a framework of meaning in which things become meaningful and understandable. Therefore, it is objectified, abstracted, idealized and formalized on essence. In contrast, the old ontology lacks a reflection of science as projection. It sees the scientific world (nature) as an immediate givenness and holds a naturalism attitude concerning the object of knowledge. Hence, it precisely keeps its ignorance of ontology of science.

From the standpoint of ontology of science, the scientific field (the framework of understanding and meaning) shares human beings' inherent finiteness and historicity. It can only be justified in a certain historical situation, and the justification cannot be given by science itself. In contrast, science cannot transcend the realities given by intentionality in its world and it is also incapable to take the place of the world it projected. Consequently, on ontological level science is futile. Ontology of science itself is not science but philosophy; moreover, it is a kind of philosophy with a critical attitude on science.

To sum up, ontology of science investigates the possible conditions and limitations of science per se as man's existential mode, while the one-sidedness of the orthodox philosophy of science is its neglect of ontological (existential) study and its non-critical attitude on the ontological foundation of natural science.

References

Chalmers, A. F. [1978] *What Is This Thing Called Science? An Assessment of the Nature and Status of Science and its Methods* (Open University Press, Buckingham).
Feyerabend, P. K. [1970] "Philosophy of science: A subject with a great past," in *Historical and Philosophical Perspectives of Science*, ed. Stuewer, R. (University of Minnesota Press, Minneapolis) pp. 172-183.
Harry, R. [1989] *The Philosophies of Science: An Introductory Survey* (Oxford University Press, Oxford).
Heidegger, M. [1962] *Time and Being*, trans. Macquarrie, J. & Robinson, E. (Harper & Row, New York).
Koyré, A. [1973] "My fields and projects," in Koyré, A., *Etudes d'histoire de la pensee scientifique* (Gallimard, Paris).
Losse, J. [1980] *A Historical Introduction to the Philosophy of Science*, 4th ed. (Oxford University Press, Oxford).
Spiegelberg, H. [1965] *The Phenomenological Movement: A Historical Introduction* (Academic Publishers, Kluwer).
Toulmin, S. [1973] "Philosophy of science," *Encyclopedia Britannica* (Encyclopedia Britannica, Chicago).

4

The Predicament of Scientific Culture in Ancient China

Hong-Sheng Wang

There were plenty of technical knowledge in ancient Chinese civilization, but no independently existing scientific-culture system. The ancient Chinese, while moving along their own tradition, had not step onto the road of modern science and technology. This is the Needham Question, which was being attributed to a cultural problem by Max Weber and then to a cultural context in Confucianism by Mou Zong-San. But more importantly, the fact is that the political power in ancient Chinese society had a function of controlling and limiting the development of scholastic culture. This phenomenon appeared not just in ancient China but also in China's modernization process since 1840.

4.1 Introduction

It was in about 1938 when Joseph Needham (1900-1995) first had the idea of writing a treatise on the history of science, scientific thoughts and technology in the Chinese culture that he regarded the essential problem as: Why modern science had not been developed in China but in Europe? [Needham, 1998]. This is the well-known Needham Question. In fact, what should also be mentioned is that he and many others had tried to answer the question for more than half a century afterwards. Their achievements are summarized in the seven volumes of *Science and Civilization in China* [Needham et al, 1954-2004].

In a wider scope, many Chinese and foreign scholars had also discussed a similar issue; some even did that before Needham (see [Liu & Wang, 2002]). Moreover, I myself had proposed a *super*-Needham

Question: Since 1840, the year of the Opium War between China and United Kingdom, it has been 172 years passed and why China has still not become a super power with advanced science and technology? And, there is even a *future*-Needham Question: Could China become a super power with advanced science and technology in 2040, 200 years after the Opium War? [Wang, 1997]. While the Needham Question pays attention to China's past, the super version is about the recent developments of China; the future version, the future of China. In fact, the three questions could be answered with the same cultural and historical accounts. Thus, it would be meaningful to analyze the Chinese scientific culture from a holistic social angle and examine scientifically the Chinese civilization itself.

Accordingly, the ancient Chinese scientific culture should be probed first by finding out whether scientific spirit existed or not in the ancient Chinese civilization. If the answer is negative the reason should be investigated. For a broad understanding, the issue will be discussed not only from the viewpoint of science but also from that of civilization. In this regard, some important ideas of the Westerners, including Max Weber (1864-1920) and others would be discussed. Especially, the ideas on Chinese civilization expounded by the Chinese philosopher Mou Zong-San (1909-1995) should be mentioned. At the end, we will arrive at the conclusion that the traditional Chinese political culture had prohibited the growth of scientific culture in ancient China. In other words, *the domination of political culture had impeded the growth of scientific culture in ancient Chinese civilization.*

Before discussing the predicament of ancient Chinese scientific culture, some conceptual explanation about science ought to be given. What is science? In the new discipline Science Matters (scimat) introduced by Lui Lam [2008], science is defined as humans' seeking of knowledge about Nature without appealing to God or superstitions, and Nature includes all the nonhuman systems and the human system (see also [Lam, 2014]). Thus, all the topics covered in the humanities and social sciences are included in scimat. In particular, he provides a *new* answer to the Needham Question which says that the ancient Chinese is more concerned about the complex system of human society while modern science originated from the study of simple systems. That is, the

Chinese focused on the "wrong" topics as the development of science is concerned [Lam, 2008, pp. 31-32]. However, why this happened is a cultural matter. Thus, in order to probe properly the Needham Question, the relationship between science and civilization should be considered.

Generally speaking, science represents rationality in human culture and civilization; broadly speaking, technology is the hardware of scientific culture. The contents of science consist of mainly a variety of different forms of knowledge; its essence is rationality that reflects a particular objective attitude of humans towards nature, society and culture. In another sense, science is the most vibrant culture in modern civilization. The most obvious difference between modern and ancient civilizations is the degree of development of science. A nation without the growth of scientific culture is not a modern nation. A civilization without the comprehensive development of science cannot be considered a perfect modern civilization.

4.2 Scientific Knowledge Was Plentiful in Ancient Chinese Civilization but Was Not Treasured Up

Historically, science is not a precise term but could be understood from different cultural dimensions. When humans began to make tools, domesticate animals, cultivate plants and manufacture pottery and bronze, science would have germinated, but in a form of experience, not theory [Mason, 1962]. The experience had been built into different forms of knowledge. The knowledge was not the fire stolen by Prometheus from Zeus but the precipitation of human reasoning.

The four ancient civilizations of Egypt, Babylon, India and China all had sprouted a large amount of scientific and technological knowledge, which represents the rational parts of their culture [Dampier, 1958]. These rational cultural elements were not pure knowledge, but blended in with religion and superstition. With the advance of civilization, these rational cultural elements in the form of knowledge had gradually broken away from religion and superstition and assumed its independent form. For example, Euclid's geometry and Archimedean physics are pure knowledge of ancient rational cultural elements with its independent form of science. Generally, in these four ancient civilizations, there were plenty of scientific and technological knowledge but only the ancient Greece civilization contains the scientific spirit. The ancient Chinese civilization does contain plenty of scientific and technological knowledge, but no scientific spirit [Weber, 1920 (1995)]. Consequently, the history of science in ancient China is not like a river that has run successively, but is more like a wasted road lying with some outstanding milestones.

In fact, there were plenty of knowledge elements spreading in many disciplines in ancient Chinese civilization. These knowledge elements show the rational side of Chinese ancient civilization, but it did not constitute an independent cultural system of science. For example, in the Spring and Autumn period (770-476 BC), China was in a time of all schools of thought with each vying for attention. Among them the Mohists had studied pinhole imaging, convex mirror imaging, leverage, force and torque, buoyancy, mechanical motion, deflection of materials, geometry, etc. [Wu, 1992]. These are all basic issues in optics and mechanics. Unfortunately, there had not been successor of Mohists and the rational tendency had not been passed on and developed. Another example was that Qu Yuan (340-278 BC), a famous poet and politician of Chu, a state in the Zhou Dynasty (1046-221 BC), asked many cosmological questions in his poems that could only be answered by modern cosmology.[1] But the real reason that Qu asking these questions is due to his political misfortune which has nothing to do with his curiosity of probing the cosmological mysteries liked what Aristotle (384-322 BC) did.

---

[1] For example, Qu Yuan's poem, *Tian Wen* (*Ask the Heaven*): Of the Beginning of old, Who spoke the tale? When above and below were not yet formed, Who was there to question? When dark and bright were obscured, Who could distinguish? When matter was inchoate, How was it perceived? Brightest bright and darkest dark, What was made form only these? Yin and Yang blend and mix, What was the root, what transformed? The circular and nine tiered Heaven, Who enclosed and surveyed it? Just was how this achieved, Who originally made it? ... What does Heaven tread upon? How are the Twelve partitioned? The sun and moon are how coupled? How are the patterned stars ranged? Emerging by the boiling canyon, Arrive at the vale of night. Form light until dark, Is a pass of how many miles? What virtue moves the moon to flourish after death? What good is it to keep a rabbit in the gut? ... [Field, 1984]

Zhang Heng (78-139) of Han Dynasty researched lunar eclipse; he found that moonlight was just the reflection of sunlight on the moon and lunar eclipse was caused by the shadowing of the Sun by Earth [Du, 1992, p. 77]. Zhang also invented a seismograph, but the instrument had been lost by unknown reason. In history, the nine bronze vessels (Jiu Ding) from the three ancient dynasties—Xia (2100-1600 BC), Shang (1600-1046 BC) and Zhou—symbolizing the king's right of ruling, had been lost at the end of the Warring States period (476-221 BC) which, by its name, was full of wars between the states. For a different reason, the famous calligraphy work by Wang Xi-Zhi (303-361), *Lanting Xu* (*Preface to Orchid Pavilion*), had been lost because Li Shi-Min (599-649), Emperor Taizong of Tang Dynasty, liked it so much that the calligraphy was buried with him. Interestingly, the loss of Zhang Heng's seismograph was due to a political reason. The Chinese believed in the induction between Heaven's will and human affairs. Accordingly, earthquake was considered a warning given by Heaven when the emperor's work is below expectation. Therefore, the emperor himself disliked the invention of the seismograph because the information of an earthquake would make the emperor look bad in the eyes of his people. Thus, for various reasons it is almost certain that many of the precious material objects from ancient China would be lost.

After the Han Dynasty, Liu Hui (c. 225-295) (Fig. 4.1), a mathematician of Wei, one of three states in the Three Kingdoms period, showed his ability in cyclotomy [Du, 1992, p. 161]. Zu Chong-Zhi (429-500) (Fig. 4.1), an ancient mathematician in the Northern and Southern Dynasties (420-589), had calculated π to be between 3.1415926 and 3.1415927, a world record at his time [Du, 1992, p. 225]. Zu and his son, Zu Heng, had also derived the formula of a ball's volume. According to Zu Chong-Zhi's *Da Ming Li* (*Ming Dynasty Calendar*), there are 365.24281481 days in a year, almost equal to the modern number of 365.2422. These were really great mathematicians who demonstrated a rational scientific spirit. But as a whole, in traditional Chinese culture, all the mathematical achievements, even these great achievements like Zu Chon-Zhi's, were still being considered as technical skills of no importance when compared with Confucian theory.

Fig. 4.1. Two ancient Chinese scientists, Liu Hui (right) and Zu Chong-Zhi (left).

Shen Kuo (1031-1095), author of *Meng Xi Bi Tan* (*Brush Talks from Dream Brook*), lived in Song Dynasty and was a knowledgeable man. The book was an encyclopedia of the time and had been written by Shen after his disappointed officialdom. In the book, Shen had even noticed the magnetic declination of the earth. But like most other scholars, he spent most part of his life and career in officialdom. Similarly, *Xi Yuan Ji Lu* (*Collected Cases of Injustice Rectifying*) by Song Ci (1186-1249) was a masterpiece of forensic medicine. This book was a summary of his professional experiences as a judge, which had great impact on later generations, especially on the later Chinese textbooks of forensic medicine [Jia, 1984, pp. 237-238]. But unfortunately, in ancient China, study of human anatomy had not been advocated, or even been prohibited. Thus, Song's research had not initiated new development of human anatomy in China like what had been done by Andreas Vesalius (1514-1564) in Europe [Mason, 1962, pp. 216-217].

During the era of Jin and Yuan Dynasties, when the Renaissance was brewing in Italy, there had also been some highly talented people in China. For example, Li Ye (1192-1279), a mathematician who had studied the technique of solving equations called Tian Yuan Shu [Du, 1992], had reached very high level in his research. Furthermore, Li had deep perception of the relationship between nature and mathematics. In the preface of his book *Ce Yuan Hai Jing* (*The Sea Mirror for Measuring Circle*) (Fig. 4.2), he writes, "It is difficult to reach the infinite limits of numeral science, but it is possible to get some new findings. Why are things like this? It is sure that the numeral relationships hide in the unseen world. In fact, the manifestation of the numeral relationships have not only mathematical meaning in themselves, but also in the principle of nature" [Qian, 1964]. This book was finished in 1248 but has not been published in 1279 when Li passed away. He appreciated the book so much that when he was at his deathbed, he said to his son, "You could burn all my books but not this one. It is not only a book of mathematics but a record of all the results of my great efforts of thinking in the field. The future generations will be able to understand my thinking." [Du, 1992, p. 627]. Li Shan-Lan (1811-1882) even acclaimed the book as the number one Chinese book in mathematics [Li, (1279) 1985, p. 1].

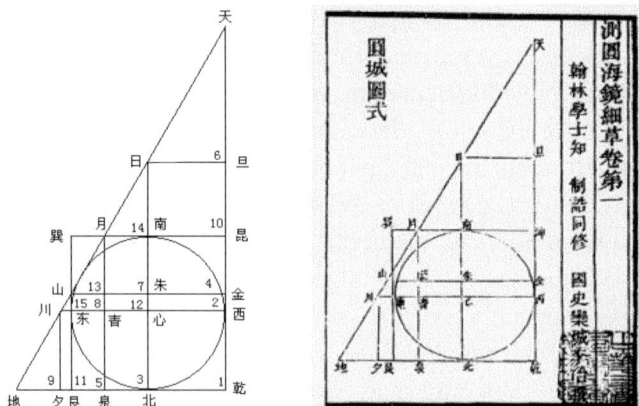

Fig. 4.2. A Figure in Li Ye's *Ce Yuan Hai Jing* (*The Sea Mirror for Measuring Circle*).

Guo Shou-Jing (1231-1316), the scientist who helped the Mongolian Emperor Kublai (1215-1294) with his talents of science, especially in the construction of water conservancy facilities and calendar reformulation, was an eminent astronomer at that time. The reformulated Shou Shi Li (Time Service Calendar) by Guo and others in 1276 had been used for more than 360 years afterward as the most accurate calendar of the world in that time. In Ming Dynasty, Li Shi-Zhen (1518-1593) had finished the

Ben Cao Gang Mu (Compendium of Materia Medica).[2] The title of the book is simple and unadorned, intended to display the scientific spirit and being totally different from all the former pharmaceutical classics such as the *Shen Nong Ben Cao Jing (Agriculture God Compendium)* which invokes God's name in strengthening the trustworthiness of doctors.[3] In Ming Dynasty, Song Ying-Xing (1587-1666), a man who failed the imperial examination, had written a book titled *Tian Gong Kai Wu (Exploring Nature with Divine Craftsmanship)* which is regarded as a handicraft-technology encyclopedia of that time.[4] Xu Xia-Ke (1587-1641), who had studied geography by traveling, had finished his famous *Xu Xia Ke You Ji (Xu Xia-Ke's Traveling Diary)* that includes plenty of knowledge with modern scientific value.[5] For example, he recorded the geographic appearances such as the karst landscape in details. He even examined the causes in forming certain geographic environments.

Yet, during the Yuan (1271-1368) and Ming (1368-1644) Dynasties, China's political and social conditions were quite different from that in Europe. In general, the development of science and technology in China was not sustainable. Most of the scientific results obtained were shelved afterward. Almost all the achievements of science and technology had been intermittent innovations, which did not appear continuously.

---

[2] The book is to correct the errors in ancient medical books. In total, it contains the description of 1,892 drugs, 11,096 prescriptions and 1,160 beautiful illustrations. It took 29 years for Li Shi-Zhen to finish the book which was printed in 1596, three years after his death.

[3] It is the earliest Chinese pharmaceutical book. The unknown authors titled their book with the name Shen Nong, which means Agriculture God in Chinese, to stress the authority of the book. It describes 365 drugs in total, including 252 from plants, 67 from animals and 46 from minerals. It also explains the essence of the traditional Chinese pharmacy. The finishing time of the book is a disputed issue. Some scholars say that it had been written during the Qin Dynasty (221-206 BC) or the Han Dynasty (206 BC-220), or even earlier in the Warring States period. Anyway, the original edition does not existing; the earliest edition in existence is the one edited by Lu Fu in 1616.

[4] It was printed in 1637. The author emphasizes the coordination between human techniques and natural forces.

[5] Afterward, Xu's diary was edited by Ji Meng-Liang, a private tutor of Xu's family, and by the author's son, Li Jie-Li. The book was further revised and updated by some private collectors. Since its first publication in 1642, the Diary has been reprinted in 38 editions up to 1985.

In late Ming Dynasty, some Chinese scholars began to be aware of the so-called *Western Learning* (a late Qing Dynasty term for Western natural and social sciences). The earliest collaboration of Sino-Western scholars was that between Xu Guang-Qi (1562-1633) and Matteo Ricci (1552-1610) (Fig. 4.3) who together had translated the first six volumes of Euclid's *The Elements*. Xu understood and mastered the Western science of the time. Referring to *The Elements* he says, "all the people in the world should learn geometrical principles and this will be a reality one hundred years later" [Du, 1992, p. 896]. But the other nine volumes of *The Elements* were not translated and published until the 19th century by Li Shan-Lan, a Chinese mathematician, and Alexander Wylie (1815-1887) and Joseph Edkins (1823-1905), two British, in late Qing's Westernization Movement. Note that Li was born 249 years after Xu Guang-Qi's birth! With the exception of these examples, there had been no major development of science and technology for a long time because China had largely closed its door during these two and a half centuries from late Ming to late Qing.

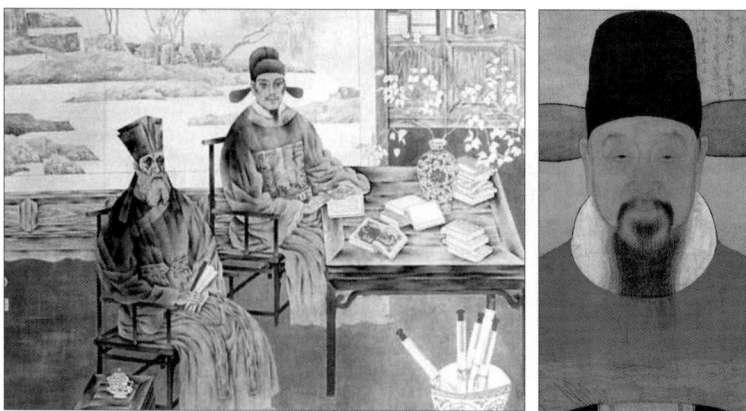

Fig. 4.3. *Left*: Xu Guang-Qi and Matteo Ricci. *Right*: Xu Guang-Qi.

4.3 Reasons for the Absence of Scientific Spirit in Ancient China

Reading the ancient Chinese writings on science and technology, one can see the sparkling of the lights of knowledge. The lights of knowledge, though from a long time ago, can still resonate with the rational thinking of modern people. It is a kind of bridge between the ancestors and modern Chinese in their understanding of science. It even seems that the ancient Chinese could step directly through the door onto modern science by following these examples of rational knowledge. But history did not happen like this. Why? Since the language and writings of the Chinese civilization are continuous, then why the intellectual rationality shown in knowledge displayed by scholars at different times had not been promoted generation by generation? Or, rather, why not the wealth of scientific knowledge in ancient Chinese civilization been developed into a system of scholastic culture? Why had it not been sublimated into a kind of scientific spirit in Chinese civilization? Furthermore, why the scientific knowledge and technology in ancient China always failed to develop to such a degree that they would help to change the society?

Obviously, these are just the same questions that puzzled Joseph Needham and many others. In fact, before Needham put out his Question, the German scholar Max Weber had already set up a kind of answer to these kinds of questions. In the book *The Protestant Ethic and the Spirit of Capitalism*, he compared the differences between Eastern and Western civilizations and, at the end, he attributed the European development of science and technology as well as capitalism to be a cultural issue [Weber, (1920) 1958]. Specifically, development of capitalism in the West was associated with the Puritan ethic but its development in China was hindered by the Confucianism ethic. To explain this view, Weber took a step further in his study of Chinese culture by writing another book in the same year, *Confucianism and Taoism*. The conclusion was that ancient civilizations other than that of Europe could not have originated capitalism and thus would be unlikely to produce scientific and technological revolutions [Weber, (1920) 1995].

In 1998, Jared Diamond (Fig. 4.4) pushed the issue a step further after he affirmed Graeme Lang's view [Diamond, 1998]. According to Lang, the rise of scientific inquiry in Europe developed within a peculiarly European institution, viz., autonomous universities where critical inquiry was relatively uninhibited by governmental or religious authority. That there were such universities existing in Europe was due to her political fragmentation which, in turn, was due to her geographical conditions. Lang further points out that there were no comparable institutions in China. He compares the maps of Europe and China and concludes that while it was easy for China to be unified into a big country it was equally easy for Europe to be split into many independent countries with their own political, linguistic, ethnical and cultural autonomies. Thus the reasons of why there was no Chinese revolution in science and technology have been attributed ultimately to geography and ecology [Lang, 1997a; 1997b]. But Diamond thinks a social and cultural interpretation has its limits and an explanation based on natural science is called for. As Diamond points out, such an explanation can always be challenged with a further question: Why were Europe and China different with regard to those social or cultural factors? For him, an advantage of his "natural science" accounting is that it escapes the circularity which often creeps into explanations which do not go deeper than the social or cultural differences between Europe and China [Diamond, 1998].

Fig. 4.4. *Left*: Jared Diamond. *Right*: Mou Zong-San (1909-1995).

Yet for the Chinese philosopher Mou Zong-San (Fig. 4.4), the situation was looked upon as a cultural predicament [1985, pp. 96-97]. Why the ancient Chinese had not step into the room of modern science and technology by following the development of their own knowledge? Mou's answer is that the essence of China's cultural life is to use everything for the benefit of human beings according to the Confucian ethical principle. It is a kind of ethical and political model of *benevolence (ren) system* that covers all the cultural fields. To use everything for human beings needs some knowledge of the world. But since the knowledge is for practical use and is sufficient for this purpose, there is no need to develop it further. Thus, the *intelligence (zhi) system* remains hidden in the *benevolence system* and has not developed into an independent system by itself (such as the logic system and the mathematics system). For example, Chinese ancient astronomy was to probe the mysteries of the universe by humans' intelligence and, being restrained by ethical and political ideas, had not been developed into real science.

Generally, there was plenty of technical knowledge in ancient Chinese civilization, but no independent scientific culture system existed. With the intelligence system hidden in the benevolence system, China developed its theoretical moral system in concurrence with the Confucian ideology; no independent knowledge system like that in ancient Greece emerged.

Miao Li-Tian (1917-2000) is a Chinese philosopher who contrasts the Chinese tradition of studying for the sake of application with the Hellenic spirit of studying for the sake of knowing what and why. As Miao puts it, ancient Chinese had the spirit of *learning for practice* but had not the spirit of *learning for truth* [1997]. The spirit of learning for practice is also explained clearly in *Ba Tiao Mu* (*Eight Clauses*) of *Da Xue* (*Great Learning*), Chapter 42 of *Li Ji* (*The Book of Rites*), a famous Confucian classic edited by Confucius himself.[6] According to this classic, pursuing knowledge is to attain a good faith in regulating the family and

---

[6] The *Ba Tiao Mu* (*Eight Clauses*): In ancient times, those who wished to illuminate lustrous virtue under heaven first brought order to their states. Those who wished to bring order to their states first regulated their families. Those who wished to regulate their families first cultivated their persons. Those who wished to cultivate their persons first made their hearts upright. Those who wished to make their hearts upright first made their ideas sincere. Those who wished to make their ideas sincere first extended their knowledge. Extending knowledge resides in going out to things. Only after you go out to things can your knowledge arrive. Only after your knowledge arrives can your ideas be sincere. Only after your ideas are sincere can your heart be upright. Only after your heart is upright can your person be cultivated. Only after your person is cultivated can your family be regulated. Only after your family is regulated can the state be in order. Only after the state is in order can all under heaven be at peace. [Shaughnessy, 2010, pp. 15-16]

state affairs. Thus, the former is just a means; the latter is the goal. From the perspective of modern science, learning for practice is just what applied science does. To a certain degree, learning for practice has a strong purpose and is not at the forefront of scientific discovery. In contrast, the spirit of learning for truth is close to basic scientific research and is promoted by the interests and curiosities of scientists, aiming at discovering new facts or new theories.

In short, Confucianism, the major ancient Chinese ideology, has the concept of learning for practice and the tradition of respecting the teachers, but lacks the academic spirit of Aristotle who says, "I love my teacher but love more the truth". Or, the concept of learning for truth had not been promoted in ancient Chinese civilization. From the points of view of race, writing (wenzhi) and culture, the Chinese civilization flows continuously. But it has failed to promote science and technology revolutions and has sublimated out the true scientific spirit, and so China has not developed her own way of modernization. This is an important problem in history and culture that warrants further consideration.

4.4 Political and Social Influences on Scholastic Culture

In addition to geographical reasons, the predicament of science and technology development in Chinese civilization could also be probed from the relations between different cultural elements, in particular, the relation between politics and academic culture. In general, the development of a society as a whole is no doubt the foundation of any academic and cultural development.

4.4.1 *Ancient Period*

In ancient China, the learning culture had always been controlled or led by politics. More precisely, ancient Chinese politics had controlled and limited the development of the learning culture [Wang, 2007]. Besides, Confucianism with a cultural shielding had prevented the ancient Chinese from free explorations in science [Weber, (1920) 1995].

The ancient Chinese society was an official-dominated society. In such a society, knowledge and rationality were not entirely excluded, but their cultural functions and objectives were controlled and limited by political standards from the authority. For example, astronomical research was controlled by the court and was greatly affected by politics. Even the development of agriculture, medicine and mathematics were also under the influence of the government. Also, the political authority could damage the social and cultural foundation of rational knowledge at the ideological level by, e.g., selecting or arbitrating academic issues in an authoritative manner or even by directly creating literary inquisitions in a violent manner. In the Song Dynasty (960-1279), the Emperor's attitude towards the scholar-officials was tolerance, so there had been some developments in science and technology. Although the literary inquisitions of the Ming and Qing Dynasties were due to political reasons, they had shocked the Chinese academics to an unmatched degree. The result of these literary inquisitions was described by Gong Zi-Zhen (1792-1841) in his poem *Yong Shi* (*On History*): "Scholars avoid speaking when meeting in fear of literary inquisition. All of their writings are just for getting rice and sorghum for eating." In short, in traditional Chinese civilization, knowledge and technology were to a large extent controlled by political culture. Science and technology appeared only because of their application values; academic rationality in traditional civilization is just a sort of ornament.

Of course, the Chinese academic culture in the Spring and Autumn period and the Warring States period had been prosperous. This was due to the political and ideological pluralism. Afterwards, political condition had changed since Qin and West Han (202 BC-9), and so was the development of the learning culture. According to *Introduction to Chinese Classics* by Qian Mu (1895-1990) (Fig. 4.5) [1997, pp. 64-65], a famous Chinese historian,

> Scholars contending in late Warring States period was especially strong. There were a mess of right and wrong, full of arguments. But First Emperor Qin (259-210) unified the whole country, and then the learning culture, in the same pace of political culture, also transformed into unified rigidity. Firstly, Qin's chancellor Lü Bu-Wei (?-235 BC) edited his *Lü Shi Chun Qiu* (*Encyclopedia*) for the purpose of collecting and caging all the theories, thus to converge

all the contending thoughts to a single authority. But he died without finishing it. Then Li Si (280-208 BC), Chancellor of First Emperor Qin, achieved his ambition and eliminated different schools by high-handed policy. So the preceding flourishing academic contending moved to extinction. For the fate of academic culture, the death of Lü Bu-Wei and the emerging of Li Si was a turning point. Li Si and Han Fei (280-233) both were students of Xun Zi (313-238 BC). First Emperor Qin liked Han Fei's theory very much. Li Si had backbitten Han Fei to death, but still took Han Fei's theory in flattering the First Emperor. Anyway, in Qin Dynasty, the political ideas all came out of Xun-tse and Han Fei, which was called the theory of Legalists. When various schools of thoughts emerged in the Spring and Autumn period and the Warring States period, they were ideas competing for political prominence through academic arguments, and at last they failed. In contrast, since the Qin Dynasty there had been just the ruler controlling the academic culture by political means. Two things in First Emperor Qin's time mark the beginning of such a new era: the burning of books and the burying of the literati in pits. (My translation)

Here, Qian Mu has specifically described the turning points of Chinese politics and ideology, and vividly explained how traditional politics controlled the academic culture.

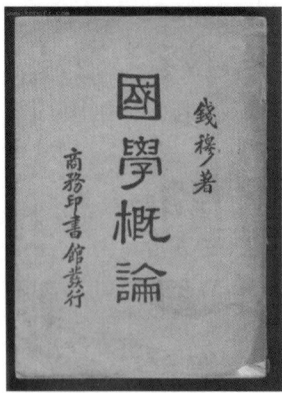 

Fig. 4.5. Qian Mu and his book *Introduction to Chinese Classics*.

Since academic contentions in the Spring and Autumn period and the Warring States periods were due to political pluralism, it is not surprising that once politics was unified in a new empire, the contentions stopped. In particular, Emperor Han Wu (156-87 BC) selected Confucianism as the orthodox official belief system and belittled all other schools of thought. In fact, what all the schools of thought in the Spring and Autumn period and the Warring State period offered are theories synthesizing politics, ethics and scholarship, which are not just about learning. Thus, what the "hundred schools of thought" of that time deliver are multiple political proposals and ideologies. Anyway, in the framework of traditional Chinese civilization, the prosperity and development of scholastic culture depend on the condition of diverse ideologies, which in turn depends on the presence of a plural political system. But the mainstream of political history in ancient China is about unity or unification, not pluralism. Of course, upon political unification different ideologies (such as Confucianism, Buddhism and Daoism) may also be found coexisting harmoniously. But this is a separate issue.

In general, the relationship between politics and science in ancient China has a harmonious aspect. In essence, ancient Chinese politics does not reject science outright but can accept a limited amount of scientific rationality and humanistic knowledge. The political system itself contains few rational elements for it is established on the old, irrational principle that "Winner be the king, Losers be bandits." In addition, after Confucianism became the orthodox of Chinese society, it also had a tendency of deviating from rational track under the influence of politics. Since scientific culture is based on knowledge and independent reasoning, it is understandable that there was no independent scientific cultural tradition in ancient China. On the contrary, if a civilization dominated by imperial power systems could accommodate an independent scientific and scholastic cultural system, it would be unreasonable. In fact, this phenomenon is not unique to the ancient Chinese civilization; it happened to most of the ancient civilizations, too [Wang, 2007, p. 209]. In ancient China, Confucianism was the main scholastic subject which, though containing academic elements, is essentially an inquiry of human values and social ideals. In fact, only in modern civilization, scholastic culture with science and technology as the core system is capable of being sustainably developed and becomes the key factor in guiding the civilization.

From a social perspective, the full development of a scientific culture based on rational thinking requires not only that the explorers freeing themselves from the shackles of political authority and ideology, but also the public going beyond their social status in pursuing the recognition and reasonable interpretation of the facts. In other words, science is a "luxury" culture. Thus, on the whole, science can appear only as a flash of the scholastic culture in most ancient civilizations; it cannot be the Sun that lights up all the civilizations. In ancient Greece, although there had born clearer concepts of science and democracy, the defects of its civilization[7] and the complexity of its social development did cause disruptions in the Greece history. The concepts of science and democracy had not been sustainable in the Greece civilization. It suggests that the science based on rational knowledge and the political philosophy based on democracy are unlikely to be the dominant cultural elements in all the ancient civilizations. This is just like the noble lily cannot grow in the barren, arid desert.

In short, the tradition of scientific culture in ancient China was not fully developed and the ancient political culture had played a role of impeding science. Confucianism as the mainstream of Chinese ideology does contain scholastic elements, but its human-value aspects are heightened into moral ideology which has the effect of overcoming scientific reasoning. All the mainstream culture systems, including political culture systems and Confucianism, do accommodate the practical side of knowledge but suppress the autonomy and independence of science. Thus, Chinese traditional politics and Confucianism had exerted restraint effects on the development of rationality needed in pure science. In the narrow space between officials and Confucians, there was little space left for scholastic culture in ancient China. Consequently, the sapling of science in ancient Chinese civilization could not grow freely and sustainably for the lack of a proper socio-cultural soil.

---

[7] According to Ruan Wei [2001], the lack of internal cohesion and a long-term internal strife were the reasons that the Greece civilization failed to grow continually.

4.4.2 Modern Period

The traditional Chinese culture had also an impact on China's modernization process from late Qing Dynasty to present. The controlling and limiting of learning culture by politics and ideology had also functioned in the modernization process; e.g., late Qing officials had insisted upon Confucianism and accepted science and technology just for practical uses. During the Republic of China and in 1945, after China's complete victory in her battle of resisting the Japanese invasion, the two major Chinese parties, Guomingdang (KMT) and Chinese Communist Party (CCP), fought against each other in a massive civil war. The social history of that period was forged by blood and fire. Then, modern Chinese scholars split into different camps according to their political and ideological beliefs. The destiny of scientific culture was just like a reed in storm at such a moment of history.

In the last 64 years of the People's Republic of China, members of CCP's collective leadership have significantly different attitudes towards scientists. For example, leaders like Zhou En-Lai (1898-1976), Zhu De (1886-1976), Chen Yi (1901-1972), Nie Rong-Zhen (1899-1992) and Deng Xiao-Ping (1904-1997) had always respected intellectuals to a certain degree and kept some scientific rationality in dealing with social affairs. For them, China needed modern science and technology and for that, the active participation of intellectuals was essential. But the traditional influence was still very strong and Mao Ze Dong (1893-1976) (Fig. 4.6) had wanted to remold the intellectuals' mind for political and ideological reasons [He, 2009, pp. 62-66]. This policy of Mao did not come from the teachings of Karl Marx and Friedrich Engel but was learned from Joseph Stalin [Gao, 2000, pp. 186-192] and was based on the traditional Chinese model of controlling science and technology through politics and ideology.[8] Even so, Chen Yin-Ke (1890-1969) (Fig. 4.6), an honest historian, should be mentioned regarding the scientific

---

[8] Science and technology here include the humanities (literature and history in particular), which, together with the social sciences, belong to the science of humans.

spirit in modern China. In the 1950s, Chen still upheld his belief of *independent spirit and freedom of thinking* in scholastic matters.[9]

During the last three decades, the relationship between science and politics had been readjusted considerably. But for the political leaders, there are still more steps to take for the full development of Chinese scientific culture, especially in comparison with the world's developed countries.

Fig. 4.6. Mao Ze-Dong (left) and Chen Yin-Ke (right). They held totally different views on the relationship between politics and learning.

4.5 Conclusion

In summary, there were plenty of scientific and technological knowledge but no scientific spirit in ancient China. The ancient Chinese could not step onto the road of modernization by following their old ways. The Needham Question is in fact about the predicament of scientific culture in ancient Chinese civilization. The predicament was caused by two major reasons. One is that in Confucianism, the intelligence system is always hidden in the benevolence system and had not developed into an independent system such as the logic system and the mathematics

---

[9] In 1956, Chen Yin-Ke was invited to Beijing from Guangzhou to be Director of the Institute of History of Chinese Middle Ages (which no longer exists) by Guo Mo-Ruo (1892-1978), President of Chinese Academy of Sciences. Chen refused the offer and replied with a letter, in which he insisted that in the study of history, one should not follow the presupposed Marxist viewpoints but should instead keep scientifically the independent spirit and freedom of thinking [Bian, 2010].

system. The other reason is that the Chinese political power system has always exerted a strong control on scholastic culture by way of ideology as part of its controlling of the whole society.

References

Bian Seng-Hui [2010] *Chronicle of Chen Yin-Ke (First Draft)* (Chung Hwa Book Co., Beijing).
Dampier, W. C. [1958] *A history of Science and Its Relations with Philosophy and Religion* (Cambridge University Press, Cambridge).
Diamond, J. M. [1998] "Peeling the Chinese onion," *Nature* **391**, 433-434.
Du Shi-Ran (ed.) [1992] *Biographies of Ancient Chinese Scientists* (Science Press, Beijing).
Field, S. [1984] *Tian Wen: A Chinese Book of Origins* (Penguin, New York).
Gao Hua [2000] *How the Red Sun Rises: An Analysis of the Rectification Movement in Yan An* (Chinese University of Hong Kong Press, Hong Kong).
Jia Jing-Tao [1984] *History of Forensic Medicine in Ancient China* (Masses Press, Beijing).
He Qin (ed.) [1999] *A History of the People's Republic of China* (Higher Education Press, Beijing).
Lam, L. [2008] "Science Matters: A unified perspective," in *Science Matters: Humanities as Complex Systems*, eds. Burguete, M. & Lam, L. (World Scientific, Singapore) pp. 1-38.
Lam, L. [2014] "About science 1: Basics—knowledge, nature, science and scimat," in *All About Science: Philosophy, History, Sociology & Communication*, eds. Burguete, M. & Lam, L. (World Scientific, Singapore).
Lang, G. [1997a] "Structural factors in the origins of modern sciences: A comparison of China and Europe," in *East Asian Cultural and Historical Perspectives*, eds. Tötösy de Zepetnek, S. & Jay, J. W. (University of Alberta Press, Edmonton) pp. 71-96.
Lang, G. [1997b] "State system and the origins of modern science: A comparison of Europe and China," *East-West Dialogue* **2**, 16-30.
Li Ye [(1279) 1985] *The Sea Mirror for Measuring Circle*, trans. Bai Shang-Shu, proofread Zhong Shan-Ji (Shandong Education Press, Jinan).
Liu Dun & Wang Yang-Zong [2002] *Chinese Science and Scientific Revolution: Selected Essays on Needham Question and Related Issues* (Liaoning Education Press, Shenyang).
Mason, S. F. [1962] *A History of the Sciences* (Macmillan, New York).
Miao Li-Tian [1997] "Learning for practice and learning for truth," *China Reading Weekly*, July 9 (Guangming Daily Press, Beijing).
Mou Zong-San [1985] *Moral Idealism*, 6th ed. (Students Press, Taipei).

Needham, J. [1998] *Sinology Series of Joseph Needham* (Tianjin People's Press, Tianjin).
Needham, J. et al [1954-2004] *Science and Civilisation in China*, Vols. 1-7 (Cambridge University Press, Cambridge).
Qian Bao-Cong (ed.) [1964] *A History of Chinese Mathematics* (Science Press, Beijing).
Qian Mu [1997] *Introduction to Chinese Classics* (The Commercial Press, Beijing).
Ruan Wei [2001] *The Manifestation of Civilizations: An Assessment of 5000 Years of Human Civilization* (Peking University Press, Beijing).
Shaughnessy, E. L. [2010] *Confucian & Taoist Wisdom* (Duncan Baird, London).
Wang Hong-Sheng [1997] "On the Needham Question and how to surpass it," *China Reading Weekly*, Nov. 12 (Guangming Daily Press, Beijing).
Wang Hong-Sheng [2007] *The Waterfalls and Gorge of China's History: The Cultural Structure and Modern Transformation of Chinese Civilization* (China Renmin University Press, Beijing).
Weber, M. [(1920) 1958] *The Protestant Ethic and the Spirit of Capitalism* (Charles Scribner's Sons, New York).
Weber, M. [(1920) 1995] *Confucianism and Daoism*, trans. Wang Rong-Fen (The Commercial Press, Beijing).
Wu Long-Hui [1992] *Mo-tse: Modern Translation in Vernacular Chinese* (China Bookstore, Beijing).

5

What Do Scientists Know!

Nigel Sanitt

Recently the largest refrigerator in the world—the £4.4 billion new instrument operating at CERN, Geneva—was inaugurated and running. The machine's purpose is to smash together high energy protons in order that scientists can learn about the world of matter, identify what the world is *really* made of, and discover the Higgs particle which confers mass on all the other particles. The Higgs boson was indeed found in 2012 and a Nobel Prize was awarded to the theorists the next year. My purpose in this article is not to denigrate this example of one of mankind's achievements, but to point out a number of problem areas in science which do not get much publicity, and which address the question of what scientists *know*.

5.1 Introduction

I am sure readers will be relieved to hear that in spite of the fact that my background is science and mathematics I am not going to quote any mathematical formulae, display any graphs nor describe in detail any physical theories about the Universe.

What I want to present in this chapter are a few myths and problems in science, which are currently happening and maybe are about to happen. I am going to introduce a few strands, which may, at first, appear unconnected, but which I hope to bring together. I am also going to introduce a model of scientific theories which incorporates two ideas—Integrationism and Problematology—which, I believe, can provide a basis for understanding how science works.

Integrationism is a theory of meaning in language put forward by Roy Harris [2005] of Oxford University. He introduced the idea of a "Language Myth" which integrationism aims to identify and argue against. The integrationist viewpoint rejects the myths that there are (1) fixed correlations between words and ideas, and (2) words substitute for objects, and we know which words stand for which objects. Meaning, for the integrationist, in science or any other area, is created at all levels of interaction and is open-ended.

The other idea I want to introduce is *Problematology*, which sees questions as the fundamental bedrock of thought, and in particular, of science. The genesis of this idea goes all the way back to the ancient Greeks, but in its modern form was elucidated by Michel Meyer [1995] at the Free University of Brussels. I shall return to these two ideas later but in the meantime I want to introduce a number of problem areas in science as a series of strands, which I hope to thread together.

5.2 Main Strands

In the first strand of this chapter I want to communicate to you the rather desperate (but interesting) state that astrophysics finds itself in at the present time. Apparently four fifths of the matter that we believe comprises the Universe is made of an unknown substance that we cannot see or interact with.

Consider the following rough analogy: When academics get together and talk, they use quite a wide vocabulary and generally have a rational discussion on the matters in question. When five year old children talk to each other they similarly, in their way, have a generally rational discussion of matters that interest them, but of course their vocabulary is much more limited; their experiences and knowledge of the world are also correspondingly much less than those of adults. Imagine a meeting consisting of five year olds discussing the present state of knowledge about the Universe. I am sure such a meeting would be interesting and amusing to us. The sad fact is that when astrophysicists get together and talk about the Universe, since they are only familiar with about one fifth of what there is, their present state of knowledge and experience of the Universe is about the same as a five year old compared to an adult. In particular, some scientist's claim that we may be reaching the "end of physics" is foolish and laughable.

My second strand in this chapter is to refer to a remarkable result, which underpins the whole of mathematical physics—both classical physics and quantum physics. This result was discovered (or rather proved as it counts as a mathematical theorem) by a woman mathematician in 1915, called Emmy Noether and goes by the name of Noether's theorem. Noether was forced to flee Nazi Germany in 1933 because she was Jewish. She taught at Bryn Mawr and at Princeton until her death in 1935. Outside of mathematical physics most people have never heard of her. She has been described as probably the greatest woman mathematician who has ever lived, and has both a crater on the far side of the moon and an asteroid named after her. Her theorem has different versions and applies to certain classes of theories in physics. What it says, in essence, is that whenever you have a continuous symmetry in a law of nature you get a conserved quantity (and vice versa). So a time-translation symmetry gives conservation of energy; a space-translation symmetry gives conservation of momentum and a rotational symmetry—the world does not change if you turn your head—gives conservation of angular momentum; the theorem also applies in the quantum realm. So from calculating trajectories of rockets to the moon—or anywhere else—to the behavior of electrons and photons in a television set, it is all down to the "penetrating mathematical thinking" of Emmy Noether. By the way, "penetrating mathematical thinking" was the phrase that Einstein used about Noether's theorem.

My third strand is the Large Hadron Collider (LHC). This is a truly gigantic machine built 100 meters underground on the Swiss-French border near Geneva. In the shape of a ring, its purpose is to collide beams of protons with each other at high energy in order to investigate the structure of matter. Just to give you a small insight into what is going on: the machine is always shut down during the winter months because the cost of electricity, which is higher in the winter, would not be affordable within the near £700 million annual budget of the facility.

My fourth strand is the so-called Standard Model of particle physics. This model is not as its name implies one model but a whole class of models partly, but not wholly reliant on the operation of Noether's theorem, which I referred to in my second strand. The strength (and weakness) of the Standard Model is that it relies on such general underlying principles, such as symmetry. Strength, because so much can be gleaned from ideas of symmetry; but weakness, because symmetry only gets you so far. It is not impossible for the Standard Model to be disproved, but it is surprisingly difficult, as any new results can usually be accommodated within the hundreds of free parameters that the model generates.

My fifth strand is a question. The language of particle physics is redolent of *reocentrism*, a word which refers here to talking about everything in terms of objects rather than processes. Even the name *particle* physics implies an object-centered view of the world. My question is: why? Why is the scientific world of particle physics so imbued with the idea of particles? Although it is ironic that at the quantum level particle physicists are happy to accept that particles do not exist as discrete entities that can be pinned down, but as waves of probability.

Consider the following context: Noether's theorem talks about continuous symmetries in time and space and conserved quantities such as energy and momentum. Einstein's theories talk about the velocity of light and geometrical transformations in time and space. Gravity (Newtonian and Einsteinian), electromagnetism and the strong and weak forces of quantum mechanics all talk about forces and processes involving matter and energy. That is not reocentric at all; in fact objects as such hardly get a look in. The idea of talking about things in terms of objects rather than processes is not primary in physics but secondary. The language of science as we know it today is thus heavily skewed away from processes in favor of objects, even though most theories are about processes. Why is this the case?

I think the answer to this question probably lies more in the realm of history, culture and psychology, deep rooted in our culture right back to classical times. I would say that it is not because of modern science but rather in spite of modern science that reocentrism reigns supreme in science. In spite of their innovative and radical credentials, scientists are in fact hidebound and deeply conservative. Maybe it is a question of time, maybe in a few hundred years scientists will eventually migrate form object to process. I hope we can push this movement along a bit faster.

It is interesting here to look into the historical record. The dustbin of science is full of forgotten theories and discarded ideas. Two items of interest are the following: there used to be a controversy as to whether light was made of particles or waves. This controversy has now been settled and light is considered as both and neither. In the mid-19th century Michael Faraday, who discovered the relationship between magnetism and electricity, believed neither in the aether nor in matter as a physical substance. His idea was that the Universe was filled with invisible lines of force—matter being created where these lines intersected. Scientists still talk about lines of force but the rest of his theory has been discarded.

The second item occurred in the 1960s. The number of particles that were being discovered went through an enormous increase and some scientists began to question the idea of what a particle was. They started to coin the term "resonance." However, this term was in the end consigned to the dustbin, although it still endures in the term "resonance particle" to describe a particle which is particularly short lived. My reason for bringing up these forgotten ideas is to show that even within science the reocentric view has been questioned in the past. These ideas were not simply a question of substituting one type of object for another, but actually questioning the idea of an object—not a break with reocentrism, but a starting point—even though these ideas were eventually submerged.

My sixth strand is to ask another question: What do scientists mean by the terms *fundamental particle* or even *galaxy*? And what exactly is Mass/Energy? Equivalence is not the same as identity. When scientists ask about the origin of mass, what does the question mean?

If you ask scientists these questions they broadly fall into two camps. One section will jump straight in and quote theories, color them with formulae and insist that "it is all in the maths." The other camp—usually comprising those who are older (and presumably wiser)—will smile condescendingly as a glaze comes over their eyes while they try to dodge the questions.

When I was an undergraduate physics student many years ago there were one hundred and eighty of us in the year. I remember one question in the year-end exams was: "What is an electron?" I spoke to one of the examiners afterwards and he admitted that not a single student in the year had attempted to answer that question—I learnt a lot about physics that day!

My seventh strand is to look at analogies in quantum physics. The problem is that there is not much in our everyday experience that gives sustenance to metaphorical language in the quantum domain—maybe that is another reason why reocentrism dominates. A simple analogy is that A *is like* B, where B is something we are familiar with and A is unknown. A lot of analogy in science is where both A and B are unknown. The *myth* that scientists follow being that the "*is like*" still applies in any meaningful way.

My eighth strand is a well-known story. A traveler in Ireland is trying to reach Dublin and has lost his way. He stops to ask for directions from a farmer. The farmer says to him: "Well if I was going to Dublin, I wouldn't start from here."

What some scientists see as an 'end of physics' is in fact a cul-de-sac that science has gotten itself into, and you do not start a journey facing towards a dead end. One thing I have tried to promote is for scientists to be more aware of the philosophical aspects of their subject. My purpose is not to make scientists more culturally and academically rounded individuals, however beneficial that may be, but to encourage scientists to think more about their subject and at least get to a place from which they can start their journey.

The importance of Noether's theorem cannot be overstated, but we are approaching its 100th anniversary and the time has come, I believe, for science to wean itself off the over reliance on symmetry for its theoretical basis.

So what of the LHC? This £4.4 billion refrigerator: Will it find the missing mass in the Universe? Will it identify the Higgs particle and help scientists unlock the secrets of the basis of mass? Will it lead scientists to new and wonderful theories? In 2012 the Higgs particle was indeed confirmed experimentally by the LHC and in the next year, François Englert and Peter Higgs have won the Nobel Prize in Physics for their theoretical prediction of the Higgs Boson. However, I am in no doubt that all these things will come to pass. The cry of "more data, more data" is always the never-ending chorus that scientists chant. And let us face it: for £4.4 billion there is no way that the powers that be will allow such an investment to yield zero return. The question that I ask is how much, if any, there is an increase in scientific understanding and even what scientific understanding means.

5.3 Questions

I have tried to look at how scientific theories are structured and I conclude that networks of questions and answers (problematological networks) are a key element. My model structures scientific theories as networks of questions and answers [Sanitt, 1996; 2007], with the questions being the most important part. With the language of science describing processes rather than objects [Harris, 2005], the resultant model confronts the skeptical argument that science has no validity. In the present model the general idea of truth in science is jettisoned, because questions do not have truth values. Thus science does not progress by reaching greater truths, but progresses by integrating more questions within our theories. I set out some of the arguments as follows in the form of 15 questions and answers:

1. What is truth? Truth represents logical structure or deductive inference and therefore is not a category that can be applied en bloc in science. That is not to say that there are no specific prescribed areas of science where the concept of truth may be applicable, but these involve formal mathematics used within science and the mathematical structure of networks which form the present model.

2. What is scientific questioning? It is our human response to the world.

3. What is a scientific theory? Scientific theories are recipes, protocols, algorithms or heuristics for coping in the world.

4. Why are networks important? Directed graphs or networks are the mathematical structure for scientific questions which form part of scientific theories.

5. Who decides the questions and answers? We (scientists) ask the questions and attempt to provide answers where possible from observations and experiments.

6. What part does Nature play? Nature does not do anything; it is just our descriptive word for what there is.

7. How do we refer to things and what things are we referring to? There are no things, but processes. Our references to things are no more than verbal coat hangers.

8. What is meaning or understanding in science and how do we understand more and progress? Meaning is a process of integration. Scientific theories provide a logical template for our questions and answers. Understanding emerges within the expanding range of integration of meaning, as our theories encompass more questions.

9. How do we fit questions and answers together and form theories? The methodology of science and human intellect weaves answers and questions together.

10. Why are scientific theories useful? Theories provide a unified understanding, codified within a logical structure. Even though answers may come and go and theories are modified or abandoned, they can still conform to a mathematical structure.

11. What is causality? The general idea of causality is part of our theories. A restricted meaning of causality refers to the impossibility of going back in time.

12. What is the difference between expectation and prediction? We try to predict using our theories. Expectation is warranted prediction. When expectations are not realized we are more surprised than when predictions do not work out.

13. How do we distinguish between good and bad science? What do these terms mean? In terms of the present model, scientific theories and parts of scientific theories are distinguished. The distinguishing factor is the number of empirical questions covered by a theory compared to the number of theoretical questions raised. Bad science

is when a part of a theory, which generates too many theoretical questions, is promoted as being a whole valid structure. Good science gets better by covering more questions within a valid structure.

14. What does the present model of the structure of science achieve? The present model (1) codifies the structures of scientific theories and is useful as a meta-language for theories. (2) It delimits possible combinations of questions and answers. (3) By putting the category of questions to the forefront, it puts truth in its proper place and underscores the different categorical status that questions have from answers. (4) It provides a logical (mathematical) narrative to discuss the structure of a scientific theory.

15. Does the present model help us to create scientific theories? Beyond describing theories can the present model be used as a tool? I believe that the present model can be useful as a pointer for future work as well as a critique of existing theories. The model provides a way of talking about theories at the cognitive, meta-theory level.

5.4 Conclusion

Astrophysics and particle physics are going through a crisis period at the moment. These times are to be welcomed not feared. It is also at times like these that scientists are more amenable to new ideas, particularly from sources outside science.

Many years ago I was watching a performance of a two act play in a theatre. After the first act ended the interval seemed to go on longer than usual. Eventually, the manager appeared on stage and sheepishly inquired if there was a doctor in the house. The theatre was in Cambridge and I knew that there were several physicians in the audience. Somewhat hesitatingly one of them stood up and we were later given our money back as the rest of the performance was cancelled: the lead actor in the play had broken his arm in a fall during the interval. It was several years before I could catch up with the second act of the play. My point in relating this story is that you know things are badly wrong in a theatre when they ask if there is a doctor in the house. When scientists are open to ideas from philosophers or linguists you know that things have gone badly wrong. I hope that we now have an opportunity to do more than fix a broken arm.

References

Harris, R. [2005] *The Semantics of Science* (Continuum, London).
Meyer, M. [1995] *Of Problematology: Philosophy, Science and Language*, trans. Jamison, D with Hart, A. (University of Chicago Press, Chicago).
Sanitt, N. [1996] *Science as a Questioning Process* (IOP Publishing, Bristol).
Sanitt, N. [2007] "A mingled yarn: Problematology and science," *Revue Internationale de Philosophie* **61**, 435-449.
Sanitt, N. [2008] "The tripod of science: Communication, philosophy and education," in *Science Matters: Humanities as Complex Systems*, eds. Burguete, M. & Lam, L. (World Scientific, Singapore) pp. 121-135.

6

How to Deal with the Whole: Two Kinds of Holism in Methodology

Jin-Yang Liu

From the traditional viewpoint of holism, we know how to deal with a whole *only* when we do understand what the whole is. But from the perspective of methodology, "what is a whole" or "what kind is a whole" is not decided by the object itself in ontology but by "how to deal with the whole" in methodology. We thus need to make a paradigm shift from rigid ontological premise to methodological pragmatism, regarding the whole as a methodological hypothesis. And there are two kinds of holism in methodology: One is *constitutive holism* which regards an object as the constitutive whole (including entity, structure or function) and solves it through constitutive methods; the other is *generative holism* which regards an object as the generative process and deals with it through generative rules. However, neither kind has the strength of the other. It is necessary to think of a whole as a methodological concept in a strong sense, while taking it as an ontological reality in a weak sense.

6.1 Introduction

The term *holism* is not very old. It was firstly coined by Smuts in his book *Holism and Evolution* [1926] although discussion of holism can be traced back to ancient Eastern and Western philosophies. Holism is presented in a variety of forms such as *organism, Gestalt, collectivism, connectionism*, etc., and one may wonder whether the different forms are descriptions of a single philosophical position or there are various holisms. A common claim of different holisms is that *the whole is more than the sum of its parts*, and any theory that holds such a view can be taken as a sort of holism in a broad sense.

Holism had been regarded as a philosophical issue since long time ago. Different philosophical perspectives give rise to the diversity of holism such as ontological holism, property/relational holism, nomological holism [Healey, 2009].[1] Most debates of the philosophy of holism focus on the ontology,[2] such as Aristotle, Plato and Smuts, regarding the whole as a property, entity or fact in nature; some discuss the wholeness of theory in epistemology such as Duhem [(1906) 1991] and Quine [1956], who aroused arguments on *conformation holism* and *meaning holism*.[3]

Since the 20th century, some scientists have realized that holism can be a new methodology[4] beyond reductionism. Modern sciences (including Gestalt psychology, quantum mechanics, systems science, complexity science and cognitive science) are the main force leading to the *movement of holism* in different disciplines. For example, Gestalt psychology first introduced the concept of *whole* into the psychology in history;[5] scientists such as Bertalanffy [1968] and Auyang [1998] take

---

[1] According to Healey [2009], *Ontological Holism*: Some objects are not wholly composed of basic physical parts; *Property Holism*: Some objects have properties that are not determined by physical properties of their basic physical parts; *Nomological Holism*: Some objects obey laws that are not determined by fundamental physical laws governing the structure and behavior of their basic physical parts.

[2] In philosophy, *ontology* mainly deals with the essential characteristics of being (entity) itself, and asks questions such as what is or what exists? What kind of thing exists primarily? How are different kinds of being related to one another?

[3] In epistemology, holist holds that a thought, a hypothesis, or a theory, should not be considered in isolation without any reference to the whole it form part of. Confirmation Holism argues that the testing of scientific hypotheses should deal with the whole of theory, which cannot be reduced to single hypothesis tested by experience. Meaning Holism focuses on the holistic identification of means, claiming that the meaning of any word depends on its connections to every other expression in that language.

[4] In philosophy, *Methodology* is a general study of method in different fields of research (such as science, history, mathematics, psychology, philosophy, etc.) in order to establish acceptable ways of working to achieve truth; the methodology under study must possess a boundary so that the solution can be achieved through a finite number of steps. Different methodology means various roads or approaches to deal with such problems.

[5] The aim of applying Gestalt is "to find out which parts of nature belong as parts to functional wholes, to discover their position in these wholes, their degree of relative independence, and the articulation of larger wholes into sub-wholes" [Koffka, 1935, p. 22].

the whole as a methodological concept or a research approach; contemporary complexity science is considered as a new science of holism in the 21st century.

However, past discussions of holism are often based on a mixture of philosophical and scientific considerations. The result is a holism with the misleading conception that ontological and methodological are isomorphic to each other, resulting in much confusion in such an interdisciplinary research field. Thus, to clear up the confusion, we propose to reexamine the holism issue from a purely methodological framework by introducing two kinds of holism: *constitutive holism* and *generative holism*.[6] It will be shown that the previous forms of holisms can be grouped under these two categories.

6.2 Methodological Shift in Holism

Taking holism as a sort of methodology does not mean discarding entirely ontological considerations but rather to set these considerations within a methodological framework.

6.2.1 *A Hidden Assumption*

Generally, we often realize the wholeness of things through intuition and naturally took a whole as an ontological "entity" which is different from its parts. That there *exists* a whole is a premise of many holists. Smuts created holism to bridge fundamental gaps among matter, mind and life. He makes it clear that "holism is a specific tendency, with a definite character, and creative of all characters in the universe, and thus fruitful of results and explanations in regard to the whole course of cosmic development" [Smuts, 1926, p. 100]. According to him, holism not only shows itself as the whole but also creates the whole in the progressive

---

[6] Strictly speaking, method and methodology are not the same. The former often refers to specific methods, and the latter refers to principles to use for concrete methods. In this chapter, we define "method" in a broad sense, which refers not only to a specific method, but also to methodology.

phases of reality.[7]

However, when "a whole" is referred as some sort of ontological being, we will be in a great ontological puzzle. For instance, a table can be a whole consisting of four legs and one board, which also can be regard as a whole composed by wood and nails. Which one is the real whole of the table? The boundary of whole cannot be determined by the table itself. And for a special whole, is the whole prior or its parts? Which one is more fundamental among the whole, the parts or their relationships? These questions have become unsolved metaphysical contradictions which can be traced back to Aristotle, giving rise to more confusion and controversial arguments in the study of holism [Philips, 1976]. Moreover, the ontological view is often regarded as a strong support for methodology by holists. Looijen [2000, p. 1] sums up the two typical statements of holism in biology as follows:

1. An organism is essentially nothing but a complex set of atoms and molecules.

2. You cannot just simply reduce an organism to a sack of molecules!

Here, statement 1 is an ontological view while statement 2 is a methodological operation; statement 2 is a plausible deduction of statement 1. Here exists a hidden assumption: methodology and ontology are isomorphic—the methodological principle of dealing with an object strictly depends on the ontological condition of the object—or we know how to deal with a whole only when we do understand what the whole is.[8] As Auyang [1998, p. 53] emphasizes:

---

[7] There are six phases: (1) definite material structure/synthesis of parts in natural bodies; (2) functional structure in living bodies; (3) specific co-operative activity becomes coordinated or regulated by some marked central control; (4) conscious central control in personality and holistic groups in society; (5) central control becomes super-individual; (6) the ideal whole or holistic ideals, absolute values [Smuts, 1926, p. 106].

[8] This assumption can also be found in many discussions on reductionism. Some researchers regard reductionism as a *philosophical doctrine* in ontology and others refer reductionism as a *scientific procedure* in methodology. Agazzi [1991, p. vii] points out the difference between "reductionism" and "reduction." In this chapter, reductionism is mainly regarded as a methodological operation.

That a system is physically constructed of parts lawfully interacting with each other does not imply that all descriptions of the system are logically constructible from the laws governing the parts. That the behaviors of a system are the causal consequences of the motion of its constituents does not imply that the concepts representing the system behaviors are the mathematical consequence of the concept representing the constituent motion. The implications depend on strong assumptions about the structure of the world, the structure of theoretical reason, and the relation between the two. They demand that we conceptualize the world in such a way that the alleged logical or mathematical derivation is feasible for systems of any complexity. This strong assumption is dubious. Methodology does not recapitulate ontology.

Thus, we need to make a methodological shift of holism. From this perspective, that *there exists a whole* (ontological view) is not a necessary premise of *deal with such a whole* (methodological approach). In this sense, the whole is not a rigid entity anymore, but is a sort of methodological category or approach.

6.2.2 Basic Conditions of a Whole in Methodology

The basic concept of the "whole" in methodology is given here.

1. Boundary

From the perspective of methodology, we should first rigorously differentiate absolute holism from relative holism, which is often neglected by holists and reductionists alike. Some holists take the whole as an indivisible entity without parts; this position can be regarded as *absolute holism*. For example, Plato in *Phoedo* points out that the human soul is such a whole. Furthermore, the whole can be taken as "all" or "one"; the whole means the Universe. The India Vedic philosophy and theology, for instance, holds such a view that "when we tried to put our minds to 'one', we will inevitably be guided to identify 'one' as 'partlessness'".[9] Bohm, a physicist and philosopher, tries to integrate all

---

[9] Plotinus (205-270), quoted from [Jin, 2000, pp. 95-96].

parts of Universe into an "undivided whole" including observables and *implicate order* [Bohm, 1980, p. 188].

Since we are going to deal with holism from the methodological approach (see footnote 4) we will not deal with absolute whole in the rest of this chapter. The reason lies in that the holism will become a metaphysical monadism or a special kind of atomism if the whole is one entity that cannot be split into any parts, and we cannot ascertain the boundary of the whole in practice if the whole means the totality of the universe. As Smuts [1926, p. 100] emphasizes,

> The only definite application of the idea has been made by absolutists, who have applied the expression of 'the whole' to the all of existence, to the cosmic whole, to the tout ensemble of the universe, considered as a unit or a being. This particular use of the idea does not interest us at this stage of this inquiry. The great whole may be the ultimate terminus, but it is not the line which we are following".

Hence, the methodological approach prefers to deal with a relative whole with boundary rather than an absolute whole. This kind of holism is called *relative holism*. Accordingly, the "whole" in our terminology is a relative concept—any whole could be a part of a greater whole. In other words, absolute holism could be considered as a strong holism while relative holism, a weak one.[10]

2. Structure

A whole is not a collection of parts but must appear as a "structure", being an integration of many elements.[11] Bertalanffy made a distinction between a "system" (i.e., what we call a whole) and a "complex"[12] in order to explain clearly this important feature [1968, pp. 44-45]. For our

---

[10] For reductionism, there is also a difference between strong reductionism and weak reductionism. The former denies the connections between the whole and its parts, while the latter accepts the connections. Both of them claim that the whole should be accounted for through the parts [Liu, 2008, pp. 123-124].

[11] In this chapter, an element may not amount to a part. Only when a whole is consisted of more than one interrelated elements (forming a structure) that this group of elements is called a part.

[12] "Complex" here means a "set" or a "collection".

6 How to Deal with the Whole

purpose, the distinction between element, complex and whole is clarified in Fig. 6.1.

There are usually three types of complex as shown in Fig. 6.1. In Type I, cases (a) and (b) are distinguished by the number of elements; in both cases there is no part or whole; each one is a complex. In Type II, case (a) is made up of the same kinds of elements while case (b), two kinds of elements; each case is a complex. In Type III, case (b) has a structure (i.e., the elements are interrelated) while case (a) does not; case (b) is a complex with a structure [while case (b) in Type II is a complex without a structure]. For a complex with structure it is obvious that to understand the complex we have to understand first its structure. According to Bertalanffy, a complex with structure is a prototype of a system, which is more general than a structure [Bertalanffy, 1968, pp. 54-56].

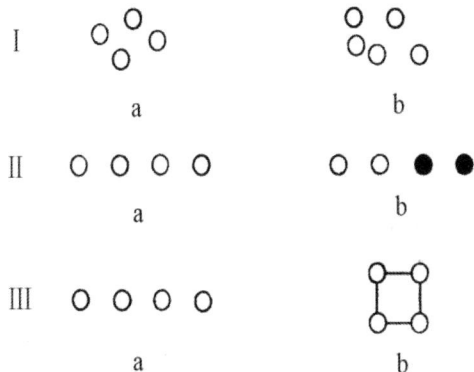

Fig. 6.1. The distinction between element, complex and whole (adapted from [Bertalanffy, 1968, p. 54]).

Note that a structure could be a physical structure such as a building or a piece of furniture, or a group of interacting relationships among components. Bertalanffy had examined exponential laws, allometric growth laws and so on in different areas such as biology, economics and sociology. In particular, in a whole consisting of parts, each of which grows with a different speed, the grow speeds have to maintain constant proportions to result in a structure. For instance, for a variety of animals, under proper assumptions, the problem can be described mathematically by a group of simultaneous differential equations resulting in the conclusion that the ratio of the metabolism rate and the weight growth rate (two competing factors) to be less than one, in consistent with empirical data [Bertalanffy, 1968, pp. 63-66].

A deeper understanding of "structure" is given by Piaget. He held that "the structure is preserved or enriched by the interplay of its transformation law, which never yield results external to it." [Piaget, 1970, p. 5]. Hence, a structure is not an entity but a transformation.

3. Non-summative

The most common understanding of a whole in holism is that the property of a whole is not the sum of its parts. The whole has properties and functions that its parts do not have. For example, life and its evolution—an emergent property—is obviously not summative. From the viewpoint of complexity science, the novelty of emergence in a complex system is beyond the simple combination of the individual properties and the functions of the parts.

The "non-summative" property triggers a methodological debate: Whether we can predict the emergence of a whole according to the properties and functions of its parts? Strong reductionism insists that a whole is the sum of its components, and a problem at a higher level can and must be solved by its components at a lower level. According to this view, we can reconstruct a phenomenon of higher level through the components of lower level.[13] Strong holism argues against such a view, holding that a whole cannot be divided into parts and we can only understand and cope with a whole through the whole itself or through

---

[13] Strong reductionism is also called "microreductionism". According to Bunge [1991, pp. 31-32], "A microreduction is an operation whereby things on a macrolevel are assumed to be either aggregates or combinations of microentities; macroproperties are assumed to result either from the mere aggregation of microproperties, or from a combination of the latter; and macroprocess are shown from microprocesses. In short, microreduction is the accounting of wholes by their parts." For example, water is composed of molecules resulting from the combination of hydrogen and oxygen atoms; social facts result from individual actions.

laws of the whole. Obviously there is a fundamental contradiction between strong reductionism and strong holism.[14]

6.3 Constitutive Holism

This contradiction can be expressed as such a question in methodology: Is it possible to study a whole only through its parts? Strong reductionists believe that it should be possible, supported by the success of modern science for more than 400 years. Nevertheless, some scientists realized the limits of reductionism, and began to consider a "whole" in science around the 20th century. As Bertalanffy emphasizes, "We are forced to deal with complexities, with 'wholes' or 'systems' in all fields of knowledge" [1968, p. 5]. To emphasize this important shift of methodological focus, from parts to the whole, we will introduce a new term, *constitutive holism*, to describe this approach.

6.3.1 *Constitutive Whole*

From the perspective of constitutive holism, the object should be regarded as a constitutive whole. Bahm calls this kind of whole "mechanical whole". He holds that the parts of a mechanical whole depend on each other in function but separate from each other in existence [1972, pp. 17-19].[15] Taking a car as example, each part of the car can be taken apart and is independent of other parts, while all the parts together constitute a whole—the car has functions that any single part of the car does not have. Obviously, mechanical whole tends to be an ontological entity while the constitutive whole includes not only such physical entities but also all kinds of composite system which can be deal with through constitutive methodology. For instance, the organization's hierarchy or a practical aim of an engineering activity is such kind of whole; both can be divided into parts from up to bottom and integrated to

---

[14] It needs to point out that there is no such a contradiction between weak reductionism and weak holism, because both of which accept connections between the whole and its parts.
[15] According to Bahm [1972], there are three kinds of whole: aggregate collection, mechanical whole and organic whole.

whole from bottom to up. Constitutive holism emphasizes that a whole is an integration of parts; one can dealt with the whole through the right division. In addition to meeting the three basic conditions of a whole in Section 6.2.2, a constitutive whole has to meet more conditions as follows.

1. Nearly decomposable

Constitutive whole should be regarded as a *nearly decomposable system* composed by more than two components; its function depends on mutual correlation between components. Nearly decomposable system was first proposed by Simon [1962, p. 474], which has two basic features: (1) in a nearly decomposable system, the *short-run*[16] behavior of each of the component subsystems is approximately independent of the short-run behavior of the other components; (2) the behavior of any one of the components depends in only an aggregate way on the behavior of the other components in the long run. "Nearly" means that we cannot decompose a whole into parts without relations otherwise the "whole" cannot be regarded as a whole.

These two features can be abstracted as: (1) *separated independence*: the existence and properties of any one component is irrelevant to other components in the whole; (2) *decomposable aggregation*: the whole is the aggregation of the functions or capabilities of the components; the function of any one component in the whole is relevant with other components. Thus, the parts in a constitutive whole are relatively independent of each other; they have their own performance and can be dismantled or replaced. For example, a laptop can be decomposed as display, processor, graphics, memory, storage, keyboard, LAN, battery, input/output and so on, any of which is independent, can be disassembled and can even be used elsewhere; however, they are integrated as a whole with more functions.

2. Stable structure

That there is a stable structure not only means that there exists a stable physical construction but also the possible existence of inner relations

---

[16] "Short-run" or "short range" means the interactions in a system only exist among near or neighborly components.

among the different components (such as quantitative operation, functional combinations and so on). In methodology, if the stable structure of a whole could not be *found*, we cannot deal with the whole. For instance, only do we make clear the structure of a machine or an organization and then decide its invariant, closed boundary, we can know how to cope with it.

3. Timeless

Whether the structure of a constitutive whole is static or dynamic, the time is reversible and is just an external parameter. Consequently, the constitutive whole is actually timeless in methodology. A constitutive entity (e.g., machine, crystal, etc.) does not change with time, and abstract structures are out of time. The stable structure of constitutive whole to some extent appears as an absolute consistency from past, present to future. This kind of time is not irreversible *real time* but is *spatial time* without direction.[17]

6.3.2 *An Example: Synthetic Microanalysis*

"Synthetic microanalysis" proposed by Auyang [1998] is an example of the constitutive holism. As shown in Fig. 6.2, the "system" can be considered as a "whole". For isolated theory, the system and constituents are not connected with each other, but there are theories to explain system and constituents, respectively. Without connection between whole and parts, strong holism can be taken for an example of system theory (ST), and strong reductionism for constituent theory (CT). For microreductionism, system is composed of constituents, and explanations for components are equal to interpretations of system. As a result, a

---

[17] In physics, time is reversible and in all laws of physics "we have found so far there does not seem to be any distinction between the past and future" [Feynman, 1965, p. 109]. However, in everyday life, time is an irreversible passage from past to future. Whitehead [1932, pp. 72-74] analyzes the spatial character of time in Newtonian physics, and Bergson thinks *duration* (acting and irreversible) as real time of life. "Real duration is that duration which gnaws on things, and leaves on them the mark of its tooth. If everything is in time, everything changes inwardly, and the same concrete reality never recurs...we do not think real time. But we live it because life transcends intellect." [Bergson, (1911) 1922, pp. 48-49]

whole becomes a collection of interacting elements, and micro explanations completely replace macro explanations. *Synthetic microanalysis* combines the comprehension of up-bottom and bottom-up together effectively with a broad theoretical framework. Micro explanations for components are restrained by macro system theory, and macro explanations of system must consider *micro mechanism* (mm) of component.

This method can be widely used in physics, economics and biology. Synthetic microanalysis does not exclude the division from the whole, but it requires us to be good at cutting in key joints, so that the up-bottom decomposition can build a proper connection between the whole and its parts. This is the key operation of constitutive whole. For example, when we analyze the whole significance of an English sentence, the right way is to divide the sentence into words and phrases, but not into letters. These words with independent meanings constitute meaning of the sentence.

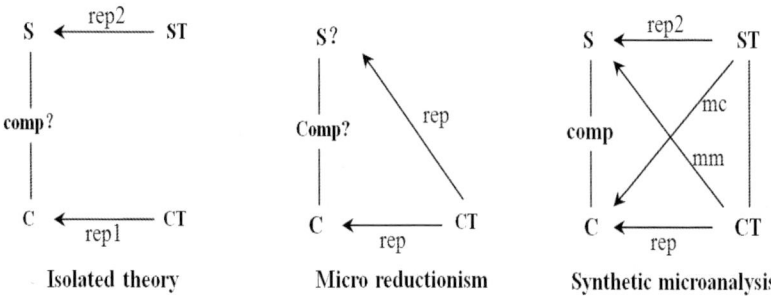

Fig. 6.2. Isolated theory, Micro Reductionism and Synthetic Microanalysis (adapted from [Auyang, 1998, p. 56]). S: system, C: constituent, comp: compose, ST: system theory, CT: constituent theory, rep: represent (rep1 is different from rep2), mc: macro constraints, mm: micro mechanisms.

6.4 Generative Holism

Different from constitutive whole's timeless features, evolutionary behaviors have distinctive temporal features such as birth, maturing, aging and death of a life; or creation, growing and expansion of an organization. Some philosophers realized the significance of evolution in overcoming the limits of mechanical worldview based on modern science. Bergson holds that the essence of evolution is due to *"élan vital"* (impetus of life) and *"duration"* [1911 (1922)].[18] Whitehead regards the *process* as a basis of the reality [1929]. Morin sharply criticizes the limits of system theory and tries to construct a real holism based on his understanding of complexity [1977]. Moreover, scientists and linguists also pay much attention to it. Bohm believes that the world is not "building blocks" but a universal flux of events and processes [1980, pp. 12-15]. Chomsky founds *"generative grammar"* in linguistics to focus on the generative procedure of natural language. What amazes him is "how a finite mechanism can construct an infinity of objects of this kind" [Chomsky, 2002]. All these discussions can be included in the consideration of a new sort of holism which we call *generative holism*.

6.4.1 An Example: Cellular Automaton

Let us begin with an example in detail as a way of introducing generative holism. Cellular automaton (CA) is our ideal example, which is a computer model of evolution, consisting of four basic components: cell, lattice, neighbors and (deterministic) local rules. In addition, "cell state" and "boundary condition of neighbors" are also important for the generation of a CA. A CA could be considered as a computer program consisting of cell spaces and its transformation rules defined in the spaces [Lam, 1998]. The time in CA is a discrete set. Wolfram made remarkable discoveries of CA in the 1980s, in the simplest one-dimensional (1D) CA. A one-dimensional CA has two advantages: (1) intuitive; cell states of each generation can be shown together in a line. (2) It is easy to be studied; the states and rules are very simple, providing easily a good indication of the working principle of CA.

---

[18] "The impetus of life, of which we are speaking, consists in a need of creation. It cannot create absolutely, because it is confronted with matter, that is to say with the movement that is the inverse of its own. But it seizes upon this matter, which is necessity itself, and strives to introduce into it the largest possible amount of indetermination and liberty" [Bergson, 1911 (1922), p. 265].

For instance, in a line of 256 cells, each cell state S has only two possible conditions {S_1, S_2} with S_i = 0 or 1; neighbor radius r = 1. According to the state of the cell and that of its direct neighboring cells on left and right, the cell state evolves. So there are 3 variables, and each variable has 2 state values. As a result, local rules will get $2^3 = 8$ different combinations of variables. Local rules can be definite as long as these 8 combination values are given. Given any 1D sequence of 0 and 1 (represented by a black or white cell, respectively), the next sequence will be created by applying the rules. Suppose the local rules are given by Table 6.1 and the initial state of the CA consists of one black cell at the center of the line. As Fig. 6.3 shows, the CA creates an evolutionary sequence, resulting in a complex structure.

Table 6.1. Local rules of a particular CA.

t	111	110	101	100	011	010	001	000
$t+1$	0	0	0	1	1	1	1	0

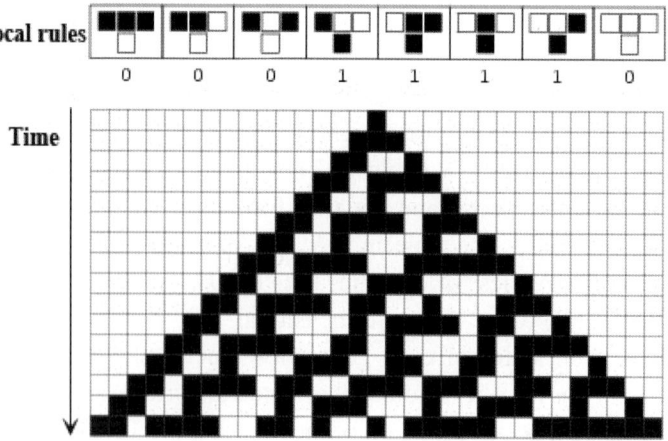

Fig. 6.3. Structure generated by the local rules in Table 6.1 of a one-dimensional cellular automaton. Each horizontal line is a generation. Source: Weinstein Eric W, "Cellular Automaton" (http://mathworld.wolfram.com/CellularAutomaton.html.)

If we remove the border and let the process continue infinitely, it will generate more new complex structures. If we set out from different initial configurations, it is sure to see completely different graphs of structures, even though the same rules are applied (Fig. 6.4). As Wolfram [1984] emphasized, "Even though the initial states are disordered, the systems organizing itself through its dynamical evolution, spontaneously generating complicated patterns". This shows that the process of generation is not a linear one and the end result is not a summation of initial values.

Fig. 6.4. Two structures with different complexity generated by the same rules [Wolfram, 1984].

Now we can think of CA from a view of holism. The first question one can ask is whether the CA is a whole? According to the three basic conditions of a whole mentioned above, we can analyze it as follows.

- *Boundary.* Any one CA is limited to a grid space, a line of finite length, even though the grid space can, in principle, be extended to infinite boundary.

- *Structure.* Any graphical configuration consisting of two or more generations of a CA can be regarded as a structure because it has more features than the cells. And the graph of each generation is one of parts which together constitute the final intricate triangular structure.

- *Non-summative.* The final intricate triangular structure is not the sum of cells, which has features that any single cell or any generation of cell do not have.

According to these analyses, there are no doubts that the CA is a whole. However, the CA is not a constitutive whole.

1. *The relations among different CA generations are not timeless.* On the contrary, each generation is the result of its former and the cause of the later. It is a typical temporal relation between generations, which do not emerge at the same time but appear as a temporal sequence of different time. Obviously, the whole is not a timeless entity.

2. *The CA is not a nearly decomposable system.* For CA, although we can make a distinction between different generations, any one generation does not exist by itself; it is a link between the previous generation and the next generation. They are closely connected with each other in time.

3. *The structure of CA is changeable*, which evolves constantly in time and each generation makes an irreversible change to the existed old configuration and then generate a new one which is usually more complex. What did amaze Wolfram was the diversity of evolutional behaviors in CA. He discovered four kinds of evolutional behaviors of CA, based on his deep research and a good number of computer experiments: stationary, periodic, chaotic and complex [Wolfram, 1984]. These spatial graphs of CA vividly demonstrate many evolutional characteristics and much complexity of the evolution process.

However, there also exist some sorts of invariance during the evolution. In the evolution process, everything is changeable, including the cell and its state, time and space, configurations of different generations and the structure, except the internal local rules. A set of given rules decide the special CA which has a definite evolutional sequence though one cannot always predict the result by the initial condition. Hence, during the whole generating process, it is not the entity or structure but the internal operational rules that remain unchanged in time, providing a self-organized evolutionary mechanism. From this perspective, the essence of evolution is a becoming mechanism based on a set of operational rules.

6.4.2 *Philosophical Discussion on "Generation"*

There are many discussions about "becoming" in Western philosophy, but it is difficult to find an exact term to describe its essence. "Organic whole" as a term of organism may partly describe evolutional relations of CA, but "organic" is apt to mislead, too. The CA is not an organic entity.

Due to a holistic idea of nature, there are plenty of relevant metaphysics thoughts in ancient Chinese philosophy such as that contained in *The Book of Changes*, *Lao-tse*, etc. For example, in *Lao-tse* (Chapter 42) one finds: "From Dao there comes one. From one there comes two. From two there comes three. From three there comes all things" [Fung, (1948) 2007, pp. 156]. According to Fung's interpretation, being is one; two and three are the beginning of the many. "Dao" itself is the "Super One" or the invariable laws of nature. Although things are ever changeable and changing, the laws that govern this change of things are not themselves changeable (pp. 156-157). Fung did not explain the meaning of "comes", which is translated from the Chinese word Sheng. Sheng has multiple meanings such as create, generate, produce and so on. We prefer to use "generate" because the expression "comes [from]" can only tell us a temporal order of things while overlooking its emerging mechanism. From this perspective, Tao is a set of rules to generate the many, and generation or generating is the essence of evolution in nature.

This view, due to interdisciplinary thinking of modern science and Chinese philosophy, is called *generativism* in China, advocated by Jin [2000], Li [2004] and others. Jin holds that generativism is a new holism different from the old "substantial holism" and "system holism".[19] "Generative" here means "creative" and "forming", which introduces a new understanding of whole as follows.

---

[19] According to our understanding, Jin & Cai's "substantial holism" [2007, pp. 2-9] is a sort of absolute holism, which holds that a whole is not a set of parts but the entity itself.

> The whole is dynamic and alive, which does not consist of any parts but only consists of itself. The whole has been a whole since its birth. There is not a conceptual relation between a whole and its parts. The process of being born is closely interacted with the forming of a whole; any kind of machine does not have such a feature... From the view of generative holism, parts are just the appearance, expression of the whole. Parts exist not only as components but as a specific expression of a whole which constantly show itself in the forms of parts. [Jin & Cai, 2007, p. 5]

Li considers generativism as a new revolution in constructivism. She wants to explain where the generative whole comes from. Enlightened by *The Book of Changes* and chaos theory, she points out that there exists an essential cause leading to the generation of a whole, which can be called the "element of generation". The "element of generation is the cause of an evolutional process, and this kind of cause is not a "material cause" but an "efficient cause" and the "final cause" in term of Aristotle's expression[20] [Li, 2004, p. 7]. No doubt, the rule of generation is the whole itself.

6.4.3 Generative Whole

The notion of generativism has increased our understanding of whole. One breakthrough for holism is that the generative whole is not an entity but a process or a set of rules (also called an "organizing principle" [Lam, 2005, p. 2322]). In the light of this, the rule of the generation is the whole itself. The basic features of a generative whole can be described as follows.

1. Nonseparability

The essence of an evolution is that it is a holistic process. It is not difficult to divide it into phases or stages which are often regarded as

---

[20] According to Aristotle, there are four kinds of cause: *material cause* (out of which things come to be), *formal cause* (what things essentially are), *efficient cause* (sources of movement and resting) and *final cause* (purpose or ends). It needs to point out that "cause" here differs from its modern use, which considers "a cause" as an agent or event exerting power and effecting a change; only "efficient cause" resembles this modern notion [Bunin & Yu, 2001, p. 151].

"parts" of the whole. However, logically speaking, they are just parts in name but not real parts of an evolutionary process. Strictly speaking, there exist no parts in a generative whole. For instance, if each generation of a CA is a part, the rule of whole should be different from the rule of parts because their features are not the same. But in fact every generation of a CA has (i.e., generated by) the same rule, and the rule of whole is the rule of generation. In a sense, "life emerges as a whole at the outset" [Kauffman, 1990, p. 309]. From the methodological perspective, generative holism does not try to divide a whole but to deal with it from the generative rule. Hence, the argument of "whole vs. part" is not fit to be used in describing evolution any more. The real reason lies in that a temporary process cannot be divided into "temporary parts" as is the case in carving a spatial entity [Oakland, 1992].

2. Invariable self-organizing mechanism

For a generative whole, everything including its constituents and structures are continuously changing with time because the number of cells and the structure of each generation are different. Only the generative mechanism based on a set of rules persist invariably. The generative mechanism is the essence of a generative whole maintaining itself through time. Different mechanisms decide different evolutional processes, representing different CAs, for example. These mechanisms are a typical of self-organizing models, as Eigen and Schuster [1977, p. 542] emphasize: "once established, it exists forever",[21] which is the reason for the order of life.

3. Open system

A living organism must be an open system since its survival depends on exchanging matter, energy and information with the outer environment. For instance, the CA cannot go on without continuous supply of information from outside. An open system is ipso facto the basis of

---

[21] According to Eigen and Schuster, a once-forever selection is a consequence of hypercyclic organization. A detailed analysis of macromolecular reproduction mechanism suggests that catalytic hypercycles are a minimums requirement for a macromolecular organization that is capable to accumulate, preserve and process genetic information.

self-organizing mechanism. According to [Bertalanffy, 1968, p. 141], the primary difference between a live organism and dead machine lies in that "the living organism is maintained in a continuous exchange of components." The open system makes this continuous exchange possible in a living organism.

4. Temporal irreversibility/history

Any generative whole is evolving in time continuously till its end, governed by (time-independent or time-depending) generative law(s). Changes of the whole in spatial configuration are a reflection of the temporal process. Here, the time is real, irreversible time. The essence of a generative whole is an irreversible process of generation and evolution; the entity and structure are just the resulting manifestations of the growing process. This feature is important for methodology, i.e., unless we know the *history* of a generative whole we cannot understand it.

6.5 Comparison between the Two Approaches

Typically, constitutive holism is to "cut" in order to decompose whole into parts while generative holism is to "generate" to learn how the mechanism of a whole works.

6.5.1 *Constitutive Holism: Carve Nature at Its Joint*

Constitutive holism as a form of weak holism can be regarded as a modified version of weak reductionism, because both do not reject dividing a whole into parts. This kind of holism includes hierarchy theory, cybernetics, systems engineering, synthetic microanalysis, etc. From the perspective of constitutive holism, the whole can be understood as a set of constitutive relations based on the constituents, structure and function. The core task in constitutive holism is to deal with the relations between the whole and its parts, whose premise and conclusion are deductive.

The key operation of constitutive holism is to "carve nature at its joint" (Plato, *Phaedrus*, 265d-266a), which means to divide a whole without damaging it. This approach is quite effective in scientific and engineering research and has been used to solve many practical problems. For example, Hall introduced morphological analysis to establish a typical research model in the field of systems engineering — Three-Dimensional Morphology of Systems Engineering, which provides useful guidelines to tackle engineering problems. We can decompose a general problem or system into its basic variables, each variable becoming a dimension of a morphological box (Fig. 6.5). "When the values that each variable can assume are found, a set consisting of one value of each variable defines a solution to the problem or a species of the general system" [Hall, 1969].

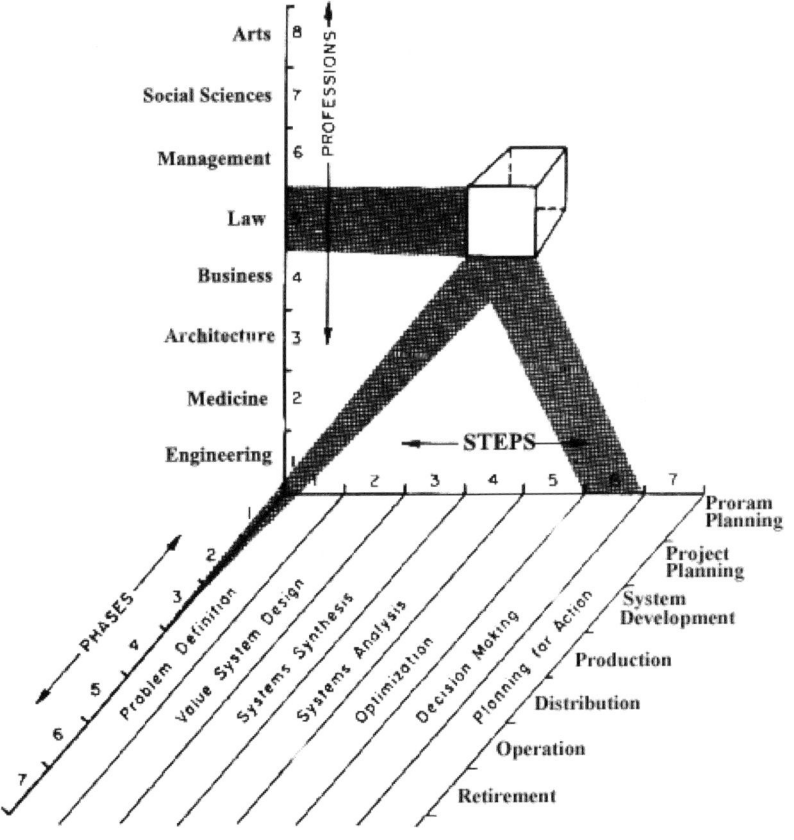

Fig. 6.5. Morphological box for systems engineering [Hall, 1969].

There are three different dimensions of an engineering system: time dimension segmented by major decision milestones (cut process/phases into a sequence of activities from inception to retirement), logic dimension (decompose the main problem into sub-problems of different levels, which can be operated in steps), and profession dimension (also called knowledge dimension, referring to the body of facts, models and procedures which define a discipline, profession or technology; integrate different knowledge to solve problems). It can be shown that the structures, phases and steps of timeless cutting are the essential operations of systems engineering.

6.5.2 *Generative Holism: How the Mechanism Works*

The idea of generative holism can be traced back to ancient philosophy. But as a scientific approach, it roots in new developments of contemporary science since the 20th century. [22] Different from the perspective of constitutive holism, in generative holism the whole is not taken as a timeless entity or structure but an evolutional process. According to complexity science, most problems of evolution are complex problems that cannot be tackled without consideration of time, such as the butterfly effect, emergence, self-organization criticality, etc. [Bak, 1996].

Thus, the core of generative holism is to find, by simulation or otherwise, the generative mechanism of a whole to learn how it works. For instance, scientists recently found that genes should not only be

---

[22] Some famous scientists such as Schrödinger, Bertalanffy and Turing took part in discussions on the problem of life in the 1940s. All of them realized the limits of reductionism and tried to give new explanations for the origin of life. Ashby described the self-organized characteristic of neural system in 1947. Influenced by it, system dynamics was created and made great developments by Forrester, Lofgren, etc. In the 1970s, more theories about self-organized evolution emerged; examples are: hypercycle theory (Eigen, 1967), dissipative structure theory (Prigogine, 1969), synergetic (Haken, 1969) and catastrophe theory (Thom, 1971). Since its founding in 1984, the Santa Fe Institute (SFI) has promoted complexity science as the approach in solving problems of evolutional emergence in different fields, which triggered more theories such as the complex adaptive system (CAS) theory, artificial life and complex network, and further developments of cellular automaton.

taken as the building blocks separated with each other but also the generative rules governed by switches within DNA which decide the final expression of genes and generate novelty in evolution. Understanding the genetic and molecular mechanisms governing the evolution of morphology has been a major challenge in biology [Carroll et al, 2008].[23] Moreover, contemporary scientists are exploring the human brain with a completely new approach. The Human Brain Project (HBP) aims "to build a completely new ICT (information communication technology) infrastructure for neuroscience, and for brain-related research in medicine and computing, catalyzing a global collaborative effort to understand the human brain and its diseases and ultimately to emulate its computational capabilities."[24] During six ICT-based research platforms of HBP, a Brain Simulation Platform will *simulate the actual working* of the brain; it integrates relevant information in unifying computer models, allowing in silicon experiments, impossible in the lab, in order to create realistic simulations of life processes. This is an enormous challenge for humans but it has showed that generative holism is necessary for such an exploration.

More extensive explorations for generative approach are carried out in the field of economics and management. For example, Bowles, in collaboration with SFI, established a new methodological paradigm, *Evolution Paradigm*, in social science which is based on evolution and behaviors in economics. In the past 100 years, neoclassical economics generally has been based on the following hypotheses: (1) Individual behaviors are established on the basis of preference, and the doer could make far-sight evaluation on the behavioral consequence; (2) preference is self-interested and is decided by exogenesis; (3) social interactions

---

[23] Current science research indicates that mutations in DNA "switches" that control body-shaping genes, rather than in the genes themselves, have been a significant source of evolving differences among animals. For example, 99% of human gene could find their copies on mouse. What cause the huge diversity may be the different gene switches, whose regulation determines whether one gene could express successfully [Carroll et al, 2008].
[24] The Human Brain Project, building on the work of the Blue Brain Project, is led by Henry Markram of the Ecole Polytechnique Fédérale de Lausanne (EPFL) (www.humanbrainproject.eul, Nov. 22, 2013).

only adopt contract communication, and increasing returns could be ignored in many situations. This set of hypotheses is called Walrasian paradigm in economics. But Bowles points out that economics should shift from the old Walrasian paradigm to a new evolutionary and behavioral paradigm, which integrates multiple methods and strategies, including evolutionary game theory, population biology and analysis tools offered by behavioral simulation dynamic system [Bowles, 2004, p.479]. Table 6.2 shows the comparison between the two paradigms.

Table 6.2. Comparison between Walrasian paradigm and evolution paradigm [Bowles, 2004].

	Walrasian economics	Evolutionary social science
social interactions	complete and enforceable claims exchanged on competitive markets	direct (noncontractual) relationships in noncompetitive settings are common
technology	exogenous production functions with non-increasing returns	generalized increasing returns in both (endogenous) technology and social interactions (positive feedbacks)
updating	forward-looking individuals instantaneously update based on knowledge of entire system	backward-looking (experienced-based) individuals update using local information
outcomes	a unique stable equilibrium based on stationarity of individual actions	many equilibria; aggregate outcomes may be long term averages of non-stationary lower entities
time	comparative statics	explicit dynamics
chance	relevant only to risk-taking and insurance	essential component of evolutionary dynamics
domain	the economy as a self-contained, self-regulating entity: exogenous preferences and institutions	the economy as embedded in a larger social and ecological system: coevolving preferences and institutions
preferences	self-regarding preferences, defined over outcomes	self- and other-regarding preferences, defined over outcomes and processes
price and quantities	prices allocate resources; actors are not quantity constrained	quantity constraints; wealth-dependent contractual opportunities
method	reductionism (methodological individualism)	non-reductionism; selection on individual and higher order entities

Based on the discussions of Bowles and in the rest of this chapter, a detailed comparison between constitutive holism and generative holism can be finally drawn, as shown in Table 6.3.

Table 6.3. Comparision between constitutive holism and generative holism.

Item	Constitutive holism	Generative holism
object	timeless entity and structure	evolutional, irreversible process
property	wholeness	potentiality
relation	spatial causality (material cause)	temporal causality (final cause, efficient cause)
system	close system	open system
level	hierarchical division	inter-hierarchy modeling
operation	cut a whole into its parts from up-bottom, and construct whole from bottom-up	no division between a whole and its parts, but explore generative mechanism of evolution
examples	systems analysis, systems engineering, synthetic microanalysis, structural design	artificial life, systems dynamics, systems thinking, CAS theory, complex network

6.6 Conclusion

It is necessary to shift the discussion of a whole from the focus of ontological view to methodological hypothesis. Only then, holism can exceed the limits due to traditional views. The main points of this chapter can be summed up as follows.

1. Absolute holism is just a metaphysical view which cannot be applied in any kind of methodology. Relative holism is a fundamental premise of any kind of holism in methodology.

2. The limit of the old ontological perspective lies in that the whole or the wholeness of an object is the internal property, entity, function or the object itself, which cannot ascertain what the real whole of an object is, leading to an unsolved internal contradiction in theory.

3. Methodology and ontology are not isomorphic to each other. That there exits a *whole* is just a *methodological hypothesis* for holism.

Whether an object is a whole or what kind of whole it is not decided by the object itself but by the practical aim of the method.

4. *Constitutive holism* and *generative holism* are two main kinds of holism in methodology. The same object could be analyzed from different views of methodological whole. For an organization, its organizational structure could be analyzed in a constitutive view, while its operational process could be understood from the perspective of generative holism. However, neither kind has the strength of the other.

5. The methodological perspective should not be taken as relativism but rather as pragmatism. There also exist some ontological conditions affecting the application of methods in practice. A constitutive whole avails to be dealt with by constitutive holism, while generative holism can be a more effective approach for a generative whole. In fact, it is not difficult to objectively distinguish between constitutive whole and generative whole, entity and mechanism, reality and process. Rescher [1998, p. 16] calls such a dependence on ontology an "ontological repercussion".

Finally, it should be emphasized that a whole may not be an ontological "thing" but *can be* a methodological "way", based on either the constitutive holism or the generative holism.

References

Agazzi, E. (ed.) [1991] *The Problem of Reductionism in Science* (Kluwer Academic, Boston).
Auyang, S. Y. [1998] *Foundations of Complex-System Theories* (Cambridge University Press, Cambridge).
Bahm, A. J. [1972] "Wholes and parts," *The Southwestern Journal of Philosophy* **3**(1), 17-22.
Bak, P. [1996] *How Nature Works: The Science of Self-Organized Criticality* (Copernicus, New York).
Bergson, H. [(1911) 1922] *Creative Evolution*, trans. Mitchell, A. (Macmillan, New York).
Bertalanffy, L. v. [1968] *General System Theory* (George Braziller, New York).
Bohm, D. [1980] *Wholeness and the Implicate Order* (Routledge, New York).

Bowles, S. [2004] *Microeconomics: Behavior, Institutions, and Evolution* (Princeton University Press, Princeton).
Bunge, M. [1991] "The power and limits of reduction," in *The Problem of Reductionism in Science*, ed. Agazzi, E. (Kluwer Academic, Boston) pp. 31-49.
Bunin, N. & Yu Ji-Yuan [2001] *Dictionary of Western Philosophy: English-Chinese* (People's Publishing House, Beijing).
Carroll, S. B., Prud'homme, B. & Gompel, N. [2008] "Regulating evolution," *Scientific American*, May, 60-67.
Chomsky, N. [2002] *On Nature and Language* (Cambridge University Press).
Duhem, P. [(1906) 1991] *The Aim and Structure of Physical Theory*, trans. Wiener, P. P. (Princeton University Press).
Eigen, M. & Schuster, P. [1977] "The hypercycle: A principle of natural self-organization," *Die Naturwissenschaften* **64**, 541-565.
Feynman, R. [1965] *The Character of Physical Law* (MIT Press, Cambridge, MA).
Fung You-Lan [(1948) 2007] *A Short History of Chinese Philosophy*, bilingual edition, trans. Zhao Fu-San (Tianjing Academy of Social Sciences Press, Tianjing).
Hall, A. D. [1969] "Three-dimensional morphology of systems engineering," *IEEE Transactions on Systems Science and Cybernetics* **5**(2), 156–160.
Healey, R. [2009] "Holism and nonseparability in physics," *The Stanford Encyclopedia of Philosophy* (Spring 2009 Edition), ed. Zalta, E. N., URL = <http://plato.stanford.edu/archives/spr2009/entries/physics-holism/> (June 16, 2012).
Jin Wu-Lun [2000] *Philosophy of Generativism* (Hebei University Press, Baoding).
Jin Wu-Lun & Cai Lun [2007] "A new understanding of holism," *Journal of Renmin University of China*, No. 3, pp. 2-9.
Kauffman, S. A. [1990] "The science of complexity and 'origins of order'," in *PSA: Proceedings of the Biennial Meeting of the Philosophy of Science Association, Vol. 1990; Vol. Two: Symposia and Invited Papers*, pp. 299-322.
Koffka, B. Y. K. [1935] *Principles of Gestalt Psychology* (Harcourt, Brace and Company, New York)
Lam, L. [1998] *Nonlinear Physics for Beginners: Fractals, Chaos, Solitons, Pattern Formation, Cellular Automata and Complex Systems* (World Scientific, Singapore).
Lam, L. [2005] "Active Walks: The first twelve years (part I)," *Int. J. Bifurcation and Chaos* **15**, 2317-2348.
Li Shu-Hua [2004] "System science: From constructivism to generativism," *Journal of System Dialectics*, No. 2, pp. 5-9.
Liu Jin-Yang [2008] *Complexity: A Philosophical View* (Hunan Science and Technology Press, Changsha)

Looijen, R. C. [2000] *Holism and Reductionism in Biology and Ecology: The Mutual Dependence of Higher and Lower Research Programmes* (Kluwer Academic, Boston).
Morin, E. [1977] *La Méthode, tome 1: La Nature de la Nature*, Chinese ed. trans. Wu Hong-Miao & Feng Xue-Jun (Peking University Press, 2002).
Oakland, L. N. [1992] "Temporal passage and temporal parts," *Noûs* **26**(1) 79-84.
Philips, D. C. [1976] *Holistic Thought in Social Science* (Stanford University Press, Stanford).
Piaget, J. [1970] *Structuralism*, trans. Maschler, C. (Basic Books, New York).
Quine, W. V. O. [1951] "Two dogmas of empiricism," *The Philosophical Review* **60**, 20-43.
Rescher, N. [1998] *Complexity: A Philosophical Overview* (Transaction Publishers, New Brunswick, NJ).
Simon, H. A. [1962] "The architecture of complexity," in *Proceedings of the American Philosophical Society*, Vol. 106, No. 6 (Dec. 12, 1962) p. 474.
Smuts, J. C. [1926] *Holism and Evolution* (Macmillan, New York).
Whitehead, A. N. [1929] *Process and Reality* (Cambridge University Press, Cambridge).
Wolfram, S. [1984] "Complex systems theory" (http://www.stephenwolfram.com/publications/articles/ca/84-complex/2/text.html).

Part II
History of Science

7

Helicobacter: The Ease and Difficulty of a New Discovery

Robin Warren

The story of helicobacter is summarized. The story illustrates the ease and difficulty of a new discovery.

7.1 Introduction

Before the 1970s, well fixed specimens of gastric mucosa were rare. Then the flexible endoscope was introduced, enabling well-fixed small biopsies from the stomach to be made. Gastric histology and pathology were clearly demonstrated. Whitehead accurately described it in 1972.

In June 1979 I was examining a gastric biopsy showing chronic inflammation and the active change. Over the next two years I collected numerous similar cases. In 1981 I met Barry Marshal and we completed a clinico-pathological study of 100 outpatients referred for gastroscopy. There was little relation between the infection and the patients' symptoms. Peptic ulcers, particularly duodenal ulcers, were very closely related to the infection. We cultured *Helicobacter pylori*. In 1986, with Marshall et al, I studied and confirmed the effect of eradication of *H pylori* on the recurrence of duodenal ulcer. Our result overturned the over-one-hundred-years-old belief that bacteria do not grow in the stomach.

In the following, the story of these developments is elaborated.

7.2 Before 1979

Before 1970, well-fixed specimens of gastric mucosa were rarely seen in clinical practice. Biopsies, taken with the rigid gastroscope or the suction method, were very uncommon. Gastrectomy specimens are clamped at each end, with the contents inside. They fix slowly from the outside. Meanwhile the mucosa autolyzes and any organisms disappear. Autopsy specimens are even worse. Most of the specimens were taken to remove tumors or ulcers and pathology descriptions were centered on this rather than the fine histology of the mucosa. If it was described at all, gastritis was given names such as "superficial" or "atrophic", which showed little real relationship to the histology.

Since the early days of medical bacteriology, over one hundred years ago, it was taught that bacteria do not grow in the stomach. When I was a student, this was taken as so obvious as to barely rate a mention. As my knowledge of medicine and then pathology increased, I found there were exceptions. Organisms, usually yeast or fungus, often grow in the necrotic debris in ulcers or tumors. Unusual infections sometimes do involve the gastric wall; once I saw tuberculosis. Bacteria, floating above the mucus layer on the epithelium, are often seen in gastric biopsies. They appear to be mixed varieties, probably just passing through, dead, or contaminants.

The introduction of the flexible endoscope changed all this. It enabled gastroenterologists to biopsy many of their patients. Small biopsies, placed immediately into formalin, fixed well. Instead of rare, these became some of our most frequent biopsies. Whitehead accurately described them in 1972. He described "active" changes, which become important in my story. These changes were common and they provided a specific feature to diagnose. His pictures show intraepithelial polymorph infiltration in the necks of the gastric glands and a remarkable distortion of the foveolar (surface) epithelium, with marked cell irregularity.

Whitehead devised a classification based on the features he actually saw and described. Most of these features are mentioned in the diagnosis. This allows any associations between histology and other clinical features to be noted. I was very impressed with Whitehead's work. I used his classification, with minor alterations, and the pathology of the stomach suddenly seemed to make sense. The diagnosis described in one short line the features actually seen.

Microbiological stains are excellent for staining bacteria in smears, especially from a clean culture. However, histology shows a complex mass of tissue structures that also stain. To see bacteria, it is necessary to contrast them with the tissue. Gram positive organisms and, with care, acid fast organisms can be seen in tissue sections. Warthin-Starry silver stain of tissue shows spirochaetes (in syphilitic chancres) and bipolar Donovan bodies (the Gram negative bacilli in granuloma inguinale). I was interested in microbiological stains, and after seeing several cases of granuloma inguinale in which the bacteria were clearly visible with the silver stain, I was experimenting with this stain on other Gram negative organisms, with variable success.

7.3 The Breakthrough

My adventure with Helicobacter began in June 1979. I was a young pathologist when the high quality gastric biopsies became frequent. I had an interest in gastric pathology, based on Whitehead's work and, in particular, his description of active gastritis. I was interested in bacterial stains. A routine biopsy showed severe active chronic gastritis (Fig. 7.1). I thought I saw bacteria growing on the surface.

The epithelium showed gross cobblestone change, very similar to Whitehead's description. Nuclei were out of alignment. Mucus secretion showed a marked patchy reduction. Focal intraepithelial polymerphonuclear leucocytes were present. There were numerous lymphocytes and plasma cells in the stroma. A thin blue line was visible on the surface, which on high power I thought consisted of numerous bacteria. My colleagues could not see them, so I stained them with the Warthin-Starry silver stain and numerous bacteria were visible at low power, for all to see.

I took tissue from the wax block used for standard histology and obtained the electron microscopy. The images were of good quality and showed the bacteria well. There were small curved bacilli closely applied to the surface. Some were attached to microvilli. The cells bulged out.

Mucus secretion was reduced. Bacteria were between the bulging tops of the cells.

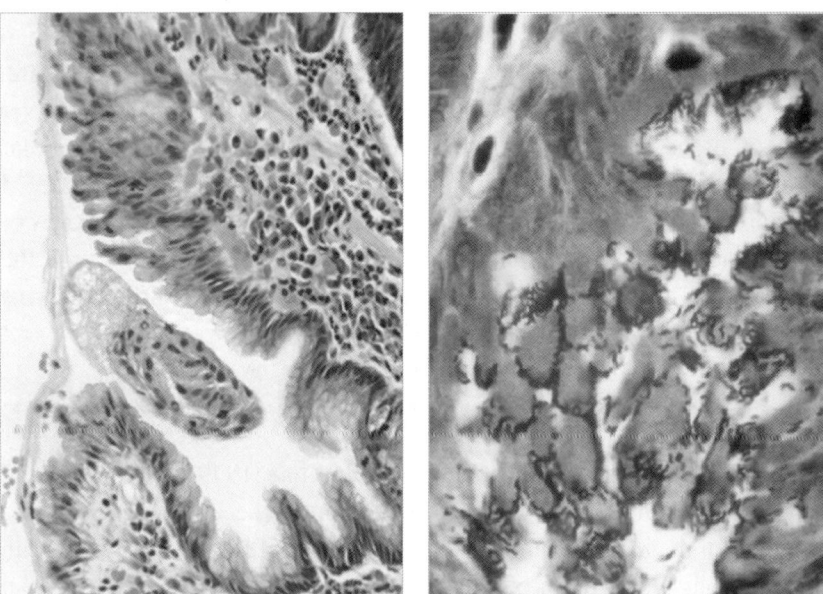

Fig 7.1. My first case. The mucosa (left) shows gross cobblestone change. High-power view with the silver stain (right) shows numerous curved bacilli on the distorted epithelium.

Finally, my colleagues believed the bacteria were there. However, they doubted their importance, and challenged me to find any more cases. I tried and, to my surprise, I found them in quite a significant number of biopsies. The number increased with experience. Many cases showed only mild pathology, but the basic changes were still present. Eventually I was finding them in about a third of the gastric biopsies.

I was unable to convince the clinicians of the importance of the organisms. Generally, they did not believe they were there at all. "Everybody knew the stomach was sterile". Gastritis was not considered to be of much significance anyway. Most thought that if the bacteria were there, they were just secondary to the gastritis. The histology suggested the opposite to me, but it was hard to prove. Another common question was "If they are there, why has not anyone described them before". At that stage I did not know why I had not seen them, let alone anyone else.

It has become apparent over the years that gastric bacteria have been described many times over the last 100 years [Marshall, 2002]. However, these descriptions were not generally known. Most of them were either veterinary biopsies or from research animals, which provided well fixed specimens without regard for "patient" well-being. Most descriptions were looked on as peculiarities, of no particular importance, even by their authors. The apparent absence of any previous report was given to me as one of the main reasons why they could not be there at all.

I worked in a laboratory, without patient contact. Although they were better than they had been, the gastric biopsies we received were usually taken from visible lesions, often ulcers, to diagnose or exclude carcinoma. As a result, the histology showed the effects of the nearby lesion. I needed biopsies from apparently intact antral mucosa, to show the effects of the bacteria. The idea of taking gastric biopsies for culture was considered ludicrous. The patient's well-being was the prime consideration.

7.4 Marshall and I

After two years I had collected many cases and was almost ready to publish my findings. Then Barry Marshall, the new gastroenterology registrar, strode into my room and demanded to see my work. He had been told to find a research project, and since he did not like the one suggested, his superiors sent him to me. He was the first person to show any interest in my work, so I showed him. He did not seem impressed at first, but he agreed to send me a series of biopsies from apparently normal gastric antrum, to see if the same findings were present. He soon became more enthusiastic, and I finally had a clinical collaborator.

In 1982, we obtained biopsies for culture and histology from 100 consecutive outpatients sent for gastroscopy. Most complained of peptic symptoms or pain, so this could not be investigated. They all completed a detailed clinical protocol that listed every symptom Barry could think of.

The results were totally unexpected. First, the bacteria were not related to any significant symptoms, only bad breath and burping. Then the gastroscopy reports showed that the gastric infection was most closely related to duodenal ulcer. At first, no bacteria were cultured. Finally, plates incubated over the Easter holiday showed a culture of a new type of bacteria, not described previously.

I sent a letter to the Lancet in 1983, a summary of the work I had done before I met Barry [Warren, 1983]. Barry sent an accompanying letter describing our joint work. He also presented our findings at the Brussels Campylobacter conference. Martin Skirrow, who chaired the conference, was very impressed with our work.

We sent our definitive paper to the Lancet in 1984 [Marshall & Warren, 1984]. Although the editors wanted to publish, they were unable to find any reviewers who believed our findings. Our contact with Skirrow became crucial here. We told him of our trouble, and he had our work repeated in his laboratory, with the same results. When he informed the Lancet, they published our paper immediately, unaltered.

I continued as a clinical pathologist, with an interest in Helicobacter. The subject rapidly expanded throughout medicine over the next decade. Original methods for diagnosis and treatment were all suggested by Barry. I was involved with the pathology from two attempts to fulfill Koch's postulates,[1] improved methods of culture and studies of duodenal ulcer.

Helicobacter patients show considerable variation. I was involved with these early examples.

- Barry gave himself a severe acute gastritis, to the disgust of his wife, in an attempt to fulfill Koch's postulates.

- Arthur Morris, in New Zealand, gave himself chronic gastritis and took years to cure it.

---

[1] Robert Koch was a German microbiologist in the late 19th century, who stated that "a bacterium can be said to be the cause of a disease if it is obtained in a pure culture from a patient with the disease, the infection of a previously well animal with this culture results in the same disease and the same bacteria are cultured again from this animal".

- My wife developed arthritis and as soon as she took NSAIDs she developed severe epigastric pain. Stopping the NSAIDs reversed this. And again. I sent her to Barry, who found Helicobacter, treated it and she was able to take the NAISD. Don't take it for granted that NSAIDs are the only guilty party.

- Most patients are symptomless. I was an example of this. After she was treated, my wife complained I had bad breath. I was positive for *H pylori* and after treatment marital bliss returned.

A later study of duodenal ulcers in 1986 confirmed that the ulcers could almost be viewed as distal pyloric rather than as true duodenal [Marshall et al, 1988]. A large amount of histological material from this study shows conclusively that, as well as preventing recurrence of ulcers, eradication of *H pylori* rapidly reverses the changes of active gastritis (Fig. 7.2). This strongly suggests that the bacteria cause the gastritis, not the reverse, as was often suggested to me.

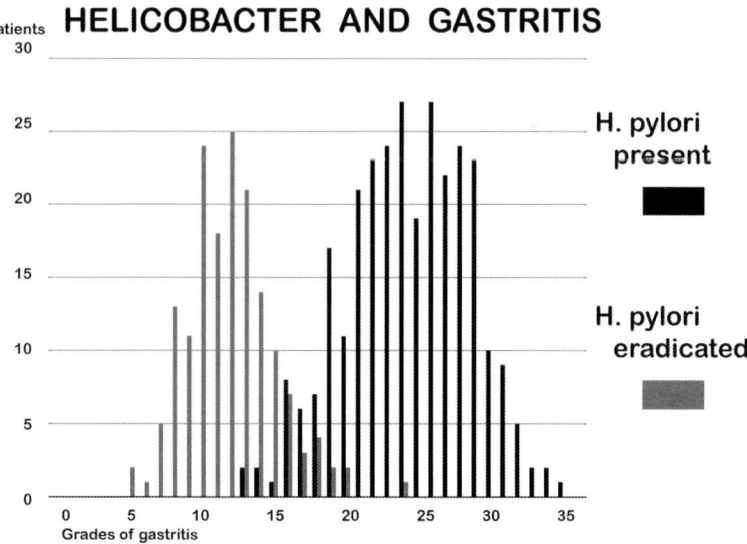

Fig. 7.2. Histogram, comparing gastritis before and after eradication of *H pylori*. The normal range is (0 – 14), in the absence of pre-existing disease or infection.

7.5 Conclusion

Now, the importance of Helicobacter is generally recognized, particularly with regard to duodenal ulcer. As a pathologist, I am disappointed that active gastritis is not considered worthy of treatment. Unfortunately, it does not cause many symptoms and nobody is interested. Only pathologists see it. So overall, we now know that Helicobacter had been seen and largely ignored for over 100 years. I saw them and linked them with active gastritis 25 years ago. Barry Marshall and I cultured the bacteria and linked them to duodenal ulcer. In various different ways over the next few years we proved these relationships.

References

Marshall, B. J. (ed.) [2002] *Helicobacter Pioneers: Firsthand Accounts from the Scientists Who Discovered Helicobacters, 1892-1982* (Blackwell, Melbourne).

Marshall, B. J., Goodwin, C. S., Warren, J. R., Murray, R., Blincow, E. D., Blackbourn, S. J., Phillips, M., Waters, T. E. & Sanderson, C. R. [1988] "Prospective double blind trial of duodenal ulcer relapse after eradication of *Campylobacter pylori*," *Lancet*, ii: 1437-1442.

Marshall, B. J. & Warren, J. R. [1984] "Unidentified curved bacilli in the stomach of patients with gastritis and peptic ulceration," *Lancet*, i: 1311-1315.

Warren, J. R. [1983] "Unidentified curved bacilli on gastric epithelium in active chronic gastritis," *Lancet* (letter), i: 1273.

Editor's Note: In 2005 Robin Warren, together with Barry Marshall, was awarded the Nobel Prize in Physiology or Medicine for the "discovery of the bacterium *Helicobacter pylori* and its role in gastritis and peptic ulcer disease".

8

Science in Victorian Era: New Observations on Two Old Theses

Dun Liu

Based on the science conditions in Victorian era (1837-1901) in England, we point out the deficiency of Friedrick Engel's thesis concerning the three great scientific discoveries in the 19th century. Moreover, we question Mitsutomo Yuasa's thesis on the rule governing the shift of the world's scientific center, which has been popular since the 1960s. Both theses were widely accepted in China.

8.1 Introduction

Due to the influence of special political and historical environments, *Dialectics of Nature* by the German philosopher and revolutionary Friedrich Engels (1820-1895), a manuscript discussing natural philosophy, was considered as a classic in directing the scientific research in China (as well the former USSR and some Eastern European countries) [Engels, 1971]. In these places, the book was also a must read among researchers in history of science or philosophy of science. Moreover, "Dialectics of Nature" has been developed into an academic discipline, enjoying a position similar to that of history of science, philosophy of science or sociology of science in the West [Gong, 2005]. In particular, Engels' thesis about "the three great scientific discoveries in the 19th century", presented in a "Reading Note", was widely accepted by the scholars in these countries. While asserting Engels' philosophical insight on the significance of important scientific discoveries, in this chapter, we question his thesis mentioned above and give some supplementary comments on the thesis' shortcomings.

Similarly, we question another thesis which is also widely circulated among the Chinese academics. In our opinion, the rule governing the shift of the world's scientific center proposed by Mitsutomo Yuasa (1909-2005), a Japanese historian of science, in 1962 needs to be refined [Yuasa, 1962]. For instance, in the late 19th century, England was still the world's leader in science; her greatest intellectual heroes are represented by life science's Charles Darwin (1809-1882) and physical science's James Clerk Maxwell (1831-1879). In fact, within a certain period in modern history, it is not just possible but indeed a historical fact that two or more science centers could exist simultaneously in the world.

8.2 What Engels Missed Out

In reality, *Dialectics of Nature* is an unfinished work, which comprises of ten essays and 171 notes written by Engels from 1873-1886. It was published in 1925 by USSR's national press with two languages, German and Russian, in one book. The full translation appeared in 1932 in China. After People's Republic of China was established in 1949, the book's translation was directed by the Chinese Communist Party's Central Translation and Editing Bureau. The most popular version is *Dialectics of Nature* published by People's Press in 1971 [Engels, 1971; Gong, 2005]. It has been read by China's every university student. "The three great scientific discoveries in the 19th century" proposed in the book appeared frequently in various examinations. According to Engels, the three great discoveries are: cell theory (1838, M. Schleiden; 1839, T. Schwann), law of energy conservation and transformation (1842, J. R. Meyer; 1847, J. Joule and L. A. Colding), and biological evolution theory (1859, C. Darwin) [Engels, 1971, pp. 175-177].

As a politician not formally trained in science, Engels' recognition of these discoveries is admirable since the three discoveries are indeed humans' top achievements in understanding nature's rules up to that time, two of which belong to life science and the third belongs to physical science. The question then is: Is there anything else that was left out? Or, regarding its influence on humans' understanding of the workings of the world is there any other scientific achievement in the late 19th century that is comparable to these three discoveries?

In fact, within biology, there is a very important discovery which, though not as influential as Darwin's evolution theory, is at least more important than the cell theory pioneered by Matthias Schleiden, Theodor Schwann, and others. That is, the understanding of some hereditary rules of life initiated by Gregor Mendel (1822-1884), which could be considered as the beginning of an ongoing revolution in the life sciences. From 1856-1863, Mendel carried out a series of pea hybridization experiments in a monastery for eight years and thus discovered two fundamental laws governing biological heredity. He finished his report in 1865 and published it in 1866. The significance of his discovery was not noticed until and after 1900 when his work was repeated by others and his contribution recognized by the heredity researchers [Orel, 1984].

To the public, Mendel was just an unknown monk living in a small cloister; it is not surprising that Engels (and others) knew nothing about him or his work. Engels himself is also a great man who received neither a university education nor rigorous scientific training. But he paid a lot of attention to the newest developments in the natural sciences and innovations in technology of the 19th century. And he gave an incisive summary of the significance of "the three great scientific discoveries" based on his views in speculative philosophy.

Here let us turn to the physical sciences. One of the great developments was finished in England in the late 19th century. Michael Faraday (1791-1867) in 1831 discovered the electromagnetic induction phenomenon. He made his discovery through his experiments and explained it by using the intuitive concept of lines of force. His successor, James Maxwell, was an intellectual hero in the Victorian era (1837-1901). He published in 1865 his masterpiece *A Dynamical Theory of the Electromagnetic Field*. The latter gives a completely mathematical description of Faraday's concept of lines of force and expresses the relationships between electricity and magnetism through a beautiful set of differential equations [Tolstoy, 1981].

The final description of the unified theory of electromagnetic fields resulted from the perfect combination of the experimental tradition expounded by Francis Bacon (1561-1626) and the describing-nature-by-mathematics tradition represented by Galileo Galilei (1564-1642). It signified the transition of the age of steam power to the age of electricity and led to the subsequent understanding of the temporal-spatial structure as well as the quantization of electromagnetism. Maxwell is a "high mountain" lying between classical physics and the new physics (quantum physics and relativity) of the 20th century. The Maxwell's equations could be called the "Divina Commedia" of physics. If one has to pick a single representative physicist that lived between the years of Issac Newton (1642-1727) and of Albert Einstein (1879-1955), I would give my vote to Maxwell. The time that his electromagnetic theory and his concept of "fields" being recognized by the scientists and verified by experiments happens to fall into the period that Engels wrote his *Dialectics of Nature*.

If there is anything missing in Engels' list of 19th century's important scientific discoveries, it will be the great achievement represented by the understanding of the connections between electricity and magnetism, beginning with Faraday and ending with Maxwell.[1]

8.3 Questioning Yuasa's Thesis

Mitsutomo Yuasa, a Japanese historian of science from Kobe University, proposed in 1962 a theory explaining the shift of "center of scientific activity". His work is based on the research of John Bernal (1901-1971) of England and the use of scientometrics, with data taken from *Science and Technology Chronological Table* and other literature. Its main argument is: The world's scientific center in the 16th century, after the Renaissance, was in Italy, the motherland of Galileo. It moved in the 17th century to England, the stage of the early Industrial Revolution, the Royal Society and Isaac Newton; and then to France, from Enlightenment to the French Revolution. The center then shifted to

---

[1] In Engels' time, the periodic table discovered by the Russian chemist, D. I. Mendeleyev (1834-1907), is of equal significance to that in Engels' list of three great discoveries. Engels did mention the period table as a brilliant example of Hegel's "quantitative change leading to qualitative change" [Engels, 1971, p. 51]. Since this chapter focuses on Victorian science, this topic will not be discussed further.

Germany from 1810 to 1920; and since the end of World War I, American science has assumed the top position in the world.

However, in the Victorian era there were plenty of bright stars in England's scientific "sky". During the mid- and late-19th century England (including Ireland which had not yet become independent) produced a large number of excellent scientists and scientific results in the areas of mathematics and natural sciences. Based on the literature [Lightman, 2004], some representative examples of these scientists and their works are list here.

Mathematics in the Victorian era excelled in logic and algebra, represented by George Peacock (1791-1858), Charles Babbage (1792-1871),[2] Augustus de Morgan (1806-1871), George Boole (1815-1864),[3] Charles Lutwidge Dodgson (1832-1898) and John Venn (1834-1923).[4] This school continued and led to the pioneering work in the foundation of mathematics by Alfred North Whitehead (1861-1947). There were also excellent works done during this period in the development of calculus and algebra, represented by George Green (1793-1841), James Joseph Sylvester (1814-1897) and Arthur Cayley (1821-1895). Moreover, the quaternion of William Rowan Hamilton (1805-1865), biquaternions of William Kingdo Cliffod (1845-1879), vector analysis of Oliver Heaviside (1850-1925), etc. provided important mathematic tools in the development of mathematical physics.

In physics, after the foundation of classical physics was laid down by Newton, thermodynamics and electromagnetism became the mainstream in the 19th century. The former is represented by James Prescott Joule (1818-1889) and William Thomson (Lord Kelvin, 1824-1907); the latter, Faraday and Maxwell mentioned above. In the exploration of the micro world, we have the electrooptic and magnetooptic effects discovered by John Kerr (1824-1907), the cathode-ray tube by William Crookes (1832-1919), and the electron by Joseph John Thomson (1856-1940). Furthermore, fluid dynamics by George

---

[2] He also designed the differential machine, the early version of modern computers.
[3] This name reminds us of the famous Boolean algebra, the foundation in the design of basic elements in modern computers.
[4] This name reminds us of the Venn Diagram, an essential element in formal logic and set theory.

Gabriel Stokes (1819-1903), colloid physics by John Tyndall (1820-1893), optics and statistical physics by John William Strutt (Third Baron Rayleigh, 1842-1919), scale contraction effect (Lorentz-Fitzgerald contraction hypothesis) by George Francis Fitzgerald (1851-1901), and so on are all important works.

In chemistry, John Dalton (1766-1844) established modern theory of atoms. Following the tradition of Humphry Davy (1778-1829),[5] William Ramsay (1852-1916) discovered a number of new elements (inert elements: helium, neon, argon, xenon, radon, etc.).

In life sciences, apart from the evolution theory proposed by Darwin and Alfred Russel Wallace (1823-1913), there were Brownian motion due to Robert Brown (1773-1858); and biometry and eugenics pioneered by Francis Galton (1822-1911), Darwin's cousin. Galto's successors, Karl Pearson (1857-1936) and Walter Frank Raphael Weldon (1860-1906), applied systematically statistics in biology; they not only perfected the foundation of biometry but also contributed importantly to modern mathematical statistics. Moreover, the works on comparative anatomy and paleontology by Richard Owen (1804-1892);[6] zoology and botany by the father and son, William Jackson Hooker (1785-1865) and Joseph Dalton Hooker (1817-1911); and the great discoveries in surgery sterilization by Joseph Lister (1827-1912) all are important events in the history of science.

The greatest geologist in the Victorian era was Charles Lyell (1797-1875). His earth-evolving theory based on gradualism expounded in *Principles of Geology* (1830-1833) inspired Darwin's evolution theory in biology. The most eminent astronomers were the Herschel family; *Outlines of Astronomy* (1849) by John Frederick William Herschel (1792-1871) was translated into Chinese in 1867, which became the first book that helped the Chinese to understand Western astronomy. Furthermore, the prediction and discovery of Neptune by John Couch Adams (1819-1892) was a great victory for Newtonian mechanics.[7]

---

[5] Davy was the most influential figure for Victorian chemists because of his discovery of a series of chemical elements (chlorine, potassium, sodium, calcium, magnesium, barium and boron) although he died before the reign of Queen Victoria.
[6] He coined the word dinosaur.
[7] The French astronomer Urbain Le Verrier (1811-1877) shared the credit with Adams.

In particular, the Victorian era produced Darwin and Maxwell, the superstars whose works singularly influenced science for a few hundred years. Their importance could be understood in terms of the grading system of visible stars proposed by the Victorian astronomer, Norman Robert Pogson (1829-1891), who suggested in 1856 to make a so-called Pogson's Ratio. According to Pogson, the dimmest visible stars are given the grade of 6 while the brightest ones, grade 1; thus, the brightness of the former is only 1% of that of the latter, while a first magnitude star is $100^{1/5}$ or about 2.512 times as bright as a second magnitude star.[8] The intelligence "ruler" proposed by the Russian physicist Lev Davidovich Landau (1908-1968) is even larger in its divisions. He thinks the contribution of first-rate physicists is more than ten times larger than that of second-rate physicists, and the super-first-rate Einstein is far beyond anything the second- or third-rate physicists could contribute. In any case, both Darwin and Maxwell could be rated as superstars in the scientific "sky", with the brightness far beyond those of grade 1 in Pogson's Ratio.

Victorian era's England was a unique superpower in her time. The empire's overall strength was number one in the world, with a great economy, a stable society and a liberal political system. The scientific stars were numerous and innovations in science and technology aplenty. With the incomparable Darwin and Maxwell (the latter the successor of Newton and forerunner of Einstein) it is hard to say that England was not the world's scientific center in the Victorian era, in contrary to what Yuasa's thesis says.

8.4 Conclusion

Engel was mainly a political figure. In spite of his great attainments in philosophy and his deep interests in the contemporarily development of natural sciences, his writings in a single essay should not be taken to be the basis of a conclusion in the study of scientific history. Similarly, his unfinished essays in *Dialectics of Nature* should not be used as the "map" in the theoretical studies in science.

---

[8] http://en.wikipedia.org/wiki/Norman_Robert_Pogson (Nov. 30, 2013).

Yuasa's thesis was based on numeric statistics. Starting with several reference-type books, he rated the scientific and technological results obtained in several countries during the same period. His work is a good example in scientific metrology, reflecting the trend in the shift of world's scientific centers. However, the database of scientific metrology available in his time was incomplete. His rating was based more on quantity than on quality, and therefore is too rough by today's standard.[9]

As in any scientific inquiry, research in history of science has to be based on empirical materials. Only through comprehensive analysis of the data done seriously and not through the sayings coming from the authorities or the books that one can reach reliable conclusions.

References

Engels, F. [1971] *Dialectics of Nature*, Chinese trans. (People's Press, Beijing).
Gong Yu-Zhi [2005] *Dialectics of Nature in China* (Peking University Press, Beijing).
Lightman, B. (ed.) [2004] *The Dictionary of Nineteenth-Century British Scientists*, Vols. 1-4 (Thoemmes Continuum, Bristol).
Orel, V. [1984] *Mendel* (Oxford University Press, London).
Tolstoy, I. [1981] *James Klerk Maxwell* (University of Chicago Press, Chicago).
Yuang, Jiang-Yang [2005] "A re-examination of the trend about shift of scientific center," *Science & Culture Review* **2**(2), 60-75.
Yuasa, M. [1962] "Center of scientific activity: Its shift from the 16th to the 20th century," *Japanese Studies in the History of Science* **1**(1), 57-75.

---

[9] The aim of this chapter is not to give a complete evaluation of Yuasa's thesis. To show that the scientific development in the 19th century does not conform to his conclusion, it is sufficient to point out one instance. Since Germany as a unified national country existed only after 1870 then how can one says that Germany was the world's scientific center from 1810 to 1820? It seems that the starting year 1810 is related to the emergence of research universities. Was Yuasa influenced by some biased concepts when he worked in his statistical analysis? See [Yuan, 2005].

9

Medical Studies in Portugal around 1911

Maria Burguete

This chapter describes how medical sciences at University of Coimbra evolved over the course of the 19th century, and in the process highlights the relationship between the Faculty of Medicine and Faculty of Philosophy. In the mid-19th century the two Faculties were the scene of an effort to reform their teaching methods, on the basis of relationships they established with a number of prestigious European scientific institutions. Their research concentrated on the biological, physiological and chemical foundations of life. The creation of laboratories of experimental physiology, histology, toxicology and pathological anatomy was the result of the reorganization of the medical faculty at Coimbra University from 1866-1872, according to the following paradigm replacement: the superficial look at disease was replaced by the study of the inner body and an attempt to understand the symptoms, giving rise to a new paradigm of medical practice—evidence-based medicine. In this chapter, we give an overview of this process which enhanced the European influence upon the development of medical studies in Portugal.

9.1 Introduction

The evolution of medical sciences during the 19th century in Europe can be seen as a movie where Portugal played an important role. Why? Because medical sciences could only evolved when the necessary conditions were created; this means the creation of universities with its own laboratories and scientific apparatuses as well as a scientific staff with a university practice based on research, in which the research school or institution played a crucial role.

When we look to the emergence of universities in Europe we realize that Portugal, with University of Coimbra created in 1290 by King D. Dinis in Portugal, was the fifth country in Europe with a university including Medical Studies. This was only possible after the signature of an Agreement in 1288 when the conflicts among King D. Dinis and the bishops (Avignon pope authority) came to an end, in spite of the solid structure of a centenary church.

Curiously, during the whole course of the 19th century Portugal was more "European" than it is now despite the European Community adhesion since 1986. Therefore, we will present the influence of Europe on the development of medical sciences in Portugal, particularly concerning Coimbra University and the evolution of medical studies.

9.2 First European Universities

In Europe the domination of political power in university studies side by side with religious influence is the most relevant character in the development of universities. At the Middle Age we have religious influence linked to the emergence of universities all over the catholic Europe; later on this paradigm is replaced by the increasing relevance of the kings' Absolutism.

So the King became the source of all development related to university, in Portugal and other European countries such as Spain, France, United Kingdom, Italy, Germany and Austria. Therefore, the first European university was in Bologna, Italy (1087); the second, Oxford, UK (1214) followed by Paris, France (1215), Montpellier (1289) and the Coimbra University (1290), Portugal. In Germany the first university emerged in Heidelberg in 1385 and three years later in Cologne. This was the beginning of the exodus from the University of Paris to these German universities to escape from the Avignon pope authority. Following the German example, King James I created St. Andrews (1410), the first Scottish university, followed by Glasgow (1450) and Aberdeen (1494). Finally, King Christian I, dominating the whole Scandinavian countries (Sweden, Norway and Denmark), founded the first university in Uppsala (1477) and then in Copenhagen (1478).

During the 13th century the great intellectual development accredited to universitary studies resulted from theological reflection, including the development of Dialectic and Aristotle Logic. It was around 1400 that Theology became an independent discipline, creating two separate worlds: the world of faith and the world of reasoning, liberating "natural sciences" from theological concerns.

Coimbra University, the fifth European university created in Europe, naturally followed the pattern of other European universities, especially the French and German model. The Pombalin Reform of Coimbra University introduced by Marquis of Pombal in 1772 [Rodrigues, 2007] created the Faculty of Philosophy and Faculty of Mathematics, with close relationship to the Faculty of Medicine, which was modernized at the same time [Rasteiro, 1999]. (A Faculty consisted of several Departments, called College in some other countries.) This reform was marked by a number of truly audacious measures, the most important of which included: separating philosophical/natural studies from theological studies for the first time in Portugal; fully introducing the ideals of the Scientific Revolution at the curricular level; attaching value to the laboratory teaching of the physical/chemical and natural sciences; and a revamped way of teaching medicine, with more lab sessions. The study plans of the three faculties (philosophy, mathematics and medicine) were organized articulately, with complementary scientific and pedagogical aspects [Burgucte, 2010a].

9.3 Emergence of Laboratory Teaching at Coimbra University

Excellent conditions were established for the new lab teaching with the creation of the Laboratories of Experimental Physics [Carvalho, 1978] and of Natural History; the construction of a modern building for teaching and research in chemistry—the Chemical Laboratory (Laboratório Chymico), which is now the university's Science Museum;[1] the creation of the Botanical Garden, an Anatomy Theatre in the new Faculty of Medicine, and a new Teaching Hospital and Pharmaceutical

---

[1] See: Catalog of *Museu da Ciência*: *Luz e Matéria* (2006), Universidade de Coimbra, Coimbra, Portugal.

Dispensary [Pita, 1996]. Of particular interest in this period was the emphasis with experimentation, which was clearly expressed in the introduction of lab sessions in physics, chemistry, natural sciences and medicine.

However, after the fall of Marquis of Pombal the spirit of reform gradually evaporated. The beginning of the 19th century was a period of great decadence in medical studies, not only in the lecture halls but also in the establishments that had been created to give students lessons in experiments. From 1850 onwards there was a significant development in the teaching of both the physical/chemical sciences and the medical sciences at the Faculty of Philosophy and Faculty of Medicine—a process that is evident from the collections of the Coimbra University's libraries, archives and museums [Burguete, 2010a]. In 1834 the medical faculty was temporarily moved to the Colégio de São Jerónimo building, which was empty following the departure of the monks as a result of the Law on the Abolition of the Religious Orders. This 17th century building was then transformed into the new Faculty of Medicine Hospital (Fig. 9.1).

Medical studies at the medicine faculty existed since 1290. However, it was only from 1863 on with Costa Simões' teaching of general histology and physiology that this faculty emerged as a Faculty of Medical (Medicine) while teaching medicine as Science of Health, taking the experimental feature into account.

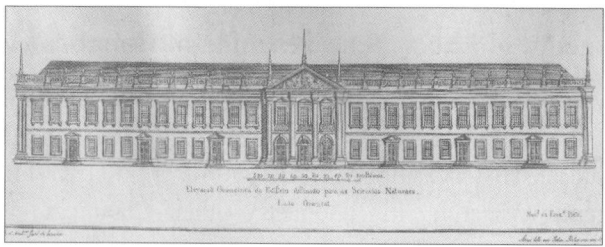

Fig. 9.1. Faculty of Medicine at Colégio de Jesus, Coimbra University.

9.4 Faculty of Medicine at Coimbra University (1863–1892)

Coimbra University's Faculty of Medicine and related developments from 1863 to 1892 are described here.

9.4.1 Faculty of Medicine (1863–1872)

Medical studies were again expanded between 1863 and 1872, resulting from a number of academic visits undertook by medical professors to a variety of university centers around Europe. Particularly significant were the travels of António Augusto da Costa Simões (1819-1903) (Fig. 9.2) who, together with the preparer (like a technician) Inácio Rodrigues da Costa Duarte, visited various medical establishments in France, Belgium, England and Germany. These trips permitted the introduction of new ideas and led to the modernization of the laboratories in Coimbra. The main purpose of the modernization was to make it possible to establish new modules with experimental nature, such as Histology and General Physiology, and Pathological Anatomy, which were introduced as part of an expansion of medical studies laid down in a Legal Charter dated May 26, 1863. Shortly afterwards, in July the same year, the study plan was changed to include these two modules.

Fig. 9.2. António Augusto da Costa Simões, Marck Athias, Kurtz Jacobsohn and Egas Moniz (left to right).

9.4.2 Medical Laboratories and Scientific Travellers

On August 18, 1864 Costa Simões and Costa Duarte began a scientific visit to various schools in Germany (Bonn, Wurzburg, Heidelberg, Munich, Gottingen and Berlin), France (Paris) and Switzerland (Zurich), with the goal of developing the histology and physiology modules. The report on this trip clearly shows the work they did and how much they learned [Costa Simões, 1866a].

Costa Simões studied experimental physiology and histology at the University of Paris, noting the excellence of their animal histological preparations in spite of the lack of human histological preparations. Similar studies were also performed at Belgian, German and Holland universities, including a report with the best scientific instrumentation found in the laboratories of experimental physiology at Liege, Utrecht and Berlin. Costa Simões developed relationships with some of the most prominent scientific researchers all over the world. Among others, the physiologist Emil Du-Bois Reymond (1818-1896), the pathologist Rudolf Virchow (1821-1902) and Hermann von Helmholtz (1821-1894) belonged to the circle of students around the Berlin anatomist and physiologist Johannes Müller (1801-1858), a significant figure who helped to usher in the dawn of the natural scientific era.

Costa Simões also attended some lab courses at Ghent University where Richard Boddaert (at that time one of the best in Europe) was appointed pathological anatomy professor (1863-1892) once he became a specialist in physiological sciences in 1862. He also researched and compared the teaching methods of histology and experimental physiology at other universities in Belgium (Brussels, Louvain and Liege) and Holland (Amsterdam, Rotterdam, Leyden and Utrecht), and made a list of the main apparatuses he found in the laboratories of experimental physiology at Ghent, Liege and Utrecht.

Later on, Costa Simões also visited medical establishments in Munich, Gottingen, Berlin and Vienna, and commented on the similarities in the overall way in which histology and experimental physiology studies were organized at the German universities. He particularly praised the Berlin Histology Laboratory at Berlin University (1810), describing it as a model establishment (Fig. 9.3). Costa Simões [1866a; 1866b] gave a detailed description of them all and thus found himself in possession of information about the most modern apparatuses. He studied their mechanisms and acquired the skills needed to carry out some more sophisticated experiments. After performing several experimental works, Costa Simões decided to adopt the German model from Berlin University, to set up the Laboratory of Experimental Histology and Physiology equipped with scientific instruments from Berlin, Breslau, Vienna, Munich and Paris. Some of them are described in Gerhardt Catalogue, still existing at the medical faculty of Coimbra University.

Costa Simões surveyed the teaching of experimental histology and physiology at the University of Paris. He noticed that although there were some excellent preparers for general microscopy (who worked on small animals, delicate plants and elegant crystallisations), they did not prepare any experiments for human histology. Therefore, in October 1866 the Experimental Physiology Laboratory was established at Coimbra University's medical faculty, with the preparations of human histology being one of the main purposes. These preparations of human tissues were quite a novelty at the time even for the best laboratories such as the Laboratory of Paris University.

Fig. 9.3. Faculty of Medicine, Berlin University.

Costa Simões' knowledge also resulted from the experimental lectures he attended at several European laboratories, in particular, the Berlin School of Medicine [Burguete, 2010b]. A few details about the medical faculty of Berlin University will be presented below in order to understand why the Berlin school was the model chosen by Coimbra's Faculty of Medicine.

9.4.3 The Berlin School of Medicine

The great Berlin university hospital, the Charité, had been doing research in life sciences for 300 years, since King Frederick I (1657-1713). A plague house had been built outside the town walls to the northwest side of Berlin to fight the terrible 1709 epidemic plague that moved towards Prussia's borders from northeast Europe. Three years later, in 1713, Berlin received an anatomical theatre where, under the care of the Royal Society of Sciences, military "barber surgeons"[2], surgeons, midwives and students from outside Berlin received free tuition. It offered space for 167 students at 6 levels. Yet a good army also needed good military physicians who were experienced in surgery and internal medicine. Therefore, anatomical and surgical instruction in the anatomical theatre was extended to include the subjects of pathology and therapy, physics, chemistry, mathematics, botany and pharmacy. However, only in October 1810, the medical faculty of Berlin University—the Berlin Frederick William University and today the Humboldt University—was founded. The founding dean of the medical faculty was the royal physician Christoph Wilhelm Hufeland (1726-1836) while Wilhelm von Humboldt (1767-1835) was the intellectual father of the new university [Burguete, 2010a].

9.4.4 The Dawn of the Natural Science Era

Philosophical approaches became less relevant and research concentrated on the biological, physiological, chemical and biochemical foundations of life. In 1861 the medical curriculum and examination guidelines at Prussian universities were reformed and "tentamen philosophicum" was replaced by a new "tentamen physicum". Medical students today are still required to study this subject after their first semester. Since then, questions about the humanities have occurred rarely; the content is primarily about anatomy, physiology, physics and biochemistry. Curiously a similar reform occurred in Coimbra's medical curriculum but only later in 1872 at the centenary celebration of Coimbra

---

[2] Barber surgeons are "surgeons" with insufficient experience. Before the medical studies were reformulated, there were only "barber surgeons" and no "surgeons".

University's reform [Mirabeau, 1872]. On this occasion, the medical faculty was again reminded of the need for new modules in order to develop a number of scientific branches (such as cutaneous diseases) that had since gained in importance [Burguete, 2010a].

As mentioned above, before this happened and in 1864-1865, Costa Simões made his scientific trip around Europe. He attended several classes and studied the developments of medical courses in the most developed countries. The names connected with this natural scientific era were already mentioned: Emil Du Bois Reymond, Rudolf Virchow and Hermann von Helmholtz. By measuring electrical current with a string galvanometer, von Helmholtz was able to show that human beings can produce electricity simply by tensing their muscles. Now we can perfectly understand why Costa Simões adopted the model of Berlin's medical school and why he developed the experimental works executed by remarkable scientists such as von Helmholtz, Reymond, Claude Bernard, and Theodor Schwann who was known as the father of physiology.

Upon Costa Simões' return, in October 1866, the Laboratory of Experimental Physiology based on the German model of the Berlin School was ready to start at Coimbra's Faculty of Medicine. The collection of scientific instruments of this laboratory consisted mainly of instruments recommended to him while visiting Berlin, Bonn, Breslau, Vienna and Paris. The program on the subject of histology and general physiology was based on the lectures given by Costa Simões from 1861 to 1864. The experimental study of the general physiology of organic systems within this program is limited to both the muscular and the nervous system. The old cupboard in Fig. 9.4 shows some scientific instruments and the textbook of Histology and Physiology (catalogue) used at the physiology laboratory beginning with the academic year of 1878/1879. Among these scientific instruments we have those used for the study of general physiology of the muscular system. The description of the twelve microscopes available at the Laboratory of Histology in 1873 is given on page 7 of the catalogue.

Fig. 9.4. An old collection of instruments of physiology and histology at Coimbra University

9.4.5 Coimbra Microscopes Collection (1748–1872)

From the Coimbra microscopes collection [Burguete, 2010a], four instruments—two Nachets (a small and a large monocular model) together with one Hartnack small model and one Mechet—were acquired by Costa Simões after his trip. All the rest already existed in the laboratory of the medicine faculty. The most reliable microscope from this collection was the Smith, Beck & Beck (1861) which was recognized in 1861 by William Carpenter, President of the Microscopic Society of London: "in point of general excellence of workmanship, this scientific instrument microscope Smith, Beck & Beck cannot be surpassed" [Santos Viegas, 1862].

9.4.6 Faculty of Medicine (1872–1892)

The period in which the scope of medical studies was broadened as a result of the international contacts made by Costa Simões was followed by one of the most notable eras of transformation in the history of medicine at Coimbra, which enabled the progress in medical sciences accomplished by the university in the late 19th century. That is, the subjects that medical students took over a period of three years at the Faculty of Philosophy was replaced by one year of more appropriate study at the Faculty of Medicine, shortening medical courses from eight to six years. The organization of the medical curriculum in 1872 took its idea from Costa Simões' reform proposal of 1866 which recommended that medical students study ancillary sciences such as the physical/chemical sciences and those related to natural history, given by the Faculty of Philosophy, after incorporating them into the Faculty of Medicine [Mirabeau, 1872]. Coimbra thus witnessed the beginning of the study of the specialist clinical fields, which placed the faculty in a favourable position to broaden its program of practical studies.

The conditions available for lab courses improved constantly over the second half of the 19th century. At the end of the 19th century, besides seeking to adapt to the scientific developments of the day, the teaching of the physical/chemical sciences continued to bear in mind the training needs of future medical students. The progress achieved in the natural sciences made such a reform indispensable if one were to *place the teaching entrusted to the said faculty in harmony with the present state of those sciences* [Freire, 1884; Mirabeau, 1892]. A reform that responded in full to the demands of modern education would require not only an almost complete makeover of the existing legislation, but also a substantial increase in costs as a result of the creation of new modules and the outfitting of laboratories and departments that would be indispensable for the lab study of the natural sciences. The Faculty Council's proposal thus represented a transitional reform, in which the Faculty of Philosophy would be divided into two sections: physical/chemical sciences and historical/natural sciences.

The chemistry training of future medical students was a constant concern. Some of the main measures took account of the specific needs for training in the fields of chemical analysis and biological chemistry. Chemical analysis was proving so complex in terms of its different branches that it required the creation of a special module or course. Experience had shown that the existing chemical analysis course did not respond to the demands of an up-to-date training. It was becoming necessary to create an obligatory lab course in the subject. Albeit accompanied by experiments that could be conducted in the ordinary classroom, oral presentations were no substitute for the students doing the work in the laboratory themselves, because they did not give a clear idea of the processes involved.

Biological chemistry was proving indispensable to the study of physiology and pathology, and becoming absolutely necessary for future medical students. The Faculty Council's solution was to include these topics in the organic chemistry module, thereby transitionally filling in a gap in the curriculum.

9.5 Medical Science after 1911: The Lisbon Case

When medicine as a science of health was launched as an experimental science as a result of the reorganization of the medical faculty at Coimbra University in 1866-1872, conditions were created for the emergence of scientific research within the medical field. However it was not until 1911 that Coimbra's first Faculty of Sciences was created by combining the Faculty of Philosophy and the Faculty of Mathematics. The founding of Faculty of Medicine in Lisbon also happened at 1911. In this case the evolution of medical sciences in Lisbon after 1911 followed the professional development of three persons—Marck Athias (1875-1946), Kurtz Jacobsohn (1904-1991) and Egas Moniz (1874-1955)—who were deeply involved in the study of medicine (Fig. 9.2).

Marck Athias was born in the Madeira Island and studied in Portugal and France. He started to work in University of Lisbon in 1897. The research program developed by Athias' school started with nervous histophysiology in 1897, an area in which Athias made the first scientific contributions in 1895 and continued until 1915. This area expanded to include general histophysiology from 1905 onwards, branching into physiology and physiological chemistry from 1911 and into histopathology from 1923.

Kurt Jacobsohn, considered the father of biochemistry in Portugal, was born and educated in Berlin. Since 1929 and for the rest of his life, he worked at the Bento da Rocha Cabral Institute (RCI) in Lisbon. The Institute, the Portuguese version of the Rockefeller University devoted to the study of medicine, was established in 1921. Jacobsohn became a Portuguese citzen in 1935, and was a member of the various scientific societies created by the research school of Marck Athias.

Egas Moniz, born in Avanca, Portugal, studied at Coimbra University and later in Bordeaux and Paris. He became professor at Coimbra in 1902; nine years later he became the new chair in neurology at University of Lisbon where he joined the "Generation of 1911", a group of researchers that included Athias [Burguete, 2015]. He also worked as a physician in the Hospital of Santa Maria, Lisbon. In 1903 Moniz, served as a Deputy in the Portuguese parliament until 1917 when he became Ambassador to Spain. Later in 1917 he was appointed Minister for Foreign Affairs and one year later he was the president of the Portuguese Delegation at the Paris Peace Conference.

Moniz have done several important works in the area of neurobiology in RCI which culminated in 1949 while sharing with Walter Rudolph the Nobel prize in Medicine and Physiology for their discovery of brain angiography and prefrontal leucotomy. The RCI was the first top quality international research center in Portugal where scientific research within the areas of medicine, biochemistry and physiological chemistry could be developed.

9.6 Important Achievements

The important achievements are summarized below.

1. The scientific and pedagogical relationships between the Faculty of Philosophy and Faculty of Medicine became stronger over the course of the 19th century; bridges between the two fields of knowledge grew up in ways that helped to develop both of them. This was a period in which the need to bring the physical and chemical sciences into the explanation of the phenomena linked to the human body and health became clearer.

2. Medical studies stopped looking at the human body from a solely external point of view, and moved towards a more in-depth knowledge of what happens inside it—a process made possible through the use of the latest teachings in physics and chemistry. The experimental learning of the medical sciences provided by the acquisition and use of the appropriate instruments made it possible to contribute to the enrichment of the basic sciences.

3. With usefully conceived and artistically crafted experimental instruments, physiologists attempted to fathom the biological regularity

of life especially through experiments on animals. Only when masterful use of the instruments had been achieved, could isolated natural phenomena be exposed and the knowledge made clear.

4. In the middle of the 19th century, the laboratory took its place next to the pathologist's dissection hall as a central location for medical research. Young researchers transferred the latest insights from biology, physics and chemistry to understand the human body. The study of medicine started a new paradigm of medicine practice—evidence-based medicine, i.e., understanding the symptoms by the study of the inner body leaving behind the superficial look of disease. Their aim was to perform pure science and not getting lost in philosophical speculations about the essence and meaning of life. Accordingly, organic life was joined indivisibly with the solid structures and liquid parts of the body. Using special experimental devices, the researchers wanted to find, define, measure, note and evaluate the mechanisms of life—in the healthy as well as the sick—that can be perceived with the senses. These standards reflect the expectation that objective models of the functioning of the human body could be developed in order to derive sustainable diagnostic and therapeutic measures.

5. Spaces, laboratories and instrumentation were implemented and developed as a result of Costa Simões' involvement concerning scientific trips. From these trips an interesting collection of scientific instruments merged at the Faculty of Medicine at the University of Coimbra, which were acquired from several European cities (Paris, Berlin, Liege and Vienna). And implementation of new ideas and practices in teaching and research allowed teaching of experiments.

6. The program of Athias' research school brought about two important changes: (1) It consolidated in Portuguese physiological research an approach at the cell level and not only at the level of the organ; with physiological chemistry, a transition from the cellular to the molecular level was announced, within the scope of physiology. (2) Athias and his disciples upheld a set of ideals, some of them of positivist inspiration, which were characteristic of republican ideals and advocated a model of university and scientific research inspired by the German university reform, which started in Berlin in 1809.

7. Besides upholding a university practice based on research, in which the research school played a crucial role, they also defended scientific dissemination and communication in society in general, the promotion of a scientific culture among researchers as a form of counterbalancing an excessive specialization, the scientific interchange between people and institutions and, finally, the creation of scientific societies and specialized publications.

In this context, Athias involved himself in the creation of two scientific societies directed towards biomedical sciences. In 1907, together with Celestino da Costa and Abel Salazar, he founded the Sociedade Portuguesa de Ciências Naturais, and in 1920, the Sociedade Portuguesa de Biologia. Additionally, he controlled the editorial activity of these societies, viz., the corresponding magazines which were created to disseminate the original ideas or works coming out of the laboratories or institutions resulting from the experimental procedures. It was precisely through these dissemination channels of knowledge that Athias and his school not only "internationalized" Portuguese medicine during the first half of the 20th century, but also created in the local scientific community the need for original works with international standards.

8. During his 50 years of scientific activity in Portugal, Kurt Jacobsohn created a research school of biochemistry at the Bento da Rocha Cabral Institute. He gathered together medicine and chemistry students and developed an innovative research program on enzymology which earned him international recognition. He also consolidated the scientific tradition inaugurated by his group, which provided the basis for the institutionalization of biochemistry at the Faculty of Sciences at University of Lisbon.

9.7 Conclusion

United by a common ideal of university and scientific research, Marck Athias, Costa Simões, Egas Moniz and later on Kurt Jacobson, formed a close-knit group with its own identity, the so-called "Generation of 1911". As soon as Athias started his scientific activity in Portugal, he surrounded himself with young people interested in pursuing a scientific career. He created conditions for them to realize their dreams. The institutions he established stood side by side with other European institutions of renowned prestige.

In short, Costa Simões, Mark Athias, Egas Moniz and Kurt Jacobsohn were the "pivot" in the contextualization of experimental medicine in Portugal, in the trail of a greater "scientificity" in this area of human knowledge.

References

Burguete, M. [2010a] *Medical Studies at Coimbra in the XIX Century* (Lap Lambert Publishing, Saarbrücken, Germany).
Burguete, M. [2010b] "Laboratories at the Faculty of Medicine of the University of Coimbra in the XIX century," *Scientific Research and Essays* **5**(12), 1402-1417.
Burguete, M. [2015] "Generation of 1911: A case study in Portugal" (to be published).
Carvalho, R. [1978] *História do Gabinete de Física da Universidade de Coimbra desde a sua fundação (1772) até ao jubileu do professor italiano Giovanni Antonio Dalla Bella (1790)* (Imprensa Universidade de Coimbra, Coimbra, Portugal).
Costa Simões, A. [1866a] *Relatórios de uma viagem scientifica* (Imprensa Universidade de Coimbra, Coimbra, Portugal).
Costa Simões, A.[1866b], *O Instituto, Jornal Scientífico e Literário*, vol. XIII, 152-156. [Imprensa Universidade de Coimbra, Coimbra, Portugal].
Freire, F. [1884] "Projecto de reforma da faculdade de philosophia da universidade," *O Instituto* **31**(4), 186-193; 228-240 (Imprensa Universidade Coimbra, Coimbra, Portugal).
Mirabeau, B. [1872] *Memoria histórica e comemorativa da Faculdade de Medicina nos cem annos decorridos desde a reforma da Universidade em 1772 até ao presente* (Imprensa Universidade Coimbra, Coimbra, Portugal).
Mirabeau, B. [1892] *Additamento à memoria historica e commemorativa da Faculdade de Medicina, 1872-1892* (Imprensa Universidade Coimbra, Coimbra, Portugal).
Pita, J. [1996] *Farmácia, Medicina e Saúde Pública em Portugal (1772-1836)* (Editorial Minerva, Coimbra, Portugal).
Rasteiro, A. [1999] *O ensino médico em Coimbra: 1131-2000* (Quarteto Editora, Coimbra, Portugal).
Rodrigues, M. A. [2007] *A Universidade de Coimbra: Figuras e factos da sua história,*. Vol. I. (Campo das Letras-Editores, Porto, Portugal).
Santos Viegas, A. [1867] "Viagem scientífica do Dr. António dos Santos Viegas," 1st Report, Dec. 1866-May 1867, *Diário de Lisboa* (the official daily gazette of the Portuguese Government), October 1867.

10

The Founding of the International Liquid Crystal Society

Lui Lam

The story of the founding of the International Liquid Crystal Society in 1990 is told here for the first time. The founding process lasted three years starting 1987 and is quite different from the usual case concerning other learned societies. A personal account of the why and how as well as the background and crucial events is given. It is written for those working in or interested in science, liquid crystals in particular, and for science historians.

10.1 Introduction

Liquid crystal is a state of matter intermediate between liquid and crystals. The molecules of the organic compounds that exhibit liquid crystal phases may be rodic, discotic, or bowlic in shape [Lin, 1982; 1987]. Rodic liquid crystals are the ones used in liquid crystal display (LCD) today and were discovered in 1888 by the Austro-Hungarian botanist Frederick Reinitzer (1857-1927). Since the industrial application of liquid crystals as display was proposed in the 1960s, there has been a resurrection of intense interest in these materials [Kawamoto, 2002]. The explosive commercialization of LCD televisions since 2007,[1] a $100 billion industry, makes the study of liquid crystals as a research field more important than ever.

---

[1] In 2007, LCD TVs overtook cathode-ray-tube TVs in sales worldwide for the first time (en.wikipedia.org/wiki/LCD_television, April 10, 2013).

An important landmark in the history of liquid crystals (LC) is the establishment of the Liquid Crystal Institute (LCI) in 1965 by Glenn Brown (1915-1995), an American chemist, at Kent State University, Ohio.[2] The Institute has been and remains an important driving force in LC's study and application. In August the same year, the first official International Liquid Crystal Conference (ILCC) was held at LCI, with 50 scientists worldwide and 42 papers presented.[3] Three years later in 1968, the 2nd ILCC was hosted by Kent, too, and since then, it has become a biennial series.[4]

An international "Planning and Steering Committee (PSC)" was formed to oversee this series. More precisely, it is the "PSC for ILCCs" because its only function was to decide, at each ILCC, who would organize the next ILCC. The PSC was disbanded in 1990 in Vancouver, Canada, when the International Liquid Crystal Society (ILCS) was established during the 13th ILCC held there. It was the Conference Committee (chaired by the author, Fig. 10.1) within the ILCS that took over the function of the PSC.

The ILCS differs from the PSC in many ways; the most important difference is that the ILCS is an open organization that is owned by the whole LC community while the PSC was essentially a "gentlemen's club." In any case, 23 years later, the ILCS remains the only game in town. According to the ILCS' official website, "Since 1990, the ILCS has attracted nearly 900 members in 43 countries and territories on six continents. The Society also serves as an umbrella organization for regional societies established in recent years."[5] By all standards, it is a success story.

But then something odd happens. Apart from the implied statement that the ILCS first came into being in 1990, there is no mention of its pre-history on this website. For example: (1) Where did the ILCS come

---

[2] www.lcinet.kent.edu/index.php (April 17, 2013).
[3] The number 50 comes from J. William Doane's "Glenn H. Brown, Honorary Chairman, 11th International Liquid Crystal Conference," *Scientific Sessions and Abstracts Book of the 11th ILCC*, June 30-July 4, 1986, Berkeley, California. The number 42 comes from the same Abstracts Book.
[4] Up to now, ILCC has been held in Kent five times (1965, 1968, 1972, 1976 and 1996).
[5] www.ilcsoc.org/ILCS/aboutILCC.html (April 22, 2013).

from? (2) Who initiated it? (3) Why the initiation? (4) Who saw it through? (5) How was it established? (5) Why it was 1990 and not sooner or later? Answers to these questions could neither be found in the first issue of *Liquid Crystals Today* (LCT), ILCS' official publication (Fig. 10.2). Seven years later, the then President of ILCS, Atsuo Fukuda, put in what he knew: "The International Liquid Crystal Society (ILCS) replaced the Planning and Steering Committee (PSC) for ILCCs in 1990" [Fukuda, 1998], which was a (small) step forward but did not help in answering the questions above.

International Liquid Crystal Society
Conference Committee

Chairman: Lui Lam
            Department of Physics, San Jose State University
            San Jose, CA 95192, USA

Phone: (408)924-5261   Fax: (408)9244815
Bitnet: luilam@calstate   Telex: 171 171 UD

                                                  May 2, 1991
Prof., Dr. Sci. Viktor V. Titov
NIOPIK,
Ul. B. Sadovaya, 1/4.
103787 Moscow, GSP-3
USSR

Dear Viktor:

Thank you very much for your Christmas card which I received in Jan. 1991. I am very glad to learn that Dr. Yuri Molchanov of Leningrad State Univ. in ready to organize the 15th Int. Liquid Crystal Conf. in 1994. Sorry I was not able to write to you earlier, but I just finished running a conf. here and now have more time to attend to this matter.

Enclosed here is a copy of the letter that I send to Dr. Molchanov. Hope you will continue to help in organizing the conf. and keep in touch.

Best regards.

Sincerely,

Lui Lam

Lui Lam
Professor

Fig. 10.1. Letter from Lui Lam, Chair of Conference Committee, ILCS, to Viktor Titov (1935-1966), May 2, 1991. Seven months later, on December 26, 1991, the USSR was dissolved; five years later, Titov's life was abruptly terminated [Dunmur & Sluckin, 2011, pp. 231-234]. Up to now, neither USSR nor Russia has hosted an ILCC.

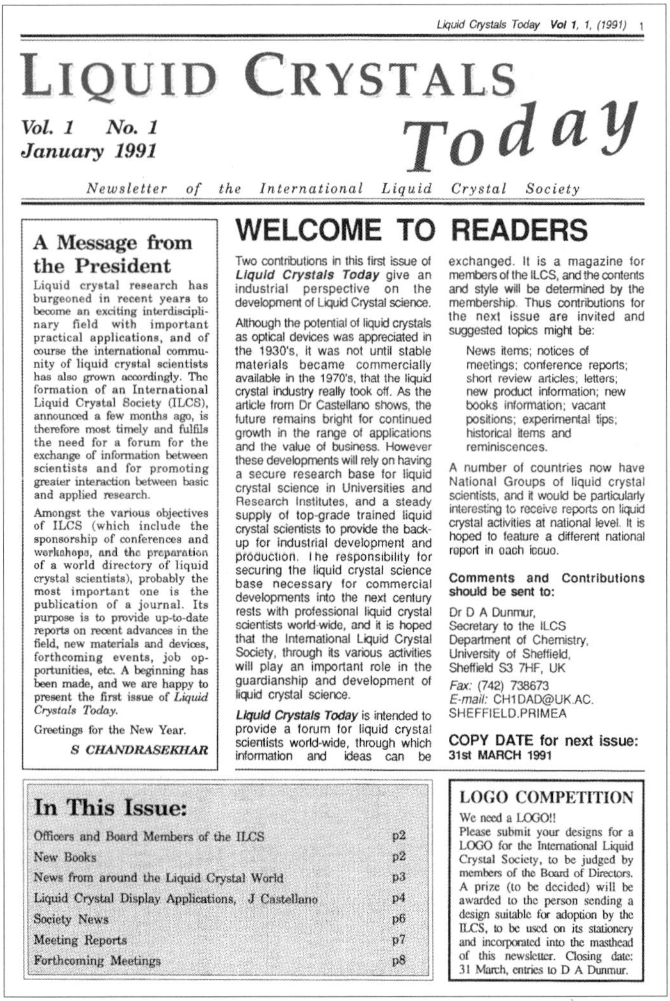

Liquid Crystals Today **Vol 1**, 1, (1991) 1

LIQUID CRYSTALS *Today*

Vol. 1 No. 1
January 1991

Newsletter of the International Liquid Crystal Society

A Message from the President

Liquid crystal research has burgeoned in recent years to become an exciting interdisciplinary field with important practical applications, and of course the international community of liquid crystal scientists has also grown accordingly. The formation of an International Liquid Crystal Society (ILCS), announced a few months ago, is therefore most timely and fulfils the need for a forum for the exchange of information between scientists and for promoting greater interaction between basic and applied research.

Amongst the various objectives of ILCS (which include the sponsorship of conferences and workshops, and the preparation of a world directory of liquid crystal scientists), probably the most important one is the publication of a journal. Its purpose is to provide up-to-date reports on recent advances in the field, new materials and devices, forthcoming events, job opportunities, etc. A beginning has been made, and we are happy to present the first issue of *Liquid Crystals Today.*

Greetings for the New Year.

S CHANDRASEKHAR

WELCOME TO READERS

Two contributions in this first issue of ***Liquid Crystals Today*** give an industrial perspective on the development of Liquid Crystal science.

Although the potential of liquid crystals as optical devices was appreciated in the 1930's, it was not until stable materials became commercially available in the 1970's, that the liquid crystal industry really took off. As the article from Dr Castellano shows, the future remains bright for continued growth in the range of applications and the value of business. However these developments will rely on having a secure research base for liquid crystal science in Universities and Research Institutes, and a steady supply of top-grade trained liquid crystal scientists to provide the back-up for industrial development and production. The responsibility for securing the liquid crystal science base necessary for commercial developments into the next century rests with professional liquid crystal scientists world-wide, and it is hoped that the International Liquid Crystal Society, through its various activities will play an important role in the guardianship and development of liquid crystal science.

Liquid Crystals Today is intended to provide a forum for liquid crystal scientists world-wide, through which information and ideas can be exchanged. It is a magazine for members of the ILCS, and the contents and style will be determined by the membership. Thus contributions for the next issue are invited and suggested topics might be:

News items; notices of meetings; conference reports; short review articles; letters; new product information; new books information; vacant positions; experimental tips; historical items and reminiscences.

A number of countries now have National Groups of liquid crystal scientists, and it would be particularly interesting to receive reports on liquid crystal activities at national level. It is hoped to feature a different national report in each issue.

Comments and Contributions should be sent to:

Dr D A Dunmur,
Secretary to the ILCS
Department of Chemistry,
University of Sheffield,
Sheffield S3 7HF, UK
Fax: (742) 738673
E-mail: CH1DAD@UK.AC.SHEFFIELD.PRIMEA

COPY DATE for next issue: 31st MARCH 1991

In This Issue:

LOGO COMPETITION

We need a LOGO!!
Please submit your designs for a LOGO for the International Liquid Crystal Society, to be judged by members of the Board of Directors. A prize (to be decided) will be awarded to the person sending a design suitable for adoption by the ILCS, to be used on its stationery and incorporated into the masthead of this newsletter. Closing date: 31 March, entries to D A Dunmur.

Fig. 10.2. First issue of *Liquid Crystals Today*, Vol. 1, No. 1, January 1991.

More strangely, in the recent book by David Dunmur and Tim Sluckin [2011] on the *detailed* history of LC, the term of Internal Liquid Crystal Society (or ILCS) is completely absent from the Timeline (1854-1997), the Index and the rest of the book, even though the LCI at Kent and ILCC are prominently mentioned. Note that Dunmur was the Secretary of ILCS and the Editor of LCT in the early years starting from the beginning.

What happened? How did the ILCS come about? Obviously, it is a story never told.[6] The first step in clearing the mystery is by providing an answer to Question 2: Lui Lam is the one who initiated the ILCS. But to answer the other questions and provide the context of the happenings, we have to go deeper and reach back earlier and ask: Why Lam? Where did he come from?

10.2 My Involvement in Liquid Crystals

I was born in mainland China. I lived in Hong Kong (1949-1965) where I went through grade schools and earned my BS, with first class honors, from the University of Hong Kong.[7] I received my MS from the University of British Columbia (UBC), Canada; and PhD from Columbia University, New York City, with thesis done at Bell Labs.[8] I had never worked on LC before I became the postdoc of Melvin Lax (1922-2002) at City College of City University of New York (CCNY), in 1972.[9]

10.2.1 *In the West (1972–1977)*

Neither he nor I knew anything about the subject when Melvin Lax asked me to work on LC. Mel was one of the founders of quantum optics and was known to be a very thorough researcher on any topic he set his eye on. He just published a 37-page long paper on a new formulation of the

---

[6] This is not strictly true. The fact that the ILCS was initiated by Lui Lam is reported briefly in [Lam, 2005a, p. 2318; 2005b, p. 529].
[7] I was absolutely a normal kid. I run away from school only once. One day in Shantou, I eluded the guard at the gate of my kindergarten during a recess, run 15 minutes home and hided myself by a sofa. I was three then and my mom convinced me that it was not such a good idea.
[8] My doctoral work on Compton profile resulted in the Lam-Platzman Correction Term [Callaway & March, 1984; Papanicolaou et al, 1991; Blass et al, 1995] and the Lam-Platzman Theorem [Bauer, 1983].
[9] While at Columbia and CCNY, I helped to found and build up the Chinatown Food Co-op, a mass organization with a political agenda [Kuo, 1977; Lam, 2010]. It was at the Food Co-op that I developed my organizational skills which later became handy in founding the ILCS. The Co-op members included Jean Quan (now Mayor of Oakland, CA) and Peter Kwong (now Professor of Asian American Studies at Hunter College, CUNY).

electrodynamics of crystals starting from a Lagrangian [Lax & Nelson, 1971; Nelson, 1979]. He saw a recent paper from Harvard's Paul Martin[10] that claims that liquid crystals are crystals [Martin et al, 1970] (which turned out to be wrong) and so Mel thought that he, actually I, could do better using his superior Lagrangian theory.

And so I started reading LC papers and that was two years before the LC book of Pierre-Gilles de Gennes (1932-2007) [1974] came out. I had no clue to proceed. In summer of 1973, I attended the Les Houches Summer School on molecular fluids. De Gennes gave a series of lectures on nematodynamics based on his book manuscript. Unfortunately, it was delivered in French and I learned nothing. But I did get something out of this school: I got to know Roland Ribotta from Orsay because we shared the same room; and convinced Michel Mirkovitch to translate a short manuscript of mine on Compton profile, from English to French, and got it published in *Physics Letters A* (which was never quoted by anyone even though it was a good paper) [Lam, 1973]. Before we left, we were given a few chapters from de Gennes' book, which, luckily, were written in English.

Back to New York, I checked every formula in the chapters and found a serious mistake in de Gennes' derivation of the Frank free energy. I wrote to de Gennes and he sent back a postcard, acknowledging the mistake but saying that the mistake had been found by someone else ahead of me. And so there was this footnote in the book acknowledging the help of someone (not me) in correcting a mistake in the manuscript.

And I still had no clue to proceed. The next summer, I went away to tour China for seven weeks [Lam, 2014] with full pay from Mel. The breakthrough came one day when, by chance, I came across an article in CUNY's library that uses a dissipation function to formulate irreversible thermodynamics in continuum mechanics.[11] I ended up using Lax's

---

[10] In the late 1960s and early 1970s two theorists, Pierre-Gilles de Gennes at Orsay and Paul Martin at Harvard, helped to bring liquid crystals to the mainstream of physics research. Both worked closely with their experimental colleagues. The Orsay team beat the Harvard team by a huge margin. De Gennes was awarded the Nobel Prize in 1991 partly for his LC works.

[11] Personal computers has been available sine 1976 (the year Apple was founded) but before 1990—the year the World Wide Web was born, the way to do a literature search,

Lagrangian to take care of the reversible processes and a dissipation function for the irreversible (dissipative) processes. It preserves the beauty and simplicity of Lax's theory in that everything, including the necessary space-time symmetries and the material symmetries, are built into the Lagrangian and the dissipation from the very beginning. The conservations laws and reciprocal relations emerge naturally and consistently from the equations of motion.

I left CUNY for Universitaire Instelling Antwerpen, Belgium, in 1975 without writing any paper for or with Mel. He was not displeased at all and even offered to find me a job at Los Alamos National Laboratory, which I declined, thinking that, wrongly, the job required a US citizenship. I worked on semiconductors and structural phase transitions in Belgium and then on superionic conductors at Universität des Saarlandes, West Germany, the following year. I was anticipating returning to China any time soon and so in 1977, I wrote up four papers summing up my dissipation-function formulation of the thermo- and hydro-dynamics of molecular liquids and solids [Lam, 1977a; 1977b; 1977c; Lam & Lax, 1978]. One of these four papers is specifically on LCs [Lam, 1977c],[12] which is my one and only one LC paper before I returned to settle in China [Lam, 2014].

10.2.2 In China (1978–1983)

In January of 1978 I returned to China to join the revolution. I was assigned to do physics at the Institute of Physics (IoP), Chinese Academy of Sciences, in Beijing. I could work on any topic. The fact that I ended

---

if you do not know what you are looking for, is to go to a good library to flip through *Chemical Abstracts* and walk semi-randomly between the long rows of shelves housing books and journals. It takes time, like treasure hunting without a map. But it is also a serene and spiritual experience, like immersing yourself in a quiet church.

[12] In [Lam, 1977c], in one stroke, we rederive the Frank free energy and the Ericksen-Leslie equation for both nematics and cholesterics *plus* the Parodi and other relations among the transport coefficients (which are nothing but the Onsager reciprocal relations). The dissipation-function approach is later used for biaxial nematics [Das & Schwartz, 1994]. It remains, in fact, the best and simplest approach in handling any new thermoviscous solids with microstructures and the hydrodynamics of any new molecular fluids (including liquid crystals).

up doing LC in China was purely incidental, which was related to Woo Chia-Wei's stay at the IoP in the last few months of my first year there [Lam, 2014].

I first worked on phase transitions [Lin, 1979; Shu & Lin, 1984a; 1984b][13] and then proposed a new kind of LC called *bowlic* [Lin, 1982; Lam, 1994] while doing propagating solitons [Lin et al, 1982; Lam & Shu, 1992]. My first Chinese LC paper, "Microscopic theory of first-order phase transitions in liquid crystals" [Lin, 1978] explains Ronald Dong's experimental results on the critical exponent of correlation length. But as the (future) ILCS is concerned, what is equally important is the LC people I met and the friends I made during these six years.

1. The USA visit (1979)

From February to May, 1979, I was a visiting scholar at Woo Chia-Wei's Physics Department, Northwestern University, Evanston. The work I did there was published in *Physical Review Letters* (PRL), my fourth LC paper and the second one published in the West [Lin, 1979]. Historically, this is the first LC paper and the first one in PRL ever came out from China [Lam, 2010]. By going beyond the mean-field approximation, among other things, it shows that the gap exponent in the nematic-isotropic transition is temperature dependent, clarifying the experimental results of Keyes and Shane related to tricritical points.

2. The Bangalore conference (1979)

Late in 1979, I received suddenly an invitation from Sivaramakrishna Chandrasekhar (1930-2004) to his LC conference in Bangalore, December 3-8, 1979.[14] I, together with a colleague from IoP and another one from Peking University, attended this conference [Lin et al, 1980]; I was housed in the guest house at Raman Research Institute. Like every visitor who visits India for the first time I was culturally awakened. This

---

[13] LIN Lei is my Chinese name in pinyin; LAM Lui are the same Chinese characters in Cantonese which I picked up in Hong Kong.

[14] I thought that invitation came from the attention of my PRL paper; Chandra, the name Chandrasekhar asked me to call him, told me years later that I was recommended by Ron Shen of Berkeley. The historical significance of this Bangalore conference is that it is the first LC conference outside of China ever attended by a delegation from mainland China.

conference was held less than two years after the discovery of discotics by Chandra's group. It was attended by a Who's Who list of liquid crystalists [15] (except Pierre-Gilles de Gennes who, a speaker at the previous conference in 1973, was no longer active in LC). Apart from Chandra, the people I met include Patricia Cladis,[16] Dietrich Demus, Christian Destrade, Adrian de Vries, Elisabeth Dubois-Violette, Charles Frank (1911-1998), Wolfgang Helfrich, Jerzy Janik, Shunshuke Kobayashi, Sven Lagerwall, Alan Leadbettter, Anne Levelut, Jacques Prost and Alfred Saupe (1925-2008).

I presented a summary of my work on the critical properties of nematic-isotropic transitions [Lin, 1980].[17] It was at this conference, while looking up to the ceiling, that I got the idea for the three-dimensional bowlic LCs [Lam, 1994].

3. My first Orsay visit and the Kyoto conference (1980)

In next year's summer, upon the invitation of Roland Ribotta, I spent a month at CNRS' Laboratoire de Physique des Solides at Université de Paris-Sud, Orsay. The lab was established by André Guinier (1911-2000), Jacques Friedel and Raimond Castaing (1921-1998) in 1959; de Gennes spent 10 years (1961-1971) there and did his LC works. I was received warmly by Friedel in his office. Before I left, I got Guinier's permission to translate his popular-science book *Structure of Materials* into Chinese [Lin et al, 1985], and finalized the arrangement for an Orsay LC group to visit China.[18]

---

[15] In this chapter a person working in LC is referred as a *liquid crystalist*, or a *LCer*, pronounced LC-er.

[16] Patricia Cladis, born in Shanghai, joined Bell Labs in 1972, the year I left. We met each other for the first time in Bangalore.

[17] I opened my talk with a remark from Chairman Mao (Ze-Dong) (1893-1976) on dialectics that says every positive thing when left too long, becomes its opposite, referring the need to go beyond the Landau-de Gennes mean-field model. The talk was closed with a quote from Rabindranath Tagore (1861-1941), India's Nobelist in literature, "Truth seems to come with the final word; and the final word gives birth to its next" [Lin, 1980].

[18] In September 1980, the Orsay's "gang of four" (Jean Charvolin, Georges Durand, Maurice Kléman and Roland Ribotta), led by Ribotta and accompanied by me, toured Beijing, Shanghai and Wuhan. Liquid crystal research in China was so far behind then,

Right after I returned to Beijing, I led a delegation of several scientists to attend the 8th ILCC in Kyoto, Japan, June 30-July 4, 1980— the first ILCC conference ever attended by mainland Chinese. As it was the case, scientists from Taiwan were listed as coming from Republic of China in the program. By my understanding, we were prohibited to show up in any conferences under such circumstances, a rule self-imposed by the Chinese government [Xiong, 2013]. The deadlock was resolved after I told Shunshuke Kobayashi, general secretary of the conference, that he only had to make an announcement at the beginning of the conference that there was a "mistake" in the program.[19]

It was at this Kyoto conference that the PSC announced that they decided to open its door a little bit on two accounts: (1) Invite the LC community to nominate candidates to the PSC; and (2) each PSC member will be allowed to serve a maximum of eight years, starting from 1980. At the end of the conference, Woo Chia-Wei told me that he had nominated me to the PSC.

4. Founding the Chinese Liquid Crystal Society (1980)

China's LC research began in 1970 amidst the Cultural Revolution [Lin, 1983]. On July 18, 1980, the Chinese Liquid Crystal Society (CLCS) was founded by Zhao Jian-An, Ruan Liang, Xie Yu-Zhang and I. I served as the Vice-President and Secretary General from 1980 to 1983.[20]

At the Kyoto conference, Lagerwall and Chandra expressed their wish to visit us in Beijing. The visa was rapidly granted and they arrived in time to join the founding ceremony of the CLCS. Lagerwall brought along a LC switch device he just invented and he kindly let us buy it from him.

Afterward, I wrote to de Vries at Kent about the existence of the CLCS. He replied with these words:

---

that these four world-renowned LCers ended up giving popular-science talks everywhere we went. But the tour was very exciting.

[19] I learned firsthand the Japanese culture through the protracted negotiations with Kobayashi, who would say "yes" whether he agreed with me or not.

[20] See *Membership List of Chinese Liquid Crystal Society (First Term, 1980-1984)*, p. 10.

Congratulations with the formation of the Chinese Liquid Crystal Society, and with your election as Secretary General and Vice President. To my knowledge, too, this is the first such society in the world. Maybe more countries will follow now, and maybe the International Planning and Steering Committee will then also be reorganized in a more formal way. I think it is time for that.

De Vries wrote this on September 4, 1980, long before anyone had the idea of ILCS in mind.

5. Martin Gordon's visit (1981)

Martin Gordon, Chairman of Gordon & Breach that published *Molecular Crystals and Liquid Crystals* (MCLC), ventured into China early, in 1981. He invited me to serve as an associate editor of the journal (which I did, 1981-1993) and asked me how he could help. I asked for a copy machine which he promised but never delivered. But I did write up a review on China's LC research for his journal, in which my 1982 prediction for bowlics is mentioned [Lin, 1983].

6. My second Orsay visit and Bangalore's ILCC (1982)

In August 1982, I gave a review on solitons in liquid crystals at the "Solitons '82" conference at the Riccarton campus of Heriot-Watt University outside of Edinburgh, UK, marking Scott Russell's centenary discovery of the solitons.[21] On November 1, 1982, my LC soliton paper came out in PRL [Lin et al, 1982; Lam, 2014][22] while I was spending two months at Orsay, working with Ribotta on LC's electroconvection pattern formation [Ribotta et al, 1986]. On my way back to Beijing, I attended the 9th ILCC, December 6-10, in Bangalore.[23] Kobayashi

---

[21] I was in the small boat in the Union Canal that tried to recreate the solitary wave John Scott Russell (1808-1882) first observed in 1834. The failed "experiment" was reported in *The Scotsman*, August 26, 1982, and by Ellbeck and Scott [1982].

[22] Historically, this is the first paper in LC literature that contains the word soliton in the paper's *title*.

[23] The 9th ILCC was originally scheduled to be held in Kraków, Poland. But in December 1981, Poland's communist government declared martial law and delegalized the opposition Solidarity and interned its key members (http://en.wikipedia.org/wiki/1982_demonstrations_in_Poland, Sept. 1, 2013). My French LC colleagues led the petition to move the ILCC out of Poland. Bangalore became the replacement in the last minute and

greeted me at the Raman Research Institute and congratulated me for being nominated to the PSC. However, my formal admission to the PSC had to wait until 1984 (Table 10.1).

Table 10.1. Members of the Planning and Steering Committee (PSC), 1980-1990. Data collected from the program book of each ILLC; the Chair of PSC is marked in grey cell. Since 1980, no member served more than a maximum of eight years (marked by five x)—a rule set by PSC in 1980, with two exceptions. Glenn Brown, the founder of Kent's LCI and the ILCC series, after serving as PSC's Chair for many years, was elevated to honorary chair after Kyoto. The other exception, Chandra, resulted from PSC breaking its own rule (see Section 10.3, Act II).

	Name	Country	1980	1982	1984	1986	1988	1990
		ILLC	8th	9th	10th	11th	12th	13th
	Number of PSC members		19	20	20	20	21	20
1	Ambrose, E. J.	UK	x	x				
2	Baur, G.	FRG						x
3	Blinov, Lev	USSR			x	x	x	x
4	Bouligand, Y.	France	x	x	x			
5	Brown, Glenn	USA	x	x	x	x	x	x
6	Chandrasekhar, S.	India	x	x	x	x	x	x
7	Chistyakov, I.	USSR	x	x				
8	Clark, Noel	USA					x	x
9	Demus, Dietrich	DR				x	x	x
10	Doane, William	USA					x	x
11	Durand, Georges	France			x	x	x	x
12	Figueiredo-Neto, Antonio	Brasil				x	x	x
13	Friberg, S. H.	USA			x	x		
14	Fukuda, Atsuo	Japan					x	x
15	de Gennes, Pierre-Gilles	France	x	x				
16	Gray, George	UK	x	x	x	x	x	
17	Gerristsma, C. J.	Netherlands	x	x	x			

that was why the 9th ILCC was held in winter instead of summer. Subsequently, Lech Wałęsa, cofounder of Solidarity, won the Nobel Peace Prize (1983) and served as President of Poland (1990-1995). Personally, to my regret, my third chance to visit Bangalore in 1986, as an invited speaker at the Second Asia Pacific Physics Conference, did not materialize due to insufficient time in applying for a visa.

#	Name	Country					
18	Hosemann, R.	FRG	x	x			
19	Janik, Jerzy	Poland		x	x	x	x
20	de Jeu, Wim	Netherlands				x	x
21	Kahn, Fredic	USA	x	x	x	x	
22	Kelker, H.	FRG	x	x	x		
23	Kobayashi, Shunshuke	Japan	x	x	x	x	
24	Lagerwall, Sven	Sweden		x	x	x	x
25	Lam, Lui (Lin, Lei)	USA (China)			x	x	x
					x		
26	Leadbettter, Alan	UK		x	x	x	x
27	Lister, J. David	USA		x	x	x	x
28	Meier, Gerhard	FRG	x	x	x	x	x
29	Okano, K.	Japan				x	x
30	Onogi, S.	Japan	x	x	x		
31	Porter, R. S.	USA	x	x			
32	Prost, Jacques	France			x	x	x
33	Rustichelli, Franco	Italy		x	x	x	x
34	Sackmann, Horst	DDR/GDR	x	x	x		
35	Saupe, Alfred	USA	x	x			
36	Skoulios, A.	France	x	x			
37	Stegemeyer, Horst	FRG			x	x	x
38	de Vries, Adrian	USA	x	x	x	x	

10.2.3 In New York (1984–1987)

I left China at end of 1983 for family reasons [Lam, 2014]. From 1984-1987, I continued to work in LC at City University of New York. I visited Beijing every summer and set up a LC group at Nanjing Normal University, resulting in ten papers on two-dimensional solitons and viscous fingering in LC disc cells (1986-1989).

In 1985, I attended the 6th Liquid Crystal Conference of Socialist Countries, August 26-30, Halle (Saale), German Democratic Republic, which was chaired by the legendary Horst Sackmann (1921-1993) [Dunmur & Sluckin, 2011, pp. 167-172]. In Halle, I learned from Christian Destrade that bowlics had been synthesized in Europe. I stopped over at Orsay to meet Anne Levelut who just finished doing an x-ray study of these materials [Lam, 1994, p. 327]. After I informed her that I had predicted bowlics while in China, she kindly mentioned my 1982 work in her paper [Levelut et al, 1986].

In summer of 1986, the 11th ILCC was held in Berkeley, California. I gave a review on bowlics [Lin, 1987] and arranged for Xie Yu-Zhang, president of CLCS and my coauthor for two papers in MCLC (1983), to give a talk there. This review is the one often quoted in the bowlic literature (see, e.g., [Palffy-Muhoray et al, 1988; Brostow, 1990; Golubović & Wang, 1992; Xu & Swager, 1993]).

The second LC journal, *Liquid Crystals*, started in 1986; I served in the editorial board from 1986-1990. More importantly, in December 1986 I started planning the first book in the new Springer book series *Partially Ordered Systems*, which I initiated.[24] Historically, the series' first book *Solitons in Liquid Crystals*, edited by Lam and Prost [1992], is the first LC book that concentrates on a single topic. The contributors include liquid crystalists from China, France, Germany, India, Japan and USA.

In the summer of 1987, I was preparing to move to San Jose State University (SJSU) in California. But there was this LC polymer conference in Bordeaux, July 20-24, which I planned to go, partly because I wanted to sit down with Prost and write the Preface and Chapter 1 for the soliton book together.[25] As was often the case, I picked a conference (or city) I wanted to go before I have an idea for an appropriate paper.[26] I never worked on polymers before; for this conference, I forced myself to come up with a presentation on bowlic and polar LC polymers and wrote it up later [Lam, 1988a].[27]

---

[24] The Editorial Board consisted of Jean Charvolin, Wolfgang Helfrich and Lui Lam; the Advisory Board: David Lister, David Nelson and Martin Schadt. With the consent of Springer, I elevated myself to the series' editor-in-chief in 1999.

[25] As it turned out, Prost drove us to a nude beach outside of Bordeaux and he wrote his part lying down nude on the sand while I was absorbing the scenery and the fresh air.

[26] I met de Gennes the second time, 14 years later, at this conference. The third and last time I saw him was in Varenna, Italy, during the Enrico Fermi Summer School on The Physics of Complex Systems, July 9-19, 1996, organized by Francesco Mallamace and Eugene Stanley.

[27] Bowlic polymers predicted in this paper were synthesized later in USA by Zeng Er-Man [2001]. It took 13 years, while the prediction of bowlic monomers took only three years to be confirmed in Europe [Lam, 1994]. Eventually, the word "bowl" or "bowlic"

Afterward, I visited Lajos Bata[28] in Budapest before traveling by car to Pardubice, Czechoslovakia, to attend the 7th Liquid Crystal Conference of Socialist Countries, August 31-September 4, 1987. At the conference I gave a review talk on bowlics and was presented with a red plastic bowl as a souvenir (Fig. 10.3).

Fig. 10.3. Two presents I received in 1987. *Left*: A red plastic bowl from the LC socialists in Pardubice (with the characters "OPP ROCKYCANY 0,5" on the bottom). *Right*: A liquid crystal thermometer from Kent's ILC. I am responsible for the imperfection at the left upper corner, resulted from the thermometer slipping from my fingers and falling to the floor in San Jose.

At this point, I had worked in LC for 15 years, published 56 LC papers, co-founded the CLCS, invented a new kind of LC called bowlic, pioneered the study of propagating LC solitons in shear flow, set up a Springer series that includes LC books, sit on the editorial board of the only two LC journals, and was a current member of the PSC—the power center of the LC profession. And I knew nearly everybody in LC, East and West. Following the example of de Gennes, my time to phase out of LC research was overdue. Besides, the job at San Jose required me to

---

appears in the title of LC papers (such as [Cometti et al, 1990; Xu & Swager, 1993; Arcioni et al, 1995; Mehta & Uma, 1998; Imamura et al, 1999; Dong et al, 2009]) and is recognized officially by the IUPAC [Barón & Stepto, 2002, p. 499] and formally in *Handbook of Liquid Crystals* [Demus et al, 2008]. The paper also contains a prediction of ultrahigh-T_c bowlic (or discotic) superconductors (see also [Lam, 1988b]) which was reported by *Superconductor Week* (October 19, 1987, p. 4), a newsletter newly created to report on high-T_c superconductors.

[28] Bata, Durand and Schadt were among the speakers I, member of the Preparatory and Advisory Committee, helped to invite to the Centenary Conference of Liquid Crystal Discovery, Beijing, June 27-July 1, 1988, which was organized by the CLCS.

work and publish with undergraduates and that would be in nonlinear physics [Lam, 1998, pp. 328-331].

10.3 Founding the International Liquid Crystal Society (1987–1990)

On my way to the LC polymer conference in Bordeaux, I stopped over at Orsay. I was shocked to learn that Mireille had passed away. I visited Mireille's lab five years ago; she was obviously a young and bright liquid crystalist. After shock, came sadness and disappointment. The liquid crystal community was relatively small; it was almost like a family. How come I, a colleague in the same field, did not know of Mireille's death until I came to Orsay, almost by accident? It occurred to me that the liquid crystal community should be organized better and should have a publication like *Physics Today* which includes an obituary in every issue. And so, I decided to establish an international liquid crystal society.

1. Prologue (July 1987-March 1988)

I asked several people, including Chandra, at the Bordeaux conference and they *all* agreed to this idea of an ILCS. In between the Bordeaux and the Pardubice conferences, I spent 12 days driving my Honda Accord from New York to San Jose. I stopped over at Kent to met Glenn Brown, Alfred Saupe, Bill Doane and others; and at Boulder to meet Noel Clark. They all supported an ILCS, too. At Kent, the LCI presented me with a gift, a LCD thermometer (Fig. 10.3). At SJSU, I received a support letter from Doane, written in his capacity as LCI's director and dated August 10, 1987, which includes these words: "Your idea of a Society of Liquid Crystal Researchers is excellent. If we at Kent can be of help in this endeavor, please let us know." The LCers in Budapest and in Pardubice also gave their support.

To ensure the success of this project, I adopted a strategy based on Chairman Mao's two important teachings: (1) Mobilize the masses; trust the masses. (2) Encircle the city from rural areas (i.e., secure the surroundings before overcoming a stronghold). Being a ballet fan, I labeled each stage of the strategy an Act. And being employed as an untenured full professor, with a teaching load of 15 hours per week, I did not find time to act until March, 1988.

2. Act I (March 2, 1988)

In February of 1988 I gave a seminar on bowlics and refreshed my memory of the maple leaves at UBC, Vancouver. My host, Birger Bergersen, helped to solicit support for the ILCS among the Canadians. In March and April of 1988, I contacted a selected group of liquid crystalists, including all current PSC members, by mail.[29] Each mail contains four pages[30]: (1) an open letter explaining why an ILCS should be formed; (2) a page called Act I laying down my vision of the ILCS (Fig. 10.4); (3) a copy of Doane's support letter; (4) a questionnaire asking the recipient for comments and her/his choice of support or not in writing, to be returned to me no later than April 25, 1988.

In fact, the need for an ILCS was lying bare there for many years, as noted by de Vries in 1980 (see Section 10.2). The overall discontent from both developing and developed countries was twofold: (1) The PSC was run like a private club without outside overseeing; (2) the club, with all the smart people in the field, did only one thing, i.e., selecting a site for a conference every two years—and absolutely nothing else, while the LC profession in many countries were not getting its fair share of attention from their respective scientific community, and jobs for new PhDs in LC declined very fast in number.

As shown in Fig. 10.4, the plan was quite in detail. All the suggestions proposed here, including the name of the society, the organizational structure, the publication of an official "newsletter," and option 3 in connection with the existing PSC—with one exception— were indeed incorporated into the ILCS. The item "Membership Due" was written with developing countries in mind, derived from my China experience. It was motivated to give people from developing countries an

---

[29] Email was not yet available. Communications between scientists were by phone, fax or mail; mailing was the cheapest.
[30] Copy of these pages and other documents pertaining to the founding of the ILCS could be downloaded free from the website (which would be updated from time to time): www.sjsu.edu/people/lui.lam/ILCS.

equal voice and would not be discriminated simply because they could not afford to go to the biennial ILCC.

3/2/88

Founding the International Liquid Crystal Society - Act I

(The ideas suggested below are meant to induce discussion. There is nothing final about it - L.L.)

Name: International Liquid Crystal Society

Aim: 1. To serve the interest of liquid crystalists as individuals.
 2. To provide a forum for people and organizations to exchange information and ideas.
 3. To serve the need of the liquid crystal community, scientists and industries, in general.

Membership: Individuals join as members; lab or group may join as a unit; company and national liquid crystal group may join as corporation member.

Membership Due: Keep it as low as possible. Members from countries with hard currency pay in US$ (or equivalent); those from other countries pay in their own currency. Reduced rate for students.

Function: Publish a newsletter (something like the Physics Today but less formal, incorporating some features of newspapers) once every two months (or may be 3 months in the first year). Organize meetings, summer or winter schools, short courses. The newsletter may be sent to each country or district in bulk and distributed by the local chapters using the local currency collected from the membership dues.

Organization: Local chapters may be formed in each country (if there is enough membership), groups of countries, or different geographical sections within a country, upon the approval of the governing board.

Relationship with the existing International Steering and Planning Committee: There are three possibilities:
 1. completely independent of each other.
 2. incorporate the Committee under the Society.
 3. form the Society from the Committee by expanding it, and then reorganize or keep the Committee intact and put it under the Society.
 (It seems to me that 3. is most practical.)

Date of founding: August 1988 during the 12th International Liquid Crystal Conference at Freiburg, West Germany. It is the centenary discovery of liquid crystal and is a good occasion; time frame is right too.

Fig. 10.4. Act I (March 2, 1988): Initial formulation of the ILCS.

The one exception that did not go as planned was the "Date of Founding"; instead of being suggested here to be in 1988, the centenary of LC's discovery, the ILCS was founded two years later. The reason for the delay is explained in the next subsection.

3. Act II (May 25, 1988)

In May of 1988, the responses to Act I were summarized in Act II (Fig. 10.5). It was based on 61 replies from 14 countries that I received; they were all positive, with nine coming from current members of PSC and two from past members. Act II was mailed to all 61 repliers and to others. As shown in Act II, more details on the future ILCS were provided; the planned founding date remained to be in August, 1988, in Freiburg, during the 12th ILCC. The plan was to hold an organizing meeting open to all participants of the conference; with enough consent, the PSC would declare itself transformed into an ILCS; and people sign up to join. Like the founding of a new country, the bylaws (the constitution) could be drafted later. Naturally, this assumed that the PSC members were among the consenters.

At this point, the missing piece was the absence of Chandra and George Gray (1926-2013), the current and previous PSC chair, respectively (see Table 10.1), among the repliers to Act I. I waited until July 15, the deadline to respond to Act II, and then August 15, the beginning of the Freiburg conference, but there were still no words from Chandra and Gray (see Fig. 10.6). Were they opposing to the idea of an ILCS? If not, what was holding them up? The answer came at the PSC meeting in Freiburg.

On August 16, 1988, PSC held its business meeting in a room without windows. After the meeting was declared open by Chandra and before anyone had the chance to raise the question of electing the next PSC chair, as required by the rules since Chandra had reached his eight-year limit , Gray said, "Let Chandra continue to be the next Chair." Everyone was silent. I was struck dumb; I looked at Doane who sat next to me at my right and saw his jaw dropped, or so it seemed. And so Chandra began his third term as PSC chair, violating the rules he and Gray helped to write and pass just eight years ago.

5/25/88

Founding the International Liquid Crystal Society - Act II

The response to Act I (3/2/88) questionnaire is very positive. At this point there are 61 replies from 14 countries, including those from 9 present members and 2 past members of the International Planning and Steering Committee for International Liquid Crystal Conferences. Summing up the comments expressed by those who responded,

o The International Liquid Crystal Society should be founded at the 12th Int. Liq. Cryst. Conf. at Freiburg, Aug. 15-19, 1988.

o It should be organized with the existing International Planning and Steering Committee as the basis.

o There should be open discussion by those attending the Freiburg conference before the founding.

o The publication of Newsletter by the Society is desired. One suggested to have it monthly; another one for quarterly; the rest seems to content to have it bimonthly.

o The membership fee should be as low as possible. (For clarification, the membership dues from members in countries without hard currency can be collected as long as it is not transferred outside. This will be used within those countries to mail the newsletter and for functions sponsored or endorsed by the Society. For countries in which paper is a scarce material printed newsletter has to be sent in in bulk; for others a good copy can be sent and reprinted there.)

o Concern has been expressed (by one) on the autonomous of local chapters. If there is no financial support from the central office to local chapter, the local chapter should be autonomous, in contrast to the practice in the Society of Information Display, it was suggested. Please keep this issue in mind and we will discuss it at Freiburg.

o In view of the positive and extensive responses outlined above I have written to S. Chandrasekhar, Chairman of the Int. Planning and Steering Committee, and to the organizers of the Freiburg Conference requesting that the International P&S Committee to start discussion on these suggestions, and to make arrangements (time and place) during the Conference for the participants to hold meetings, respectively.

o A name list of all those who sent back their questionnaire to me as of today is enclosed here for your reference. All support the founding of the Society.

• A brief statement as an open letter to the participants of the Freiburg Conf. is prepared here and will be distributed to all during the registration period (assuming the consent of the organizers, of course). If you want to remove or add your name to the list please mail the form back to me before July 15, 1988 (the date the form should reach me). If you agree to have your name there, do not take any action. Of course, your comments are always welcome. Please copy these pages and help distribute to your colleagues.

Lui Lam

For Lui Lam, Phys. Dept., SJSU, Telephone: (408)924-5261, Fax: (408)924-1018, Telex: 171171 UD
(Note new fax number)

Fig. 10.5. Act II (May 25, 1988): A summary of responses to Act I.

10 The Founding of the International Liquid Crystal Society 229

```
                      Call for Establishing an

                   INTERNATIONAL LIQUID CRYSTAL SOCIETY

Colleagues:

The 12th International Liquid Crystal Conference Marks the centenary of the
discovery of liquid crystals. The number of people working in this field, the
interdiscipliniary character of the research on these unique and fascinating
materials, the productive collaboration between industry and university and
the importance of liquid crystals in new industries require a permanent
organization dedicated to serving the international liquid crystal community.
We invite all participants at this Conference to attend an organizational
meeting to plan the formation of an INTERNATIONAL LIQUID CRYSTAL SOCIETY.

Signed:
    BRAZIL                        FRANCE                         Dowell, Flonnie
  * Figueiredo Neto, Antonio      Charvolin, Jean                Drzaic, Paul S.
    Fujiwara, Fred Yukio        * Durand, Georges                Fishel, Derry L.
    Santos, Marcus B. L.          Gasparoux, Henry               Gelbart, William M.
                                  Hardouin, Francis              Gelerinter, Edward
    BULGARIA                      Noel, Claudine               + Kahn, Frederick J.
    Derzhanski, Alexander I.    * Prost, Jacques                 Keast, Sandy S.
    Petrov, Alexander G.                                         Kumar, Satyendra
                                  GERMAN DEMOCRATIC REPUBLIC   * Lam, Lui
    CANADA                      * Demus, Dietrich              * Litster, James David
    Bergersen, Birger                                            Mahmood, Rizwan
    Burnell, E. Elliott           HUNGARY                        McAdams, Larry R.
    Dong, Ronald Y.               Bata, Lajos                    McRuer Robert N.
    Gilson, Denis                 Buka, Agnes                    Meyer, Robert B.
    Leigh, William James                                         Neubert, Mary E.
    Tracey, Alan S.               ITALY                          Ong, Hiap Liew
                                  Rustichelli, Franco            Petschek, Rolfe George
    CHINA                                                        Rosenblatt, Charles
    Dong, Chuchuan                JAPAN                          Shen, Y.R.
    Li, Guozhen                   Kawamura, Yasuaki              Ukleja, Paul
    Liang, Ruan                 + Kobayashi, Shunsuke            Vargas-Aburto, Carlos
    Liang, Zhong Cheng                                           Vaz, Nuno A.
    Liu, Han-Ming                 NETHERLANDS                    Vora, Rasiklal A.
    Shao, Ren-Fan               * De Jeu, Wim H.                 Westerman, Philip W.
    Wang, Bin                                                    Wu, Shin-Tson
    Wang, Liang Yu                POLAND
    Wang, X. J.                   Janik, Jerzy A.                USSR
    Xi, Guangeng                                               * Blinov, Lev M.
    Xi, Hua                       SWITZERLAND                    Shibaev, Valery Petrovich
    Xie, Yu-Zhang                 Lierau, Rolf R.
    Yang, Shong Ling              Schadt, Martin                 YUGOSLAVIA
    Zhao, Jing An                                                Blinc, Robert
    Zheng, Shu                    UNITED KINGDOM                 Zeks, Bostjan
    Zhong, Guofu                  Clark, Michael George
    Zhengmin, Sun
                                  UNITED STATES
    CZECHOSLOVAKIA                Acree, William E., Jr.
    Pirkl, Slavomir               Armitage, David
                                + Brown, Glenn
                                * Clark, Noel A.
                                  Davis, Frederick
                                + De Vries, Adriaan
                                * Doane, William J.

  * Present member.) of the International Planning and Steering
                   ) Committee for International Liquid Crystal
  + Past member.   ) Conferences.
```

Fig. 10.6. The Petition: Flyer distributed before August 16, 1988, the date of the PSC meeting, at the 12th ILCC, Freiburg, There are 82 signers (1 from UK and 0 from India).

> 9/20/88
>
> The Founding of an International Liquid Crystal Society - Act III
>
> In response to the suggestion of you and many other colleagues (see Act II), the Planning & Steering Committee for International Liquid Crystal Conferences agreed to transform itself into an International Liquid Crystal Society during their business meeting in the afternoon of August 16, 1988, at Freiburg, FR Germany. A subcommittee consists of
>
> > S. Chandrasekhar (India)
> > W. Doane (USA)
> > A. Fukuda (Japan)
> > S. Lagerwall (Sweden)
> > L. Lam (USA)
>
> was approved by the Planning & Steering Committee during the meeting. The subcommittee was charged to formulate the details of the transformation and make proposals to the P&S Committee. All the above were announced by the Chairman of the P&S Committee, Dr. S. Chandrasekhar, at the end of the awarding ceremony of the 2nd Glenn Brown Award on August 17, 1988 at the 12th International Liquid Crystal Conference at Freiburg.
>
> In the subsequent meeting of the Subcommittee on August 18, 1988 at Freiburg, L. Lam was asked by the subcommittee to write the first draft of the bylaws of the upcoming International Liquid Crystal Society.
>
> In short, our effort to found the I.L.C.S. has been fruitful. From this point on, any questions or comments concerning the future I.L.C.S. should be directed to members of the Subcommittee. In matters relating to the drafting of the bylaws please send your suggestions (or a copy of your writing) directly to L. Lam [Dept. of Physics, San Jose State University, San Jose, CA 95192, USA. Tel.: (408)924-5261]. This will shorten the time of communication and enables the International Liquid Crystal Society to be born earlier.
>
> Thank you very much for your support in the past and in the future.
>
> *Lui Lam*

Fig. 10.7. Act III (September 20, 1988): Announcing PSC's transformation to ILCS, and the subcommittee in charge of formulating the transformation.

I presented the Petition signed by 82 people (Fig. 10.6) and proposed that we transformed the PSC to an ILCS. Gray asked immediately, "What will happen to the PSC members?" I answered without thinking, "They will be the Board members of this new ILCS." And so, the resolution was passed unanimously. Someone proposed a subcommittee (of five members) to formulate the details of the transformation.[31] The decisions were announced by Chandra to the conference participants the next day. On August 18, I was asked to draft the bylaws because no one else in the subcommittee wanted to do it (see Fig. 10.7).

4. Act III (September 20, 1988)

A month later, I sent out Act III (Fig. 10.7) to all the 82 signers whose name appeared on the final petition (Fig. 10.6), thanking them for their fruitful support.

5. Act IV: Writing the Bylaws (1988-1990)

Before the summer of 1990 I was untenured. Needless to say, in my first three years at SJSU, I had to work like crazy. I created and taught two new graduate courses in nonlinear physics [Lam, 1998]; at one point, I supervised nine students in research and wrote papers with them; I attended several conferences and organized one or two conferences every year.[32] My research focus was no longer in LC but in nonlinear and complex systems. During this period my LC publications are merely extensions of my previous works.

Being overloaded, I asked Hiap Liew Ong (IBM) and then Flonnie Dowell (Los Alamos) to write the bylaws; both turned me down. Finally, I asked Ong to send me a copy of the bylaws of the Society for Information Display (SID). After reading the (1987) Bylaws of SID and the Constitution and Bylaws of the American Physical Society I wrote ILCS' bylaws myself in July 1989 (Fig. 10.8). The spirit and suggestions in Acts I and II were incorporated into the bylaws. In particular, I named the official magazine *Liquid Crystals Today* and stipulated that the society will *not* publish any research journals (in deferment to the two existing journals, MCLC and *Liquid Crystals*). More importantly, the President is limited to two consecutive terms of two years each term, and has to wait two years before reelected to the same position; after

---

[31] For this reason, the open organizational meeting envisioned in the Petition never took place.
[32] The conferences include two local series: Woodward Conference [Lam & Morris, 1989] and Liquid Crystals West. James Fergason (1934-2008) and Ron Shen participated in the latter series.

consecutive four years of having one (or two) President(s) from the same country, the next President has to be elected from another country. In other words, no same person or country could have the presidency for more than four years in a row.

Finally, in the summer of 1990, everything was ready for the ILCS to be formed at the 13th ILCC. And so I had to go to Vancouver. But there was a problem. I was scheduled to present a LC paper in that conference but I did not yet have enough material to write this paper.

6. Finale: Birth of the ILCS (July 22-27, 1990)

It turned out that 1990 was my very busy year. In January I organized a "Winter School in Nonlinear Physics" at SJSU. In March, in Anaheim, California, I chaired the American Physical Society Symposium on "Instabilities and Propagating Patterns in Soft Matter Physics," which I proposed. In June I was the Director of the NATO Advanced Research Workshop on "Nonlinear Dynamical Structures in Simple and Complex Liquids," in Los Alamos. And in July I attended briefly the "Nonlinear and Chaotic Phenomena" conference in Edmonton, Canada [Lam, 1991] before flying directly to Vancouver to attend the 13th ILCC, July 22-27, at the UBC campus—my Alma Mater and the first place I lived after Hong Kong 35 years ago.

It was a few days before this Edmonton conference that we rushed through a simple experiment in our Nonlinear Physics Laboratory at SJSU, in Room 55 in the basement of the Science Building. The experiment was so simple that it could be finished in less than one second. What we did was taking a LC cell, putting in LC or oil, and applying a high enough voltage across the cell. After a flash of light, the experiment was finished and we found a complicated filamentary pattern left on the inner surfaces of the coated glass plates. I wrote up the paper [Lam et al, 1991] right before I rushed to the airport and presented it in Vancouver, which led to my invention of Active Walk, a new paradigm in complex systems [Lam, 2005a].

In Vancouver, before the ILCC began, Chandra carefully sounded me out by telling me his wish. Here was the exchange:

Chandra: "Lui, I want to be the President of the ILCS."
Lam: "OK."

10 The Founding of the International Liquid Crystal Society

Being the President never crossed my mind nor was it my wish when I started this campaign to found the ILCS. I did it for Mireille and for the LC community.

Tel: (408) 924-5261   Fax: (408) 924-1018
Bitnet: LUILAM@CALSTATE (after 8/25)

July 15, 1989

To: Subcommittee to establish Int. Liq. Cryst. Soc. (ILCS)
(S. Chandrasekhar, W. Doane, G. Durand, A. Fukuda, J.A. Janik, S. Lagerwall, L. Lam, A.J. Leadbetter)

From: Lui Lam   *Lui Lam*

Re: Drafting the bylaws of ILCS

o   I will not be able to attend the 8th LC Conf. of Socialist Countries in Krakow, Poland, Aug. 28 – Sept. 1, 1989. I cannot get away from my teaching duties at that time. Fortunately, Chandra may be able to attend that, and I hope many of you will be there. I agree with Chandra that the occasion could be used for the members to discuss the bylaws and the possibility of merging the "Western" and the Socialist Countries Conferences into a single series. I am all for this idea.

o   Enclosed please find a copy of the SID Bylaws, and the **Constitution and Bylaws of the American Physical Society** for your references.

o   Here is some skeletal form of Bylaws for the ILCS written by me. It is for discussion and as a basis of modification by the Subcommittee. I intend it to be brief and general (as compared to those of SID and APS) since we are doing it for the first time.

Bylaws of the International Liquid Crystal Society

Article 1 – NAME

The Society shall be called The International Liquid Crystal Society, hereafter called the ILCS.

Article 2 – OBJECT AND SCOPE

1. The object of this Society shall be:
   (a) To encourage the scientific, literary and eduactional advancement of liquid crystals.
   (b) To provide a forum for individuals and organizations to exchange information and ideas relating to liquid crystals.
   (c) To serve the need of the liquid crystal community, including both individual scientists and the industry.

2. The scope of the ILCS is non-national.

Article 3 – MEMBERSHIP

1. Grades and Qualifications

   (a) Student Member – An individual pursuiting an undergraduate or graduate degree.
   (b) Associate Member – An individual interested in furthering the object of the ILCS.

Fig. 10.8. Act IV (July 15, 1989): Draft of ILCS' Bylaws.

Now we needed a Vice President. Chandra and I could not come up with a candidate. And so we stood at the entrance of the conference canteen watching people going in and out, and still did not see a suitable candidate. After five minutes, the conversation went like this:

Chandra: "Lui, why don't you be the Vice President?"
Lam: "That wouldn't be appropriate because then both the President and Vice President are Asians."
Chandra: "But I never see you as an Asian."

That was probably true; I acted and talked more like a New Yorker than a Chinese. Luckily, at this point, we saw Martin Schadt approaching the canteen. That was it; we asked and Schadt kindly agreed to be the Vice President.

With the President and Vice President in place and the Bylaws approved, the PSC dissolved itself. All PSC members automatically became Board of Directors members. The new ILCS was announced on July 27 at the closing ceremony and people were invited to sign up to join (Figs. 10.9 and 10.10).[33] I assumed the Chair of the Conference Committee, with all Board members as its members. In other words, the Conference Committee was the old PSC under a new name. And that was how the ILCS was born!

10.4 After 1990

My term as ILCS' Board member ended in 1994. I stopped going to ILCC in 1996 because my research, since 1995, had completely shifted to nonlinear and complex systems [Lam, 1997; 1998].[34] But I did show up in Sendai in 2000 on my way home from Beijing to have dinner with Helfrich, Cladis and others.

---

[33] ILCS' first Officers Meeting was held July 27, 1990 at UBC; present: Chandra (President), Shadt (Vice-President), Lam, Doane (Treasurer), and Dunmur (Secretary).
[34] The fact that the Chair of the 16th ILCC at Kent refused to waive my registration fee did not help. After Active Walk [Lam, 2005a; 2006], I worked on Histophysics [Lam, 2002] and Science Matters [Lam, 2008]. The latter is a new discipline that treats human dependent matters as part of science. See also www.sjsu.edu/people/lui.lam/scimat.

ANNOUNCEMENT

International Liquid Crystal Society (ILCS)

ILCS is pleased to announce its existence, and to invite applications for membership.

ILCS is a voluntary, non-profit international organization. The objects are (i) to encourage the sicientific and educational advancement of liquid crystals; (ii) to provide a forum for individuals and organizations to exchange information and ideas relating to liquid crystals; and (iii) to serve the need of the liquid crystal community, including both individual scientists and the industry. Membership consists of student member, assocaite member, member, sustaining member and affiliate Society member. (See the bylaws for more details.)

The Board of Directors of ILCS currently consists of G. Baur (Germany), L.M. Blinov (USSR), G.H. Brown (USA), S. Chandrasekhar (India), N.A. Clark (USA), W.H. de Jeu (The Netherlands), D. Demus (Germany), J.W. Doane (USA), G. Durand (France), A.M. Figueiredo Neto (Brasil), A. Fukuda (Japan), J.A. Janik (Poland), S.T. Lagerwall (Sweden), L. Lam (USA), A.J. Leadbetter (United Kingdom), J.D. Litster (USA), K. Okano (Japan), J. Prost (France), F. Rustichelli (Italy) and H. Stegemeyer (Germany).

For further information on the ILCS please contact:
Professor S. Chandrasekhar
Raman Research Institute
Bangalore 560080   INDIA

?AX: 91 812 340492; Tel: 91-812 340122; Telex: 845 2671 RRI IN ;rams; RAMANINST, or any member of the Board of Directors.

Fig. 10.9. Finale (July 27, 1990): Announcement of the *existence* of the ILCS distributed at the 13th ILLC, Vancouver (prepared by the author in June, 1990).

```
                PRELIMINARY APPLICATION FORM FOR MEMBERSHIP

                   International Liquid Crystal Society (ILCS)


Name: _____
         First              Middle initial                 Last
Affiliation (if any): _____


Address: _____

         _____

         _____


Phone: _____   Fax: _____

Telex: _____   E-mail: _____

Highest Degree obtained: (B.S./M.S./Ph.D.) _____ year _____

                              _____ Univ.

Current Interests in Liquid Crystals: (Physics, Chemistry, Biology, Polymers,
Applications, etc.)
_____

_____


Type of Membership Applied: (Student member, associate member, member,
sustaining member, or affiliate Society member)
_____


Signature: _____   Date: _____

P.S. o The membership dues will be announced by the ILCS shortly.
     o Please return this form to the reception desk of the 13th International
       Liquid Crystal Conference during the conference, or to Prof. S.
       Chandrasekhar, Raman Research Institute, Bangalore 560080, India.
     o Suggestions and Comments:
```

Fig. 10.10. Finale (July 27, 1990): The sign-up form (together with the Announcement and the Bylaws) distributed at the 13th ILLC (prepared by the author in June, 1990).

I was in Beijing June 4, 1989 [Morrison, 1989]. After that I stopped going to Beijing for eight years. Instead, I frequented Taiwan and worked with the LCers I met at the Kyoto conference [Pan et al, 1995]. I urged them to join the ILCS. With much delay, the LC society in Taiwan was formed in 1995 and affiliated to the ILCS as the "ROC Taiwan Liquid Crystal Society" [Fukuda, 1998]; the CLCS chose to stay out. Interestingly and historically, it was not until 2004 that a petition signer (Fig. 10.6) got elected to the presidency of ILCS. He was Satyendra Kumar, from Kent.[35]

In short, the International Liquid Crystal Society is a French-inspired, Chinese-initiated and truly international mass organization. Enjoy!

References

Arcioni, A., Tarroni, R., Zannoni, C., Dalcanale, E. & Du vosel, A. [1995] "Microscopic heterogeneity in a bowlic columnar mesophase as probed with fluorescence depolarization measurements," *J. Phys. Chem.* **99**, 15981-15986.
Barón, M. & Stepto, R. F. T. [2002] "Definitions of basic terms relating to polymer liquid crystals," *Pure Appl. Chem.* **74**, 493-509.
Bauer, G. E. W. [1983] "General operator ground-state expectation values in the Hohenberg-Kohn-Sham density functional formalism," *Phys. Rev. B* **27**, 5912-5918.
Blass, C., Redinger, J., Manninen, S., Honkimäki, V., Hämäläinen & Suortti, P. [1995] "High resolution Compton scattering in Fermi surface studies: Application to FeAl," *Phys. Rev. Lett.* **75**, 1984-1987.
Brostow, W. [1990] "Properties of polymer liquid crystals: Choosing molecular structures and blending," *Polymer* **31**, 979-995.
Callaway, J. & March, N. H. [1984] "Density functional methods: Theory and applications," *Solid State Physics* **38**, 135-220.
Cometti, G., Dalcanale, E., Du vodel. A. & Levelut, A.-M. [1990] "New bowl-shaped columnar liquid crystals," *J. Chem. Soc., Chem. Commun.*, 163.
Das, P & Schwartz, W. H. [1994] "Continuum and molecular theories of biaxial nematics: Calculation of the 2-Director viscosity coefficients," *Mol. Cryst. Liq. Cryst.* **239**, 27-54.

---

[35] The first few Presidents of ILCS are: Sivaramakrishna Chandrasekhar (India, 1990-1992), Geoffrey Luckhurst (UK, 1992-1996), Atsuo Fukuda (Japan, 1996-2000) and John Goodby (UK, 2000-2004).

Demus, D., Goodby, J. W., Gray, G. W., Spiess, H. W. & Vill, V. [2008] *Handbook of Liquid Crystals* (Wiley, New York).
Dong Yan-Ming, Chen Dan-Mei, Zeng Er-Man, Hu Xiao-Lan & Zeng Zhi-Qun [2009] "Disclination and molecular director studies on bowlic columnar nematic phase using mosaic-like morphology decoration method," *Science in China Series B: Chemistry* **52**, 986-999.
Dunmur, D. & Sluckin, T. [2011] *Soap, Science, and Flat-Screen TVs* (Oxford University Press, New York).
Ellbeck, J. C. & Scott, A. C. [1982] "Solitons galore," *Phys. Bull.* **33**, 426-427.
Fukuda, A. [1998] "Message from the President, Professor Atsuo Fukuda," *Liquid Crystals Today* **8**(4), 9. (All issues of *Liquid Crystals Today* could be downloaded free from www.tandfonline.com/loi/tlcy20#.UbqVAOc3uSo.)
Gennes, de, P. G. [1974] *The Physics of Liquid Crystals* (Clarendon Press, Oxford).
Golubović, L. & Wang, Z.-G. [1992] "Anharmoic elasticity of smectic A and the Kardar-Parasi-Zhang model," *Phys. Rev. Lett.* **69**, 2535-2538.
Imamura, K., Takimiya, K., Aso, Y. & Otsubo, T. [1999] "Triphenyleno[1,12-*bcd*:4,5-*b'c'd'*:8.9-*b"c"d"*]trithiophene: The first bowl-shaped heteroaromatic," *Chem. Commun.*, 1859-1860.
Kawamoto, H. [2002] "The history of liquid-crystal displays," *Proceedings of the IEEE* **90**, 460-500.
Kuo, Chia-ling [1977] *Social and Political Change in New York's Chinatown: The Role of Voluntary Associations* (Praeger, New York).
Lam, L. [1973] "Surfaces de Fermi, profil Compton et effets a N-corps," *Phys. Lett. A* **45**, 409-410.
Lam, L. [1977a] "Dissipation functions and conservation laws of molecular liquids and solids," *Z. Physik B* **27**, 101-110.
Lam, L. [1977b] "Reciprocal relations of transport coefficients in simple materials," *Z. Physik B* **27**, 273-280.
Lam, L. [1977c] "Constraints, dissipation functions and cholesteric liquid crystals," *Z. Physik B* **27**, 349-356.
Lam, L. [1988a] "Bowlic and polar liquid crystal polymers," *Mol. Cryst. Liq. Cryst.* **155**, 531-538.
Lam, L. [1988b] "Possible liquid crystalline high T_c superconductors," in *3rd Asia Pacific Physics Conference*, eds. Chan, Y. W., Leung, A. F., Yang, C. N. & Young, K. (World Scientific, Singapore).
Lam, L. [1991] "Unsolved nonlinear problems in liquid crystals," in *Nonlinear and Chaotic Phenomena*, eds. W. Rozmus, W. & Tuszynski, J. A. (World Scientific, Singapore).
Lam, L. [1994] "Bowlics," in *Liquid Crystalline and Mesomorphic Polymers*, eds. Shibaev, V. P. & Lam, L. (Springer, New York) pp. 324-353.
Lam, L. (ed.) [1997] *Introduction to Nonlinear Physics* (Springer, New York).

Lam, L. [1998] *Nonlinear Physics for Beginners: Fractals, Chaos, Solitons, Pattern Formation, Cellular Automata and Complex Systems* (World Scientific, Singapore).
Lam, L. [2002] "Histophysics: A new discipline," *Mod. Phys. Lett. B* **16**, 1163-1176.
Lam, L. [2005a] "Active Walks: The first twelve years (Part I)," *Int. J. Bifurcation and Chaos* **15**, 2317-2348.
Lam, L. [2005b] "The origin of the International Liquid Crystal Society and Active Walks," *Physics (Wuli)* **34**, 528-533.
Lam, L. [2006] "Active Walks: The first twelve years (Part II)," *Int. J. Bifurcation and Chaos* **16**, 239-268.
Lam, L. [2008] "Science Matters: A unified perspective," in *Science Matters: Humanities as Complex Systems*, eds. Burguete, M. & Lam, L. (World Scientific, Singapore).
Lam, L. [2010] "The first 'non-government' visiting-scholar delegation in the United States of America from People's Republic of China, 1979-1981," *Science & Culture Review* **7**(2), 84-94.
Lam, L. [2014] "Solitons and revolution in China: 1978-1983," in *All About Science: Philosophy, History, Sociology and Communication*, eds. Burguete, M. & Lam, L. (World Scientific, Singapore).
Lam, L., Freimuth, R. D. & Lakkaraju, H. S. [1991] "Fractal patterns in burned Hele-Shaw cells of liquid crystals and oils," *Mol. Cryst. Liq. Cryst.* **199**, 249-255.
Lam, L. & Lax, M. [1978] "Irreversible thermodynamics of thermoviscous solids with microstructures," *Phys. Fluids* **21**, 9-17.
Lam, L. & Morris, H. C. (eds.) [1989] *Wave Phenomena* (Springer, New York).
Lam, L. & Prost, J. [1992] *Solitons in Liquid Crystals* (Springer, New York).
Lam, L. & Shu, C. Q. [1992] "Soltions in shearing liquid crystals," in *Solitons in Liquid Crystals*, eds. Lam, L. & Prost, J. (Springer, New York) pp. 51-109.
Lax, M. & Nelson, D. F. [1971] "Linear and nonlinear electrodynamics in elastic anisotropic dielectrics," *Phys. Rev. B* **4**, 3694-3731.
Levelut, A. M., Malthête, J. & Collet, A. [1986] "X-ray structural study of the mesophases of some cone-shaped molecules," *J. Phys.* (Paris) **47**, 351-357.
Lin Lei (Lam, L) [1978] "Microscopic theory of first-order phase transitions in liquid crystals," *Kexue Tongbao* **23**, 715-718.
Lin Lei [1979] "Nematic-isotropic transitions in liquid crystals," *Phys. Rev. Lett.* **43**, 1604-1607.
Lin Lei [1980] "Critical properties of nematic-isotropic transition in liquid crystals," in *Liquid Crystals*, ed. S. Chandrasekhar (Heyden, London) pp. 355-360.
Lin Lei [1982] "Liquid crystal phases and the 'dimensionality' of molecules," *Wuli (Physics)* **11**, 171-178.

Lin Lei [1983] "Liquid crystal research in China: 1970-1982," *Mol. Cryst. Liq. Cryst.* **91**, 77-91.
Lin Lei [1987] "Bowlic liquid crystals," *Mol. Cryst. Liq. Cryst.* **146**, 41-54.
Lin Lei, Liu Lin & Qiu Ju-Liang (trans.) [1985] *Structure of Materials: From Blue Sky to Plastics* (Science Press, Beijing).
Lin Lei, Shu Changqing, Shen Juelian, Lam, P. M. & Huang Yun [1982] "Soliton propagation in liquid crystals," *Phys. Rev. Lett.* **49**, 1335-1338; **52**, 2190(E).
Lin Lei, Ye Peixian & Zhou Hetian [1980] "A brief note on the International Liquid Crystal Conference, 1979, Bangalore, India," *Wuli Tongxun*, No. 2, Supplement (Institute of Physics, CAS, Beijing) pp. 45-46.
Martin, P. C., Pershan, P. S. & Swift, J. [1970] "New elastic-hydrodynamic theory of liquid crystals," *Phys. Rev. Lett.* **25**, 844-848.
Mehta, G. & Uma, R. [1998] "Oxa-bowls : Formation of exceptionally stable diozonides with novel, C–H$\cdots$O hydrogen bond directed, solid state architecture," *Chem. Commun.*, 1735-1736.
Morrison, D. (ed.) [1989] *Massacre in Beijing: China's Struggle for Democracy* (Warner Books, New York).
Nelson, D. F. [1979] *Electrical, Optic, and Acoustic Interactions in Dielectrics* (Wiley, New York).
Palffy-Muhoray, P., Lee, M. A. & Petschek, R. G. "Ferroelectric nematic liquid crystals: Realizability and molecular constraints," *Phys. Rev. Lett.* **60**, 2303-2306.
Pan Ru-Pin, Sheu Chia-Rong & Lam, L. [1995] "Dielectric breakdown patterns in thin layers of oils," *Chaos Solitons Fractals* **6**, 495-509.
Papanicolaou, N. I., Bacalis, N. C. & Papaconstantopoulos, D. A. [1991] *Handbook of Calculated Electron Momentum Distributions, Compton Profile, and X-ray Form Factors of Elemental Solids* (CRC Press, Boston).
Ribotta, R., Joets, A. & Lin Lei [1986] "Oblique roll instability in an electroconvective anisotropic fluid," *Phys. Rev. Lett.* **56**, 1595-1597; **56**, 2335 (E).
Shu Chang-Qing & Lin Lei [1984a] "Theory of homologous liquid crystals. I. Phase diagrams and the even-odd effect," *Mol. Cryst. Liq. Cryst.* **112**, 213-231.
Shu Chang-Qing & Lin Lei [1984b] "Theory of homologous liquid crystals. II. Orientation correlation functions," *Mol. Cryst. Liq. Cryst.* **112**, 233-264.
Xiong Wei-Min [2013] "China's participation in international science: The case of biochemistry, 1949-1982," *Science & Culture Review* **10**(2), 50-72.
Xu, B. & Swager, T. M. [1993] "Rigid bowlic liquid crystals based on tungsten-oxo CaliM4 1arenes: Host-Guest effects and head-to-tail organization," *J. Am. Chem. Soc.* **115**, 1159-1160.
Zeng Er-Man [2001] *Design, Synthesis and Characterization of Columnar Discotic and Bowlic Liquid Crystals*, PhD thesis (Georgia Institute of Technology, Atlanta).

Part III
Sociology of Science

11

Three Waves of Science Studies

Harry Collins

There have been two waves of science studies and an attempt is being made to establish a Third Wave. The waves are described in outline. Science Matters works within the First Wave. Along the way it is argued that the human sciences are fundamentally different from the natural and biological sciences. It seems that Science Matters has failed to recognize this.

11.1 Introduction

Science Studies, or "Social Studies of Science", are the collective names for history, sociology and philosophy of science taken together. They comprise critical analysis of science from those who are not themselves scientists. In a paper written in 2002, Collins and Evans suggest that there have been two waves of science studies and propose a third wave [Collins & Evans, 2002].[1]

---

[1] The analysis presented here is not without its critics. The 2002 paper encountered fierce criticism from the science studies community and there were signs suggesting that its authors were initially being rejected from the heartland of science studies for writing it. It was the Third Wave that was the problem and the resistance of even social scientists to being seen as merely part of a movement rather than as independent thinkers. It must also be said that the 2002 paper was written from the point of view of sociologists of science and many historians and philosophers of science, who never fully engaged with the Second Wave (see below), believe the model does not describe their world. The 2002 paper, by the way, is now the second most cited paper in the 40-year history of the journal *Social Studies of Science*. Some of the flavor of the Third Wave and how it compares with the Second Wave can be obtained from a short paper in *Nature* [Collins, 2009].

11.2 The First Wave of Science Studies

The first wave of science studies grew out of the attitude to science before and after the Second World War. In that "golden age" there was no question but that science was the pre-eminent kind of knowledge. "Science Matters" [Lam, 2008; 2011] can be seen as belonging to that golden age[2] whereas for most of us in science studies it is long past. When science is seen as unquestionably the finest kind of knowledge the job of outside commentators is simple. It is to understand, explain and, effectively, reinforce the success of the sciences. In the golden age, for professional outside commentators and public alike, a good scientific training was seen to put a person in a position to speak with authority and decisiveness in their own field, and often in other fields too. Because the sciences were thought of as esoteric as well as authoritative, it was inconceivable that decision-making in matters that involved science and technology could travel in any other direction than from the top down.

The first wave began to run into shallow academic waters in the late 1960s. An iconic contributor to its demise was Thomas Kuhn's book *The Structure of Scientific Revolutions* [1962]. Though were many other contributors to the demise of the golden age, Kuhn's book was the most widely read, being said to be the best selling academic book of the Twentieth Century. Kuhn showed that science underwent cultural revolutions—for example the change from a Newtonian to an Einsteinian worldview—in which the very meanings of experiments and the words used to describe them could change radically. If that was the case then scientific knowledge could not be so obviously superior because one never knew when it would be overtaken by a revolution and all would change. By the end of the 1970s, the golden age for science no longer existed as a serious foundation for academic analysis—though, of course, it still lives on among non-professional analysts of science. For example, it lives on among scientists themselves, among amateur analysts of science, and among policy-makers who can find responsibility lifted

---

[2] Dr. Lui Lam considers that my interpretation is incorrect. My interpretation is based on some of the writings of Dr. Lam but more particularly is the result of a two day conversation I had with him in Cardiff, UK, Nov. 24-25, 2011.

from their shoulders by a science like economics even when everyone knows its predictions are close to worthless.

11.3 The Second Wave of Science Studies

The second wave of science studies has run from the early 1970s, and continues to run today. It is often referred to as "social constructivism", although it has many labels and many variants. One important variant is the sociology of scientific knowledge (SSK). What has been shown under Wave Two is that it is necessary to draw on "extra-scientific factors" to bring about the closure of scientific and technical disputes—scientific method, experiments, observations, and theories are not enough. The problem reveals itself where there is deep controversy. For example, in such cases the standard of replicability is vitiated by the *experimenter's regress* [Collins, 1985/1992].

Suppose some experimenter or group of experimenters finds a new phenomenon—"thargs". A second group tries to repeat the observation but finds no thargs. What happens in seriously disputed science, and we see this over and over again, is that the first group says the second group did not do their experiment/observations well enough. The second group says that the first group only thought they saw thargs because they did not do *their* experiment/observation properly. Both groups can be justified in making these accusations because experiments are very hard and novel experiments are even harder and there is no way to say who did the experiment best. In the end scientists reach a consensus by referring to "extra-scientific" factors such as who has the best track record, comes from the better university, who they trust, and so on. We see that the choice of what is to count as the correct result is very much tied up with social judgments not just "scientific" judgments. Or, to put it another way, scientific judgments are always, in part, social judgments. This means that many social influences enter into the heart of scientific consensus formation.

The philosophy that underlies this approach takes it that human activity can be thought of as characterized by paradigms, or "forms-of-life". This is particularly important for any program which takes it that the methods of natural science are a suitable model for the analysis of social life. It is humans that create knowledge but it is also humans who create the objects of social life. The objects of social life are things like witches and mortgages. Because humans live life differently in different places then the world is different in different places. In one place life is governed by the existence of witches and the like, in another it is governed by the existence of mortgages and the like. Even in science these differences show up in terms of Kuhn's paradigms: at one time scientific life is governed by the idea of length and mass as conserved, at another time, as not conserved. The social scientist, and this includes the analyst of science, has to begin with the world as constructed by the "actors"—the natives livening in the society being studied. To try to explain human action in a society informed by the idea of witches and the like in terms of mortgages and the like would be nonsensical. The social scientist has, therefore, two lots of categories to take into account when analyzing the world: there are the *actors' categories*, which create the world being analyzed, and there are the analyst's second-order *technical categories* built on top of the actors' categories. The natural scientist, on contrast, has to deal only with the technical categories.

The way the world is divided, or created, by the categories found in language and culture marks a fundamental difference between natural and social science; in natural science there is only one set of categories—the analysts'—in social science there are two. And in social science, including the social analysis of science, the analysts' categories are always subservient to the actors' categories. The first move in social science has to be understanding actors' categories. We normally come to understand the world of unfamiliar actors' by living among them for a while—the participatory method as used in sociology, anthropology and ethnography. Thus, history is a very difficult enterprise because, where the actors are dead, it is hard to reconstruct their world. The natural scientist is never faced with the equivalent problem.

When it comes to the history of science, the first step has to be to create an understanding of the "paradigm" that informed the scientific actors. For these reasons, history and sociology of science are very different to natural science. History of science and sociology of science are social sciences and the world of the scientists that they describe has to be understood as a species of social life.

Findings like this, along with the underlying philosophy, led some writers to think of science as continuous with any other social activity. The way science interacted with society was reconceptualized. Much valuable rethinking took place in respect of how science operated in courts of law, schools, and policy processes such as public inquiries. For example, it used to be thought that forensic science needed to be represented in courtrooms solely by state sponsored professionals but it is now understood that there can be two sides to an argument that turns on forensic evidence. In the courtroom both defense and prosecution can have their own justifiable and conflicting interpretations even where technical issues are involved.

Wave 2 of science studies became swept up into the much larger movement known as "postmodernism". The downside of seeing science as having no features which justify its being counted as an especially reliable form of knowledge began to become evident. Analysts from the humanities—the "other side" of C. P. Snow's "two cultures"—could feel they could criticize science without first developing any deep understanding of its practices and procedures; this is a downside from the point of view of the methodology of the social sciences even if it is an "upside" from the point of view of the humanities. Science also came to be said to be just a continuation of politics by other means; within social studies of science it began to seem, then, as though one could do science by doing politics and this, in my view, has damaged science studies' own academic community. The distinction between scientists and the public was also said to be no longer clear and it was argued by some that ordinary people had just as much right as experts in determining whether, say, a vaccine was safe. This seems to have led to tragedies from the measles epidemics which have followed the opposition to mumps, measles and rubella vaccine in the UK and elsewhere, to Thabo Mbeki's decision not to distribute anti-retroviral drugs in South Africa [Nattrass, 2012].

The Second Wave of science studies was a brilliant and much needed corrective to the First Wave. Science is now understood in a way that makes much of the philosophy of science that was practiced under the First Wave, with its clear separation between theories and the "facts" upon which they depended, look like fantasy. We now know that there are no clean facts—at least not until controversies have been settled. We understand the mechanisms of science and we understand what goes on in laboratories, and in scientific decision-making in general, incomparably better than we did in the 1950s and 1960s. Whatever happens, our understanding of science can—or at least should—never be the same again; the Second Wave will not (or, at least, should not) go away. Nevertheless, when Second Wave findings are subsumed into the wider movement of postmodernism and the very idea of science begins to dissolve; this should be a cause for worry. The question is, given the Second Wave, what is science?

11.4 The Third Wave of Science Studies

The Third Wave of science studies is an attempt to explain what is special about science for the purposes of policy-making without giving up the insights of the Second Wave. Long experience seems to show us that the search for a justification of science as provider of truth is always going to fail. As a mythology this idea is fine but it just does not stand up to close examination of scientific practice, most evidently where there is deep controversy among scientists. There are no clean procedures that can settle scientific controversies to the satisfaction of all technically knowledgeable parties. The settling of scientific controversy is a matter of consensus formation and, closely examined, does not look too different to the way consensus is formed in other spheres. The Third Wave squares the circle by replacing the quest for truth with an analysis of expertise. It rests on the simple idea that it is better to give more weight to the opinions of those who, literally, *know what they are talking about*. Given a choice between the opinions of lay persons and experts in the matter of whether administering a vaccine to a population is likely to cause less harm than not administering it, the Third Wave says give more weight to the opinion of the experts. Of course, this is only the beginning. What do we mean by an expert? What kinds of experts are to be preferred? What is expertise and what kinds of expertise are there? What happens when experts disagree among themselves?

To start with the last question, even the Third Wave is nearly impotent when experts disagree. It says only that the matter should be settled among the experts. It says that the opinions of lay persons do not become elevated in importance just because the experts disagree, so it sets a boundary around those who can legitimately take part in the technical phase of a debate. Note the term "technical phase". The Third Wave stresses that the technical phase should be separated from the political phase of a technological debate in the public domain. Even when the technical phase is settled to the experts' satisfaction it does not comprise a political decision, it comprises an input to a political decision. In the end, democracy trumps all. What must not happen, however, even though it is allowed under the more postmodernist interpretations of Wave Two, is that the outcome of the technical phase is determined by political considerations. Under Wave Three, any policy that is seemingly precipitated by the outcome of the technical phase can be *overturned* by the political phase but what happens in the technical phase cannot be *determined* by the political phase. When the technical phase is overturned by the political phase this is legitimate but the process should be transparent so that voters have a chance to signal their objections.

The paradigm case where this sequence did not take place is Mbeki's claim that he would not distribute anti-retrovirals to pregnant women because they were dangerous. Mbeki was able to point to a controversy on the internet regarding the question of the safety of anti-retroviral drugs and he claimed this justified his actions. But there was no serious controversy and he should simply have stated that he was making a political choice not distribute the drugs. What he did was to use a quasi-scientific justification to disempower the democratic process.

If there was a controversy on the internet over the safety of anti-retrovirals, how can the Third Wave make the claim that there was no real scientific controversy going on? The Third Wave analysis is that this so-called controversy had long passed its "sell-by date" in the mainstream scientific literature and what Mbeki was advising his parliamentarians to look at were merely remnants found only in fringe journals or the internet. By the time Mbeki made his decision it was a sham controversy that was taking place, not a real one. Mbeki did not have the expertise to see the difference; his was a clear case of a non-expert trying to participate in the technical phase of a debate. Mbeki, literally, *did not know what he was talking about*. This is a rare example of the Third Wave being able to offer input even while some would say a scientific controversy is still going on.

What do we mean by an expert? The Third Wave puts forward a detailed analysis of the nature of expertise, building up to a "Periodic Table of Expertises" (PTE) [Collins & Evans, 2007, p 14]. The basic definition of an expert under this scheme is someone who has acquired the tacit knowledge [Collins, 2010] pertaining to the domain of expertise in question. Tacit knowledge can be acquired only through social interaction with members of a domain so, though the PTE includes categories that do not turn on such social contact, these are not real expertises but more collections of knowledge.

The highest form of such knowledge is "Primary Source Knowledge" which is acquired through reading the technical literature in the absence of social contact with the expert community. But this can be deeply misleading as the journals differ in quality and even within the top quality journals there are many valueless publications or publications that the expert community know are to be ignored. One might say that Mbeki was mistaking Primary Source Knowledge for a higher form of expertise.

To date, the most discussed category in the Periodic Table is a new one—"Interactional Expertise"—which is contrasted with "Contributory Expertise". Contributory Expertise refers to what is regularly understood as expertise and its possessor is able to make new contributions to the practice of the technical domain in question. Contributory Expertise is acquired through practice within the domain—the normal apprenticeship model. Interactional Expertise, in contrast, is acquired solely through *linguistic* interaction with members of the domain without engaging in the practice. It is argued that someone who has engaged with the linguistic discourse deeply enough will be, in principle, indistinguishable from a contributory expert in technical discussions. In other words, interactional experts ought to be capable of making technical judgments that are just as good as those made by full-blown practitioners; interactional experts, as one might say, do not just "talk the talk", they "walk the talk". That this is the case has been demonstrated with Turing-type tests known as "Imitation Games" [Collins et al, 2006; Giles, 2006; Collins & Evans, 2013]. The idea of Interactional Expertise seems initially counter-intuitive but further consideration shows that, on the contrary, it is *necessary* if we are to make sense of the world. For example, wherever there is a technically complex division of labor (as in a big scientific experiment), specialists must understand each others' domains in order to coordinate their work but they can only have gained the necessary understanding through talk; if practice was always necessary to gain technical understanding then there could be no specialists and no complex division of labor. Interactional Expertise is, quite simply, the foundation of society [Collins, 2011].

As indicated, the Third Wave has implications for understanding the relationship between democratic societies and technology. Still more ambitious is the attempt to re-establish that the *values* associated with the form-of-life of science as a whole are central to democracy. The term used is "elective modernism". Elective modernism contrasts with earlier attempts to align science with democracy in being purely normative. For example, Robert Merton [1942] attempted to justify the values of science by reference to their efficiency whereas elective modernism simply considers the values associated with science to be good in a self-evident way without need for justification. Note, then, that in contrast to "Science Matters", elective modernism does not endorse scientific method but scientific values.[3] There is nothing in common to scientific method beyond observation and/or immersion in the world and the determination to describe it with clarity.

The Third Wave program is turning out to have very wide implications. Here it has been possible to provide only a "taste". More can be discovered by web searches and/or accessing the website: www.cf.ac.uk/socsi/expertise.

---

[3] Contrary to Dr. Harry Collins' claim, Science Matters (Scimat) does *not* endorse scientific method but *does* endorse scientific values (*and* Reality Check which is not mentioned in this chapter). See [Lam, 2008; 2011] and particularly Lui Lam's two articles, "About science 1" and "About science 2", in *All About Science: Philosophy, History, Sociology & Communication*, eds. Burguete, M. & Lam, L. (World Scientific, Singapore, 2014)—*Editor.*

References

Collins, H. M. [2011] "Language and Practice," *Social Studies of Science* **41**(2), 271-300. [DOI 10.1177/0306312711399665]

Collins, H. M. [2010] *Tacit and Explicit Knowledge* (University of Chicago Press, Chicago).

Collins, H. M. [2009] "We cannot live by scepticism alone," *Nature* **458**, March, 30-31.

Collins, H. M. [1985] *Changing Order: Replication and Induction in Scientific Practice* (Sage, Beverley Hills & London). [2nd edition, 1992, University of Chicago Press, Chicago]

Collins, H. M. & Evans, R. [2007] *Rethinking Expertise* (University of Chicago Press, Chicago).

Collins, H. M. and Evans, R. [2002] "The Third Wave of Science Studies: Studies of expertise and experience," *Social Studies of Science* **32**(2), 235-296.

Collins, H. M. and Evans, R. [2013] "Quantifying the Tacit: The imitation game and social fluency," *Sociology* (in press).

Collins, H. M., Evans, R., Ribeiro, R. & Hall, M. [2006] "Experiments with Interactional Expertise," *Studies in History and Philosophy of Science* **37** A/4 [December], 656-674.

Giles, J. [2006]. "Sociologist fools physics judges," *Nature* **442**, 8.

Kuhn, T. S. [1962] *The Structure of Scientific Revolutions* (University of Chicago Press, Chicago).

Lam, L. [2008] "Science Matters: A unified perspective," in *Science Matters: Humanities as Complex Systems*, eds. Burguete, M. & Lam, L. (World Scientific, Singapore) pp. 1-38.

Lam, L. [2011] "Arts: A Science Matter," in *Arts: A Science Matter*, eds. Burguete, M. & Lam, L. (World Scientific, Singapore) pp. 1-32.

Merton, R. K. [1942] "Science and technology in a democratic order," *Journal of Legal and Political Sociology* **1**, 115-126.

Nattrass, N. [2012] *The AIDS Conspiracy: Science Fights Back* (Columbia University Press, New York).

12

Solitons and Revolution in China: 1978–1983

Lui Lam

Historically it is rare that one could do scientific research and political revolution at the same time. Such a chance was offered to me in China from 1978 to 1983. Throughout these six years, solitons (i.e., localized waves that travel without, or with slight, change in velocity and shape) were one of my major research topics at the Institute of Physics, Chinese Academy of Sciences. It was a hot topic in the physics community worldwide. In this chapter, the development of soliton research and political revolution in China experienced by the author is reported. The aim is not just to keep a record for those memorable years but also to convey the excitement of the so-called "Science Spring" of 1978, the year China's reform-and-opening up revolution began.

12.1 Introduction

It is rare that one can participate in history by doing scientific research and carrying out political revolution simultaneously. A famous example is the case of Condorcet (1743-1794), a French philosopher, mathematician and political scientist [Baker, 1975]. He held many liberal ideas and participated actively in the French Revolution (1789-1799). In 1794, after being branded a traitor and while hiding as a fugitive from French Revolution authorities he finished his masterwork, *Sketch for a Historical Picture of the Progress of the Human Mind* [Lukes & Urbinati, 2012]. Soon after that, he was arrested and died in prison, at age 50.

A more recent example is the case of the thousands of scientists working in the 1950s in the new China, trying their best to do science in the midst of a socialist revolution. These include those who returned to China before and after 1949, the year the People's Republic of China was established; a lot of them earned PhD degrees from top universities in Europe and America [Wang & Liu, 2012].

However, when I returned to China in 1978 I was unaware of Condorcet's story and the full story of the Chinese scientists did not come out yet.

12.2 Returning to China

I was born in mainland China but grew up in Hong Kong. After receiving my bachelor degree in Hong Kong, I went for graduate studies first in Vancouver, Canada, and then at Columbia University, New York City [Tsui & Lam, 2010, pp. 209-211]. It was at Columbia as a physics student that I encountered the anti-Vietnam War student movement (1968) [1] and subsequently the Chinese *Baodiao Movement* (1970). Baodiao means "Protect Diaoyutai." Diaoyutai (also called Diaoyudao or Diaoyu Islands) is the group of small islands near Taiwan that was and still is under dispute among the Chinese and the Japanese (Fig. 12.1) [The Seventies Monthly, 1971].[2]

Although I had participated actively in Baodiao and lived in Chinatown to do community work [Lam, 2010] my decision to go back and settle in China was not motivated by these experiences. Rather, it was China's "The Great Proletarian Cultural Revolution". The Cultural Revolution began in 1966, the year I arrived at Columbia. The ideals put forth by this revolution and the heroic stories coming out from China stimulated the leftists and radicalized many young people all over the world. What could be more exciting than shutting down the whole country and rebuilding its entire infrastructure, leading to a new society not just in China but in every place in the world? How romantic? And it

---

[1] http://en.wikipedia.org/wiki/Columbia_University_protests_of_1968 (Sept. 21, 2013).
[2] "China and Japan: Could Asia really go to war over these?" *The Economist*, Sept. 22, 2012.

was such a basic solution to all human problems. At Columbia, we were trained and urged to tackle important, basic problems.

Fig. 12.1. The "Protect Diaoyutai" movement. *Left*: One of the Diaoyudao Islands. *Right:* The book cover of [The Seventies Monthly, 1971]; the reprint "Hong Kong, Taiwan, Diaoyutai" (pp. 103-105) was written by the author (under the pen name Huang Shi-Zhi which is omitted in this book; "shi zhi" means "stone it" in English).

While living in Manhattan Chinatown (starting summer, 1971) I joined the editorial board of *China Daily News*, a newspaper established during the resistance-against-Japan war years by some overseas Chinese in New York who were sympathetic to the Chinese communists [Lai, 2010]. One of the founders is Tang Ming-Zhao. He went back to New China early on and returned as a member of China's first permanent representatives to the United Nation stationed in New York. On the evening he arrived, he asked his personal driver to deliver a box of Tsingdao beer to the China Daily office.

In summer of 1974, I (and three other persons) was invited to tour China for seven weeks and that was well before China opened its door to tourists. China was obviously underdeveloped but was green and unpolluted. Everywhere we went, including the scenic West Lake in Hangzhou, we were the only guests in town.

After the trip I decided to go back to China to join the revolution.[3] But it was not that easy. Zhou En-Lai (1898-1976), the premier at that time, kept telling visitors that China was not ready to welcome us back. And that was true given that the whole country was practically shut down, a condition that even the premier was not allowed to tell. We kept pressing.

My three-year, postdoctoral appointment at City College came to an end in 1975, and my doctoral mentor, Philip Platzman (1935-2012) at Bell Labs [Hamann & Isaacs, 2012], found me a job at Antwerp, Belgium. I postponed my new appointment to October 1, so that I could finish my duty as the stage manager of the Chinatown event celebrating the national day of China—a position that I held for couple of years based on the skills I picked up as a member of the Chinatown Food Co-op [Kuo, 1977].

In Belgium, as in New York, I worked with the local Chinese. From time to time, I visited the Chinese embassy in Brussels enquiring about the status of my application and received the same negative answer. After ten months in Antwerp, my job shifted to Saarbrücken, West Germany. The same story, except the Chinese embassy was in Bern, a two-hour trip from my town. Finally, something happened: Chairman Mao (Ze-Dong) (1893-1976) passed away in September 1976. And in 1977, the embassy arranged me to join a Taiwanese-student delegation from Europe to observe the national day celebration in Beijing and to find out about my application over there.

I was more than excited. To increase my chance of success, I flew back to New York, drove three hours to the Baltimore home of Ren Zhi-Gong (1906-1995)[4] and asked him to write me a referral letter. Ren was

---

[3] The trend of overseas students returning to China in the 1970s and 1980s resulting from the Baodiao Movement was first predicted in an article in *Zhongwen Yundong* (Chinese Language Movement), a hand-written magazine co-founded by the author, which published a total of three issues, all in 1971. The article, "Facing squarely Hong Kong's political future" [Lam, 1971], was reprinted by *Undergrad*, the student newspaper of University of Hong Kong, September 16, 1971, which is reproduced on p. 7 of *Impact 100*, a study celebrating HKU's achievements in its first 100 years (daaoweb.hku.hk/UserFiles/Image/publication_book/CNews/Autumn2012/Impact100.pdf, June 19, 2013).
[4] Ren Zhi-Gong, graduated from Tsinghua University (1926), MIT (1928), Harvard (1931, PhD, physics); returned to China in 1933; professor at National Southwest

in his sixties at that time. He came out from China in 1946 and had served as department chair at Johns Hopkins University. Because of his seniority he knew all the physicists in China. But more importantly, he and Wang Hao (1921-1995)[5] (Fig. 12.2) were the two famous professors who participated actively and whole heartily in Baodiao. Ren was not just a nationalist; he was an anti-imperialist. Ren handed me a sealed envelope, addressed to Qian San-Qiang (1913-1992)[6] (Fig. 12.3, right).

Fig. 12.2. Two senior professors participating heavily in Baodiao: Ren Zhi-Gong (left) and Wang Hao (right). Both earned PhD from Harvard University; Ren in 1931 and Wang in 1948.

To save money, I took a train to West Berlin, crossed the border and entered East Berlin, and flew to Beijing via Romania. In Beijing the weather was perfect. I handed the referral letter to the person welcoming us. After touring the oil fields in Daqing and other places, we returned to

---

Associated University, Kunming; came to US in 1946 and stayed; fellow of National Academy of Sciences (USA) (baike.baidu.com/view/2336180.htm, Oct. 1, 2012).

[5] Wang Hao, graduated from National Southwest Associated University, PhD from Harvard University (1948), a logician and professor at Rockefeller University [He & Wen, 2006] (also: baike.baidu.com/view/58173.htm#sub6821460, Sept. 30, 2012). He lived in New York City (NYC) and once came to a Baodiao meeting at the Chinatown loft (5th floor, 22 Catherine Street) I lived; like everybody else, he sat on the floor. On the other hand, Ren Zhi-Gong lived in Baltimore and had to drive three hours to attend the Baodiao meetings in NYC. And there were many of these meetings since NYC was the nerve center of Baodiao; I saw them often.

[6] Qian San-Qiang, graduated from Tsinghua University (1936), studied in France with Madame (Irène) Curie and her husband (1937-1940, PhD, nuclear physics), returned to China in 1948, father of China's atomic and hydrogen bombs (baike.baidu.com/view/3908.htm, Oct. 1, 2012).

Beijing just in time for the National Day celebration. We watched fireworks from the open stalls at the Tiananmen Gate, ending with a lot of firework droppings on our hair. The food of the national banquet at the Great Hall of the People was a little bit disappointing. But who cared about the food while dining at People's Hall? Being there was a treat by itself.

Fig. 12.3. Two senior physicists I associated with in Beijing: Yan Ji-Ci (left) and Qian San-Qiang (right). Qian was the protégé of Yan. Both earned PhD in Paris, France; Yan in 1927 and Qian in 1940.

My date of leaving Beijing was approaching, and suddenly I was informed that Qian San-Qiang invited me to dinner. Qian was an experimentalist in nuclear physics; as the father of China's "two bombs," atomic and hydrogen bombs, he played the roles of J. Robert Oppenheimer (1904-1961) and Edward Teller (1908-2003) combined. He was the most important Chinese scientist, well-known inside and outside of the country. The dinner was at the Donglaishun restaurant at Dongan Market (now rebuilt as New Dongan), famous for its mutton hotpot. After all the delicious food was consumed, Qian informed me that my application was approved and I was assigned to work at the Academia Sinica (called Chinese Academy of Sciences, or CAS, today). I was prepared to assume any job anyway, but out of curiosity I politely asked, "Which Institute?" Qian replied, "You will know when you return."

As I learned later, Qian honestly did not have the answer. At that point, the Institute of Physics and the Institute of Semiconductors, both under CAS, were trying their best to get me assigned to their own institute.

12.3 Arriving Beijing and the Early History of Institute of Physics

China started sending students abroad in 1872 during the Qing Dynasty. Since then, there are a total of 11 generations of overseas students, counting those going out officially or privately. Many graduated from the top universities in Europe and in America and came back to contribute significantly in the country's modernization. After 1949, there are three generations of *haigui*, the nickname for those going abroad privately and subsequently returning to mainland China voluntarily. (See Appendix 12.1.) I belong to the 9th generation going abroad and the 2nd generation of haigui.

In early January, 1978, my wife and I, together with our eight-month old daughter, stopped in Hong Kong for a few days to buy clothes and some personal items (camera, watch, and bicycle—which was shipped to Beijing), all made in China. I also bought two foreign-made items: a color TV (upon the advice of the embassy in Bern) and a HP scientific calculator (costing 150 USD).[7] We walked across a bridge at the border and entered Shenzhen; that was one of the happiest moments in my life. We took the train to Guangzhou, stayed overnight and flew to Beijing. I was informed that I was assigned to the Institute of Physics, CAS.

The history of IoP is closely linked to two names, Cai Yuan-Pei (1868-1940)[8] [Zhao, 1998] and Yan Ji-Ci (1900-1996) [Jin, 2000][9] (Fig. 12.3, left). In 1928, during the Republic of China, Cai established the government-funded National Academia Sinica (NAS, the forerunner of

---

[7] The calculator turned out to be crucial in helping me to produce China's first *Physical Review Letters* paper [Lin, 1979]; the story is given in [Lam, 2010]. Even though the bicycle was China-made, the embassy advised us to buy it in Hong Kong because bicycles in China, like many other items, were rationed.

[8] Cai Yuan-Pei, a revolutionary and educator, went to Germany at age 40 to study philosophy, psychology and aesthetics, and visited France at age 46. He was the president of Peking University, 1917-1926, and a strong advocate for liberal education and academic freedom. He resigned briefly in 1919 to protest the government's arrest of students during the May Fourth Movement. He was president of the National Academic Sinica, 1928-1940. Because of the war, he lived in Hong Kong, 1937-1940, and died there at age 72.

[9] See also: baike.baidu.com/view/46332.htm (Oct. 1, 2012).

CAS) in Nanjing *and* the National Academy of Art (now called China Academy of Art) in Hangzhou [Zhao, 1998].[10] The NAC's Institute of Physics was located in Shanghai (part of which moved to Nanjing in 1948). Next year, the privately-funded National Academy of Beiping (NAB) was established in Beiping (Beijing today). Yan headed NAB's Institute of Physics in 1931-1937 and turned it into the most important place in physics research in China, in the first half of last century [Wu, 2005, pp. 237-238].

Yan Ji-Ci graduated in 1923 from Southeast University (renamed National Central University in 1928), Nanjing, and went to Paris, France, the same year; he belonged to the 5th generation of overseas students. He was a brilliant experimentalist, obtained his PhD in 1927 from University of Paris, and became the first Chinese earning a PhD in France, which was widely reported in French and Chinese newspapers [Jin, 2000, p. 11]. Yan returned to China with fame the same year, got married, and visited Paris again for two years, 1929-1930; he worked briefly in Madam (Marie) Curie's laboratory. During his reign at NAB's Institute of Physics, he published 53 physics papers, which, with two exceptions, were all published in foreign journals [Jin, 2000, p. 92]. He trained a large number of young physicists. For example, he took in Qian San-Qiang as his assistant in 1936 and brought him to Paris the next year, to work as a graduate student under Irène Curie and her husband. In 1948 Yan became an academician of National Academia Sinica and the president of the Chinese Physical Society. In short, Yan brought back physics from Europe and developed the physics enterprise in his country, similar to what I. I. Rabi, a Nobelist at Columbia University, did for America [Rigden, 2000]. In a certain sense, *Yan Ji-Ci is the father of Chinese physics.*

Soon after the establishment of the People's Republic of China [Dietrich, 1986], in 1949, the Academia Sinica was established in Beijing, the capital. Next year, the two physics institutes (of NAS and NAB) were combined to form the Institute of Applied Physics, with Yan Ji-Ci as the Director. Shi Ru-Wei (1901-1983) became the new director in 1957 (Fig. 12.4, left). The institute changed name to Institute of

---

[10] Incidentally, in the same year, Mickey Mouse was born.

Physics (IoP) and moved to Zhongguancun, the present location, in 1958. In 1981, Shi stepped down (and passed away in 1983 [Zhao, 2005]), and Guan Wei-Yan (1928-2003)[11] [Li, 2004] succeeded as director (Fig. 12.4, right) [Sun, 2008; Zhao et al, 2008].

In the first two decades of new China, the top physicists were assigned to the secret "two bombs" project; scientific research was carried out mainly in the numerous institutes of CAS (about 100 in number presently); the universities concentrated in teaching but not in research. Many of the physics-related institutes of CAS such as the Institute of Semiconductors, was split off from IoP. When I arrived in 1978, IoP employed more than 900 people, with about 600 scientists. With its history and size, the IoP was easily the top research place in the whole country.

Fig. 12.4. The two Directors of Institute of Physics during my stay there: Shi Ru-Wei (left) and Guan Wei-Yan (right). Shi earned PhD from Yale University in 1934; Guan studied in USSR (1953-1957) and graduated from Moscow State University.

12.4 Life at Institute of Physics

Life in Zhongguancun and at Institute of Physics is described, including the "Science Spring 1978," my 1979 trip to USA as a visiting scholar, and other things.

---

[11] Guan Wei-Yan, studied low temperature physics in the 1950s in USSR returned to China in 1960, director of IoP, president of the University of Science and Technology of China, academician of CAS (1980). He passed away in Taiwan.

12.4.1 Living in Zhongguancun

I first met Hao Bai-Lin when I drove from Saarbrücken to Fribourg, Switzerland, to attend his seminar there. Hao was the one who showed me around in Zhongguancun when I first arrived at IoP. He rode his bike in front of mine and demonstrated the use of hand signals while making turns.

The theoretical physics group in IoP was disbanded during Cultural Revolution on the ground that basic science was useless. Luckily, a Theory Group was preserved in the Department of Magnetism. I worked in this group, consisting of ten persons working in theory (see Fig. 12.5) plus six in computers. I shared an office room with three other colleagues (Wang Ding-Sheng, Li Tie-Cheng and Cai Jun-Dao) on the fifth floor of the main building (Fig. 12.6). There was only one very old elevator, which was not safe enough to carry people and was used occasionally to transport heavy equipments. Even Shi Ru-Wei, the 77-year old director of IoP, had to climb the stairs to his office on the third floor. Shi received me soon after I arrived; we had a nice chat. Watching Shi's back moving slowly but surely up the stairs, like in a slowed-down video, was the most touching experience I had at IoP. The image conveyed the long history of difficulties China had to overcome to move forward and her perseverance and success in moving forward, one step at a time.

Everyone was poor; the Institute was poor. But even with money, there was nothing to buy. Everything was in shortage. There was no doufu in the market, and no white paper in the shop.[12] To write notes and do calculations, I used the back of thin and semitransparent papers meant for writing Chinese characters. Months before my arrival, IoP had to find wood to make all the furniture for my future apartment, including a bed, a desk, a chair, a wardrobe, and the bookcase shown in Fig. 12.6.

We stayed at the Overseas Chinese Hotel in downtown for three months while IoP desperately tried to find a family who was willing to exchange their apartment in Zhongguancun for the apartment assigned to

---

[12] In contrast, in the war years at National Southwest Associated University, Wu Da-You (Wu Ta-You) (1907-2000) in Kunming was able to find white papers to type up the manuscript of his book *Vibrational Spectra and Structure of Polyatomic Molecules* (1939) [Wu, 2005, p. 553].

me by CAS' headquarters which was more than one hour away by bus from IoP. Eventually, IoP succeeded; our assigned three-room apartment became a two-room apartment two-minute by bike away from IoP. But a month or so before that happened, I insisted and was allowed to start working at IoP. It was nothing but simple. It meant that the hotel kitchen had to prepare breakfast for one-person, me, at 5:30 am, six days per week. At 6:00 am, I walked across the street to the China Art Gallery; rode a bus for an hour to the Beijing Zoo; stood for half-an-hour in the freezing cold, with a number of my fellow workers, in an open truck that was sent by the IoP. One could not survive the trip without a "Lei Feng hat" (which was made of fur and covered the two ears); luckily I brought one in from Hong Kong. But there was nothing to complain about; overcoming "hardship" was quite expected in doing revolution. I joyfully look forward to work every morning, in spite of the four-hour daily trip.

Fig. 12.5. Five of my colleagues in Theory Group, Department of Magnetism, Institute of Physics. *Left to right, top to bottom*: Li Yin-Yuan, Pu Fu-Ke, Hao Bai-Lin, Yu Lu and Wang Ding-Sheng. All five became academicians of CAS. In 1978, apart from these five and the author, the other four members working in theory are: Shen Jue-Lian, Li Tie-Cheng, Feng Ke-An and Cai Jun-Dao.

Fig. 12.6. The author in his office, Room 518, main building, Institute of Physics (1978). The wooden bookcase was made by IoP specially for me before my return. The books, mostly in English, were brought back by me; note *The Feynman Lectures on Physics* in red color in second row from top. The newspaper on my desk is *People's Daily*; my iron rice bowl for lunch is next to bookcase. The switches mounted on a wooden board at the back wall controlled the lightings. They "leaked" when touched; getting electrically shocked was a daily experience for the occupants of this office.

I soon understood why the Chinese had to take a nap after the lunch. The breakfast's calories were able to support you for four hours; after a bare lunch and without a nap, you would not be able to work the afternoon. After half a year, the nutrition stored in my body from Germany waned off and I found myself getting sick from time to time. There were no private doctors. To visit the hospital, I rode a bike for 30 minutes; got a number and waited, and waited. The whole process took half a day.

There was no taxi. What if you were so sick that you could not ride a bike or take a bus to the hospital? It never happened to me, but it happened to Wu Ling-An. Wu returned to China with her parents from the United Kingdom; she spoke perfect Oxford English. In one evening, Wu had to be rushed to the hospital to give birth to a child. No problem.

Her husband borrowed a three-wheel cart from IoP, wrapped Wu's body with cotton quilt and laid her down on the wooden platform, and paddled the cart to the hospital in time. A healthy baby was born.

Almost everything was rationed, including rice, flour, cooking oil and cotton cloth; others had quotas, such as bicycles and pregnancies; still others required permission from your "work unit," such as marriages and divorces—the latter was rarely granted. The exception was soy sauce. So after we moved in to Zhongguancun, I went to buy soy sauce. I said, "I want to buy some soy sauce." The saleswoman, "Where is your (empty) bottle?" I replied, "What bottle?" It turned out that the Bern embassy did remind me to bring in a TV and a bicycle but forgot to mention the soy-sauce bottle. No bottle, no soy sauce; that was it.

All these difficulties were just inconveniences. What was really handicapping physics research at the Institute, apart from other things, was the six-day work schedule. Formally speaking, we did research Monday to Friday; Saturday was for political studies which no one was allowed to skip. On Sunday, the only free day in the week, one had to wash dirty clothes by hand and attend to family chores. There were no holidays apart from the national holidays of the Labor Day, National Day and the Chinese New Year (called Spring Festival); no Christmas because it is religious and Western, and no Mid-Autumn Festival because it is feudal. One's body was stretched to the limit; not much energy was left on Monday. Yet, all my fellow members of IoP were more than eager to work after the Cultural Revolution.

12.4.2 Science Spring 1978

Hua Guo-Feng (1921-2008) succeeded Chairman Mao in 1976. Deng Xiao-Ping (1904-1997), who studied in France (1919-1927) and belonged to the 5th generation of overseas students, resumed working in July, 1977. He oversaw post-Mao reforms in science and education. In March 18-31, 1978, the National Conference on Science was held in Beijing, in which Deng gave the speech that recognized intellectuals as part of the working class and science as the first production force [Luo,

2008].[13] As an honor and for my education, I was sent by IoP to attend the conference for half-a-day; it was not the opening day and I did not see Deng.

There was a lot to catch up. Feng Kang (1920-1993)[14] (Fig. 12.7) came to IoP and gave an introductory talk on solitons, in the only large auditorium room in the main building. That was the same room used by IoP to deliver important messages to its senior members. On that day, the room with several hundred seats was filled to capacity and every theorist in Beijing seemed to be there. The reason: Soliton was a brand new topic in China, but more importantly, the theorists were ready to roll up their sleeves and work again after theory was discredited or banned in the last ten years [Hao, 2012].

Fig. 12.7. Feng Kang and siblings. *Left to right*: Elder brother Feng Huan, Feng Kang, little brother Feng Duan, and elder sister Feng Hui [Tang et al, 2010].

---

[13] Equally important, one month later, Deng sent Xi Zhong-Xun (father of Xi Jin-Ping, current Party Chairman and Head of State) to head the Guangdong Province and create the Special Economic Zone in Shenzhen, signaling the start of socialist market economy. (See: "Editorial: The old path, evil path and bloody path after the Chinese communists' 18th National Congress," *World Journal*, Nov. 24, 2012, p. A5.)

[14] Feng Kang, BS in physics at National Central University (1944), worked at Tsinghua University and Institute of Mathematics, CAS, and in Moscow (1951-1953). In 1978 he was appointed Director of the newly founded Computing Center, CAS, until 1987 when he became the Honorary Director (en.wikipedia.org/wiki/Feng_Kang, Sept. 30, 2012). His brother, Feng Duan, was the only one among the siblings who never studied aboard. He is a physicist in condensed matters at Nanjing University and an academician of CAS, and had served as President of the Chinese Physical Society.

Fang Fu-Kang of Beijing Normal University gave a series of lectures on stochastic differential equations. We biked for 30 minutes to attend every lecture. English classes were offered at IoP. A technical member, infected with hepatitis B, was so enthusiastic to improve himself to better serve the country that he joined the class and studied very hard, ignoring the advice that someone of his condition should take a lot or rest. Pretty soon, the hepatitis progressed to liver cancer and he died before the class was finished.

I soon met Yan Ji-Ci in person. When I was a student in Hong Kong I bought his high-school physics textbook, which was superbly written. The book helped me to become the top student in class. So when I first met him it seemed that I already knew him. Shaking his hand was a historical moment for me; it was like finally making connection to the Chinese physics tradition and becoming a part of it. I met him again a few more times. The last time I saw him was in 1982 when he laid sick in Hospital 301; he survived. Hua Guo-Feng was also a "patient" in the hospital; he invited me to his room and we chatted for half-an-hour. Even though Yan was the Vice President of CAS from 1978-1981, I was unaware of his appointment; there was not yet webpage to check.[15]

I did meet Qian San-Qiang again several times. There was this time that he conducted a meeting of a few theorists, me included, to discuss the role of theoretical physics. At one point, he emphasized the need of "theory linked to *shiji* (practicality)". I intercepted, "It is also important to emphasize 'theory linked to *shiyan* (experiments)'". He, a trained experimentalist, seemed to find it refreshing since, apparently, no one mentioned this to him before. Actually we were referring to two different aspects of theory: Qian was talking about the application of theory; I, how theory should be conducted in research. Linking theory to experiments is the proven way of good research I picked up at Bell Labs (see Section 12.5 for an example). This was well recognized by Peng Huan-Wu (1915-2007) (Fig. 12.8, left), too, but Peng was not at that meeting [Peng, 2001, pp. 82-83].

---

[15] The President of CAS was Fang Yi (1916-1997), whom I did not meet personally until after I left China (in the Great Hall of the People). I was among a group of high-T_c superconductor physicists from IoP that was received by Fang.

Peng Huan-Wu graduated from Tsinghua University and spent nine years (1938-1947) in Europe, obtaining his PhD under the Nobelist Max Born in 1940. He was responsible for the theoretical aspects of China's nuclear bomb project [Peng, 2001, bookback cover], the counterpart of Hans Bethe regarding American's atomic bomb. He became the first director of the Institute of Theoretical Physics (ITP), CAS, in May 1978. He and I had several exchanges during my stay in Beijing (see Sections 12.5 and 12.6).

Fig. 12.8. Two other senior physicists I associated with in Beijing: Peng Huan-Wu (left) and Huang Kun (right). Both earned PhD in England; Peng in 1940 and Huang in 1948. The Nobelist Max Born was Peng's thesis advisor and Huang's postdoctoral supervisor.

Huang Kun (1919-2005) (Fig. 12.8, right) was very kind to me. He studied and worked in Europe from 1945-1951. After earning his PhD (1948) under the Nobelist Nevill Mott, he was a postdoc with Max Born [Chen & Yu, 2008]. It is the book *Dynamical Theory of Crystal Lattices* (1954) by Born and Huang, a classic in solid state physics that made Huang the best known Chinese physicist in the Western academic circles. He returned to China in 1951 and married his European girlfriend the next year. Perhaps because of that, Huang was not drafted into the bomb project; he remained in the academia and had the chance to make important contributions in solid state physics [Zhu, 2000].

During the Cultural Revolution, Huang Kun was the only scientist who wrote in *Hongqi*, the foremost political journal; he was deemed political incorrect after the Revolution. With the approval of Deng Xiao-Ping, Huang was appointed the Director of the Institute of Semiconductors, CAS, in 1977. He invited me to visit his institute. The reason was that he learned from my resume that I had worked in semiconductors while in Belgium and he was eager to rebuild the theory group in his institute. I met him again at the Lushan conference (see below) and other meetings. The last time I saw him was in 2002 during Yang Chen-Ning's 80th birthday banquet in Beijing. He was sitting alone at the end of a very long table. I walked up to him and said hello.

In August of 1978, the Chinese Physical Society resumed its "annual" meeting in Lushan after ten years [Wang & Yang, 2012]. Lushan is the mountain resort that Chiang Kai-Shek (1887-1975), president of Republic of China, spent his cool summers but is also the place that an important meeting was held during the Great Leap Forward years in 1959. All the important physicists showed up. I was allowed to go, too. It was a very interesting experience. And, it was at the end of this conference I learned that a new Institute of Theoretical Physics was being added to CAS.

From August to December, 1978, Woo Chia-Wei, Chair of the physics department at Northwestern University, USA, visited IoP as a guest professor. Since his research involved liquid crystals and to prepare for his visit, I was asked to give a series of lectures on liquid crystals to my colleagues. And liquid crystals became my research field during my years at IoP [Lam, 2014]. In spite of China's large population, the number of physicists was very small and it was easy for one to be counted as an "expert" on a chosen subject. Thus, with only one published paper on liquid crystals I became an expert in the field of liquid crystal theories. Near the end of his stay, Woo invited the IoP to send a delegation of physicists to his department to work as visiting scholars, with everything paid by his university except for the air tickets. This proposal was quickly approved at the highest level, presumably by Hua Guo-Feng himself [Lam, 2010].

Before 1978 ended, I published my first paper in China, "Microscopic theory of first-order phase transitions in liquid crystals" [Lin, 1978a]. This paper explains an experimental result from Canada and is the precursor to my 1979 paper [Lin, 1979] (see Section 12.4.3). In fact, it is also my first paper written in Chinese; before that I had published 22 papers in English and one in French [Lam, 1973].

12.4.3 My 1979 Trip to America as a Visiting Scholar

In December, 1978, the most important Third Plenum (of the 11th Central Committee Congress of the Communist Party of China) was held in Beijing, which officially shifted the government's focus from class struggle to economic development. This meeting is regarded as the beginning of China's reform-and-opening up revolution[16] of the last 30 something years [Tang, 1998]. A few days after this meeting, on January 1, 1979, China and USA established formal diplomatic relationship. From January 29 to February 4, 1979, Deng Xiao-Ping visited the United States.

The IoP at that time was effectively run by Guan Wei-Yan, the deputy director, due to the old age of the director, Shi Ru-Wei. The IoP picked eight members to form the delegation to Northwestern: Qian Yong-Jia, Li Tie-Cheng, Zheng Jia-Qi, Shen Jue-Lien, Wang Ding-Sheng, Cheng Bing-Ying, Gu Shi-Jie and Lin Lei [Lam, 2010]. The delegation left Beijing on February 9, 1979, five days after Deng's visit. We arrived at the university in Evanston on February 15, via Paris, New York, and Washington, D.C. While in New York, *China Daily News* interviewed me (Fig. 12.9, left).

I spent three months at Northwestern while the rest of the delegation stayed behind for two years. Before leaving, I was invited by quite a number of universities to give physics seminars, and general talks on my experience in China (sponsored by the Chinese students there). The reception was very enthusiastic.

I went back to Beijing, via Hong Kong, in June, 1979. While in Hong Kong, I was interviewed by *Wen Wei Po* (Fig. 12.9, right) and was invited to write a popular-science article in this newspaper. I also wrote up my work on liquid crystals and submitted it to *Physical Review Letters* (PRL), the foremost international (weekly) journal in the physics profession. This paper, "Nematic-isotropic transitions in liquid crystals,"

---

[16] The reform-and-opening up is called "a new and great revolution" by Xi Jin-Ping, then Head of the Central Party School, in his talk "Review and thoughts on party construction at the 30th anniversary of reform-and-opening up," presented on September 1, 2008 (http://news.xinhuanet.com/politics/2008-09/08/content_9849759.htm, July 30, 2013).

12 Solitons and Revolution in China 271

Fig. 12.9. My two interviews by Chinese newspapers at the beginning and end of the 1979 trip to USA. *Left*: "The most welcomed, unexpected guest: An interview with Lin Lei, a researcher from the Chinese Academy of Sciences visiting America," *China Daily News*, New York, first of three installments, March 1, 1979. *Right*: "From University of Hong Kong to Europe and America and to Beijing: Dr. Lin Lei talks about his path of scientific research," *Wen Wei Po*, Hong Kong, June 1979.

turns out to be the first one ever coming from mainland China that got published in this top journal [Lin, 1979]. (For more details, see [Lam, 2010].) The publicity gained through the interviews and campus tours helped to make me a well-known figure among the overseas Chinese, which later played a positive role in helping my research in solitons (see Section 12.5).

12.4.4 Other Things

I was the first returnee assigned to CAS after the Cultural Revolution. The support I received from the higher-ups and my colleagues was fantastic. I was interviewed by the Chinese magazines (see Fig. 12.10 for an example) and invited to write popular-science articles in the newspapers [Lin, 1978b].

I soon found myself being useful to China. The first instance occurred when the IoP was showing a foreign film without subtitle to all the members gathered in the canteen. I was called upon to do a translation with my broken Chinese. On another occasion, the official Xinhua News Agency phoned me out of the blue and asked me to translate an English word for them. Academically, there are two physics terms the Chinese translations of which are due to me: *jiguai xiyinzi* for "strange attractor" and *guzi* for "soliton" [Committee on Physics Terms, 1996].

Fig. 12.10. My interview in the magazine *China Construct*, "In their primes: Interviews with three young overseas returnees," February 1981.

Perhaps more importantly, in 1980, I helped to found the Chinese Liquid Crystal Society and was a Vice President and the Secretary-General. Before I showed up in Beijing, Xie Yu-Zhang (1915-2011)[17] (Fig. 12.11, left) was the only physicist working on liquid crystal theory. In the 1970s, when in his sixties Xie started working in liquid crystals to support the liquid crystal display (LCD) group at Tsinghua University [Zhao et al, 1980]. The group received funding even during the Cultural Revolution since Jiang Qing, the wife of Chairman Mao, learned that large-size LCDs would enhance political propaganda works. Unfortunately, the project never fulfilled its promise. I had published papers with Xie and his student, Ou-Yang Zhong-Can (Fig. 12.11, middle).

Fig. 12.11. Three physicists in liquid crystals. *Left to right*: Xie Yu-Zhang, Ou-Yang Zhong-Can, and Shu Chang-Qing (1984). Ou-Yang became Director of the Institute of Theoretical Physics and academician of CAS.

In January 1979, the Theory Group was separated from the Department of Magnetism and became the Theoretical Physics Section directly under IoP, like the other Departments. I was elected the

---

[17] Xie Yu-Zhang, BS and MS from Tsinghua University, associate professor at physics department of National Central University (1945-1948), studied at Vanderbilt University in the United States and earned a PhD (1948-1950) (baike.baidu.com/view/ 2711126.htm, Sept. 30, 2012). He returned to China in 1957, became professor at Tsinghua University, spent four years in jail (1968-1972) during the Cultural Revolution [Wang & Liu, 2012], assumed the first presidency of the Chinese Liquid Crystal Society, and retired in 1986.

Associate Head of this Section; Pu Fu-Ke (1930-2001),[18] the Head. Somewhat later, the Section consisted of two research scientists (equivalent to full professors), three associate scientists, four assistant scientists, and 11 graduate students.[19]

From 1979 on, I attended international conferences on liquid crystals and on solitons abroad and stayed as a visiting scholar for a few months once in US (1979) and two times in Paris (1980 and 1982). The two important occasions are the International Conference on Liquid Crystals (December 3-8, 1979) in Bangalore, organized by the Indians, and The Eighth International Conference on Liquid Crystals (June 30-July 4, 1980) in Kyoto, Japan. As international liquid crystal conferences are concerned, the former was the first one ever attended by a delegation from mainland China and the latter, the first one in the biennial series; both delegations were headed by me. In Kyoto I was nominated (and subsequently elected) to the Planning and Steering Committee that oversaw the biennial series, a crucial event that led to the forming of the International Liquid Crystal Society ten years later [Lam, 2014]. (See Section 12.5 and [Lam, 2014] for more on liquid crystals in China.)

Liquid crystal is an important member of a large group of materials called soft matter. The liquid crystal group I formed at IoP is in fact the forerunner of IoP's soft-matter laboratory established in 2001.

The signatures of any revolution are the existence of extreme risks and, very likely, the loss of human lives. In this case, there were two. First, the possible downfall of Deng Xiao-Ping accompanied by the appearance of another Cultural Revolution was like a knife hanging over our head daily (especially since Chairman Mao had urged a Cultural Revolution every ten years). Second, one could catch hepatitis B by eating in the canteen and die of liver cancer. The former did not happen, but the latter did. Liao Qiu-Zhong, a PhD in linguistics from UC Berkeley, died of hepatitis B just a few years after returning to China. Some years later the same happened to Cai Shi-Dong (1938-1996), a

---

[18] Pu Fu-Ke graduated from Tsinghua University (1952) and studied in USSR (1956-1960), finished with an associate PhD degree; worked all his life at IoP and was an academician of CAS. [See: "Academician Pu Fu-Ke passed away in Beijing," *Physics* (*Wuli*) **30**, 382 (2001).]

[19] *Biennial Report, 1978-1979*, Institute of Physics, Academia Sinica, Beijing, p. 59.

1973 returnee at IoP with a PhD in plasma physics from Princeton University.[20]

12.5 Solitons in China

Feng Kang's 1978 lecture on solitons was very exciting and it was the first lecture on solitons I ever received. However, at that time I was working on phase transition problems in liquid crystals and solitons seemed to be irrelevant to my research.

A soliton is a localized wave that travels without change of velocity and shape. It was discovered by John Scott Russell (1808-1882) in 1834. However, in many real situations this definition is relaxed a bit and the velocity and shape are allowed to change a little bit while traveling [Lam, 1992]. It was a hot topic in physics from the late 1970s to early 1980s, especially after it was widely applied to condensed matter [Bishop & Schneider, 1978].

In the summer of 1981 I went to Tsinghua University to attend a Master thesis defense given by a student of Xie Yu-Zhang. After the meeting, Zhu Guo-Zhen, a colleague of Xie, led me to his laboratory and showed me his experimental results of three lines that propagate in a liquid crystal cell. Zhu just had his manuscript on this experiment rejected by *Acta Physica Sinica*, the major physics journal of China. The reason was that the sound wave equation, a linear equation, given there was at odd with the experimental observations. In the experiment, the width of each line *narrows* as the line moves while the width of a sound wave always increases as time increases. What he needed now was an interpretation of his experimental results, which was an integral part of any experimental work and could be more challenging than doing the experiments itself.

To me, it was obvious that these lines could not be a sound wave for the reason explained above. In fact, the lines looked like solitary waves and Feng Kang's lecture came to mind.

---

[20] *Remembering Cai Shi-Dong*, published by Institute of Physics, Chinese Academy of Sciences; Asian African Association for Plasma Training; and Society for Plasma Research (Beijing, 1997).

Knowing that good theory has to go side-by-side with good experiments, better be one's own experiments if possible—a lesson I learned when doing my PhD thesis at Bell Labs—I proposed to Zhu that I would send Shu Chang-Qing (Fig. 12.11, right) to his laboratory to do more experiments. The condition was that any paper coming out of this joint effort would have Shu as the first author. He agreed. At that point in time, Shu just finished his three-year Master degree with me and was continuing with his doctoral studies.[21]

I started reading the soliton literature earnestly. The library of IoP did not have originals of the foreign journals but it did have copies.[22] By this time, IoP had one copying machine, made in China. It broke down several times per day and was managed by a young technician, Chen Jie.[23] Every time I went to the copying room at the back of the main building, I passed a cigarette to Chen. After the smoking session, I gave him the journals I wanted to be copied, and I usually got it back the next day. It helped tremendously. Moreover, a few young overseas Chinese visited Beijing and asked to see me (probably because my name was known in those circles). Before departing, they would ask me what they could do to help my work (i.e., helping China). And I got several soliton books this way.

Zhu's experiment was very simple. He put a thin plate between the two glass plates of a liquid crystal cell (which is nothing but two parallel, thin glass plates with liquid crystal in between); the plate was placed near one end of the cell and was pushed forward toward the other end for a short distance. Three dark lines were generated and moved forward [Zhu, 1982].

---

[21] Incidentally, Shu's Master thesis on phase transitions [Shu & Lin, 1984] shows that, among other things, Woo Chia-Wei had misinterpreted his own theoretical results in his first liquid crystal paper which was published in *Physical Review Letters*.
[22] China was short in foreign currencies; she bought two copies of every foreign journal and stored them in Beijing and Sichuan Province, respectively, just in case, and made copies for the major libraries. (See also [Xiong, 2012].)
[23] Chen Jie rode a motorcycle to work (while motorcycles were extremely rare in Beijing). In the 1980s he was sent by IoP to work at the Lawrence Radiation Laboratory, Berkeley. He managed to stay in the US and became the anchor of the daily Mandarin news at KTSF, a local Chinese TV station in San Francisco.

Pretty soon, I figured out what happened. When the plate is pushed forward, the molecules at the middle layers move faster than those near the cell's boundaries and a velocity distribution, called shear, is created. The shear induces the rod-like molecules, which are initially vertical, to incline forward at an angle throughout the cell. When the movement of the plate is stopped, opposite shear is generated and the molecules near the plate end will incline backward, resulting in an orientation distribution like this: two domains with opposite angles at the two ends are separated by a region of (nearly) vertical molecules. In the soliton literature, this distribution is called a "kink." As more and more molecules at the plate end reverse their orientation, the vertical-molecule region shifts forward and we have a propagating kink (Fig. 12.12). With this new understanding that the shear was what generated the solitons, I directed Shu to do the experiments systematically using pressure difference to create the shear because pressures could be controlled more precisely.

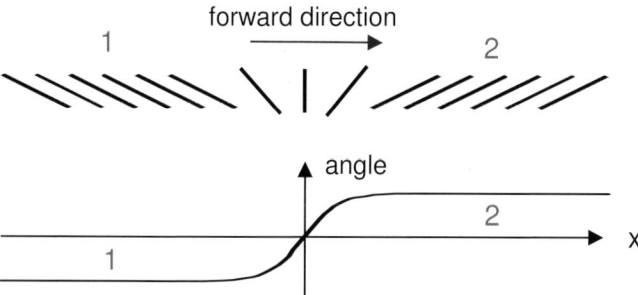

Fig. 12.12. A soliton in shearing liquid crystals. *Top*: Orientations of the rod-like molecules: a region of (nearly) vertical molecules is sandwiched between two regions of oppositely oriented molecules (labeled 1 and 2, respectively). *Bottom*: The corresponding orientation distribution: a kink soliton.

I very quickly wrote down the equation of motion: the damped sine-Gordon equation. We were excited except that we did not know how to solve it; nothing relevant could be found in the literature. I sent Shu to Feng Kang's home to ask for his advice.

Feng could not solve it either. But Shu came back deeply impressed by Feng's extremely strong memory and enormous knowledge. That no one can solve it is not surprising, even today. The problem is that solitons are solutions of particular nonlinear equations which, unlike linear equations, do not succumb to systematic analysis. Furthermore, for Feng and other mathematically inclined soliton researchers, rigorous solitons are what interest them; the damped sine-Gordon equation belongs to the domain of non-rigorous solitons [Lam, 1992, p. 11].

Luckily the soliton solutions of the equation could be figured out qualitatively and we were able to understand the major features of the experiments. This understanding and some initial experimental results obtained by Shu (e.g., a *single* dark line was generated) were briefly summarized within a general review on solitons I gave in Wuhan in 1981 [Lin, 1982a].

The review is titled "Solitons in condensed matter" (Fig. 12.13). It was presented at the First National Conference on Statistical Physics and Condensed Matter Theory, held at Huazhong Institute of Technology. Lin Pei-Wen[24] of ITP and I shared a campus dorm room throughout the conference. Incidentally, two other talks on specific soliton systems were presented at this conference: "Soliton model of polyacetylene and fractional charges" by Yu Lu (ITP), and "Solitons in ferromagnetic chains: A new example of inverse scattering method" by Pu Fu-Ke (IoP).

Pretty soon we found an analytic solution under a special condition which corresponded to the experimental situation. Calculated results from this special solution matched very well the experimental observations. To speed up the work, a small group on this project was formed, which consisted of Shu, Shen Jue-Lian and I from IoP, Huang Yun from Peking University, and Lin Pei-Wen from ITP. This group met several times at IoP.

In early 1982, we were ready to write a paper for *Physical Review Letters*. But our initial effort to submit a joint paper with Zhu describing

---

[24] Lin Pei-Wen (P. M. Lam), high school in Macau; BS, San Diego State University, CA; PhD, Washington University (Missouri); postdoc, West Germany. He returned to China end of 1980, worked at Institute of Theoretical Physics, CAS; got married and left China for West Germany with his wife in 1985. He is now a physics professor in USA.

both the theory and experiments failed. At the end, two papers were written: an experimental paper written by Zhu and a theoretical paper written by our group. Both manuscripts reached the PRL office in May, 1982; our paper arrived May 5, two days earlier than Zhu's. In those years, all outgoing official mails had to go through an appropriate office in one's affiliated institute or university for screening. Apparently, the mailing apparatus of IoP was more efficient than that of Tsinghua.

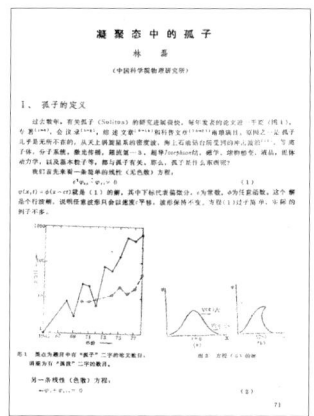

Fig. 12.13. The review, "Solitons in condensed matter" [Lin, 1982a].

Both papers were rejected. I wrote up a reply to the referees, and did the same for Zhu. Upon that, the editor changed his mind, and the two double-second PRL papers from mainland China were published in the Nov. 1, 1982 issue of PRL [Lin et al, 1982; Zhu, 1982]. Historically, in the liquid crystals literature, "Soliton popagation in liquid crystals" by Lin et al (Fig. 12.14) is the first paper with the word "soliton" appearing in the title and the first paper that talks about propagating solitons. This paper from China enabled us to lead the field for ten years, culminating in the review book *Solitons in Liquid Crystals* [Lam & Prost, 1992] (Fig. 12.15, left).[25]

---

[25] In the same year, I published the prediction of bowlic liquid crystals [Lin, 1982b] which was synthesized three years later in Europe [Lam, 1994]. I coined the terms "bowlic" and "bowlic liquid crystal" which are now recognized officially by the IUPAC and appear formally in *Handbook of Liquid Crystals*. Bowlic is one of three existing

Lin et al.[9] in which the dark lines are interpreted as solitons.

I am indebted to Yu Hao and Xu Laoli for designing and constructing the mechanical device used to push the exciter, Lin Lei for his interest in this work and his assistance in improving the English writing of this manuscript, Zhao Nanming for discussion at an earlier time, Professor J. L. Ericksen for telling me his recent view about director waves in a personal communication at my request, and finally, my teacher Professor Meng Chaoying for his encouragement.

---

[1]J. L. Fergason and G. H. Brown, J. Am. Oil Chem. Soc. 45, 120 (1968).
[2]J. L. Ericksen, J. Acoust. Soc. Am. 44, 444 (1968).
[3]F. M. Leslie, in *Advances in Liquid Crystals*, edited by G. H. Brown (Academic, New York, 1979), Vol. 4.
[4]Zhu Guozhen, J. Qinghua Daxue Xuebao 21(4), 83 (1981).
[5]Zhu Guozhen, in Proceedings of the Chinese Liquid Crystal Conference, Guilin, China, 20–25 October 1981 (to be published).
[6]Lin Lei and Shen Juelian, in Proceedings of the Chinese Liquid Crystal Conference, Guilin, China, 20–25 October 1981 (to be published).
[7]Zhu Guozhen et al., to be published.
[8]M. Born and E. Wolf, *Principles of Optics* (Pergamon, New York, 1975).
[9]Lin Lei et al., following Letter [Phys. Rev. Lett. 49, 1335 (1982)].

Soliton Propagation in Liquid Crystals

Lin Lei,[(a)] Shu Changqing, and Shen Juelian
Institute of Physics, Chinese Academy of Sciences, Beijing, China

and

P. M. Lam
Institute of Theoretical Physics, Chinese Academy of Sciences, Beijing, China

and

Huang Yun
Department of Physics, Beijing University, Beijing, China

(Received 5 May 1982)

Soliton propagation in nematic liquid crystals under shear is shown to be possible and studied theoretically. Calculations including those pertaining to the modulation of monochromatic or white light passing through such a liquid-crystal cell are presented. Recent experiments are interpreted accordingly and are in good agreement with the theory presented here.

PACS numbers: 61.30.-v, 03.40.Kf, 05.70.Ln, 47.15.-x

Solitons are important and have been found in various objects ranging from celestial bodies to laboratory systems.[1,2] However, unlike the first observation of solitons in shallow water by Scott Russell, many of the recent experimental evidences of solitons in condensed matter are indirect in nature. The experiments[3] on the ordered fluid ^{3}He are no exception. In this regard, we note that in another type of ordered fluid, viz., liquid crystal, because of the strong coupling of the director with light, it may be possible to observe the motion of the molecules and the solitons rather directly.

Discussions of solitons in liquid crystals[4] was first given by Helfrich[5] and subsequently by de Gennes,[6] Brochard,[6] and Leger.[7] In their work in nematics, the solitons (called "walls") are magnetically generated and are small in width (e.g., a few microns). Experimentally, the observation[7] of these solitons is delicate and a polarizing microscope has to be used. Recently, there have been more but still limited attention[8] paid to the role of solitons in the physics of liquid crystals.

In this Letter, we first point out and discuss a new case in liquid crystals, viz., nematics under uniform shear, in which solitons can exist and propagate. In contrast to the magnetic case[5-7]

© 1982 The American Physical Society   1335

Fig. 12.14. One of the two double-second PRL papers from mainland China, "Soliton propagation in liquid crystals" [Lin et al, 1982].

---

types of liquid crystals in the world, with important potential applications. This Chinese invention from its own backyard is not yet recognized in IoP's official showroom.

Fig. 12.15. *Left*: The world's first book on solitons in liquid crystals [Lam & Prost, 1992], which is also the starter in the Springer series "Partially Ordered Systems" founded by the author. *Right*: Book cover of *Introduction to Nonlinear Physics* [Lam, 1997] which includes two chapters on solitons, based on the review article published in China [Lin, 1982a].

Like my first PRL paper, no one at IoP paid any attention. But the reception from outside of China was quite different.[26] When the paper came out in print, I was spending two months at the Universite Paris-Sud, Orsay. I gave a seminar and solitons in liquid crystals were picked up by a group there [Madhusudana et al, 1992]. At the Chinese University of Hong Kong where I stopped over on my way home to Beijing, Yang Chen-Ning, the Nobelist of parity-nonconservation fame, invited me to give a seminar on this topic.

At a certain point after this paper, Shu and I went to talk to Peng Huan-Wu at his home. We wanted to know how instability of a wave solution is studied theoretically. He received us warmly and kindly educated us on this topic.

Overall, I published 23 papers on solitons in liquid crystals in 12 international journals. All, except for the first one, were published after I left China (see next section). Incidentally, in the last conference I

---

[26] Many years later, when I met Stanley Liu, the editor of PRL who handled my papers, at a physics conference in the United States, he still remembered my two papers, including this one.

attended before leaving, the Second National Conference on Statistical Physics and Condensed Matter Theory, Nanning, November 14-18, 1983, a paper on solitons in polyacetylene was coauthored by Huang Kun. And Shu Chang-Qing did finish his thesis "Solitons in Liquid Crystals" in April 1984, which, officially, is China's first PhD thesis in liquid crystal physics.[27]

12.6 Leaving China

Leaving China used to be quite impossible. The government took it personally, reasoning that since the revolution was such a splendor thing in the world, anyone chose to leave implied a disagreement with this assessment and should be discouraged. A married couple, Chen Ruo-Xi and Duan Shi-Yao, who returned in 1966 and worked in Nanjing, left with permission in 1974, after Zhou En-Lai, the premier, had declared to the Baodiao visitors that the official policy regarding returnees was "Welcome to return; coming and leaving are both free" [Chen, 2008]. Apparently, Zhou was already thinking of how to attract talents from abroad to strengthen the country's rebuilding once the Cultural Revolution was finished.

My wife went back to NYC for a home visit upon her father's death, taking along our daughter, in the summer of 1980. She then announced that she would not return. It was a total surprise to me; we never discussed this before (or after). In fact, it was hard to discuss anything personal when we were separated in two cities, in two continents. There was no private phone at home, and any form of communication across the border was more likely than not to be monitored. To make a phone call to her, I would ride my bike one hour to the telephone and telegraphy building in Changan Avenue, downtown. The cost was ten Yuan per minute; my monthly salary would last 20 minutes.

Three years later, her "green card" application was approved and I had to decide. There were a few returnees of my generation who already

---

[27] Unlike other countries, China's PhD thesis advisors have to be approved by the government, and I was the only one among a population of 1.1 billion certified in the field of liquid crystal physics while I was there.

left, either because they found the social reality not matching what they envisaged or because their skills were not adequately put to use, or both. (In all cases, none of them left because of low living standards.) My case was different. Since I helped to bring my daughter to the world I had the responsibility to see her grow up properly.

I resigned and left China in December, 1983. Before that, I had a private talk with Guan Wei-Yan, the then Director of IoP. After clearing up all work-related business, I asked him to appoint me as a Guest Professor so that there would be no misunderstanding for people abroad that there was bad relationship between China and I. I told him this would be good for China, and he agreed.[28]

I went to say goodbye to Peng Huan-Wu at his home. When I mentioned that I proposed IoP appointing me as a guest professor since it was good for China, he nodded his head and said, "Yes, 'good for China' is very important." Only years later I learned that "Good for China" was his motto in leading his whole life. I last talked to him at Tsinghua University in 2005 during a recession of the workshop organized to celebrate his 90th birthday, which was after he gave a long presentation on some physics topics. He looked healthy and alert; he said, without me reminding him, "You worked on liquid crystals."

Just before leaving, I was invited to have lunch with a vice president of CAS. He asked me for suggestions. The one suggestion I gave was that CAS could set up a stand at the entrance of CAS' main library, a central spot in Zhongquancun, which would post seminars coming up in the different institutes so that people could attend them freely. It was a low-tech, costless way to encourage interdisciplinary research. From what I know, that never materialized.

On the issue of leaving China, the 2012 Nobelist in Literature, Mo Yan, said on his first press conference after the announcement of the award: "One should not think that whoever leaving the country must be unpatriotic and whoever staying behind must be patriotic."[29]

---

[28] My appointment as Guest Professor of IoP, approved by CAS, was signed by Guan Wei-Yan, dated June 25, 1984.

[29] www.ktsf.com/en/nobel-winner-mo-urges-china-dissidents-freedom (Oct. 12, 2012). I think he had Gao Xing-Jian, the 2000 Nobelist in Literature, in mind when he made this remark.

12.7 Conclusion: The Missing Link

During the ten years of Cultural Revolution, normal research activities in China were practically put on hold. A whole generation of scientists was lost, resulting in a missing link—the absence of mature scientists in their sixties presently (with a few exceptions). That is why most leaders in the scientific enterprises are now in their fifties.[30] An important consequence is the lack of enough number of widely-respected senior scientists who could referee and help to maintain the high level of research in many fields of study. Counting papers is thus the alternative chosen by the administrators in academia, with disastrous results. The reason is that it takes time in writing papers and fighting the referees to get them published. Unnecessary publications thus reduce the amount of valuable time needed for doing high-quality research.

In the 1950s, China's progress in science suffered from embargos from abroad, i.e., due to external factors. Today, it suffers from *internal* and self-inflicted factors, like counting papers. The first step would be stop running the universities like running an IBM company. Maybe it is time to resume the tenure system in the top six universities.

Thirty years passed and China changed a lot. For instance, the words "soliton" and "revolution" are no longer heard. Taxis and cars are everywhere in Beijing, perhaps too many.

Appendix 12.1 A Brief History of Chinese Students Going Abroad and the Returnees

The following history is updated from [Lam, 2010]. Before 1949, there are eight generations of Chinese students *going abroad* to study, by the Western Returned Scholars Association's counting.[31]

---

[30] This was noted by Zhou Guang-Zhao, the past president of CAS, in his speech at Peng Huan-Wu's 90th birthday workshop, Tsinghua University, 2005.
[31] www.coesa.cn/info/categorymore.shtml?Cid=C01 (Mar. 20, 2009). (See also [Hong Kong Museum of History, 2003].)

1. *First generation* (1872-1875): The Qing Dynasty government sent out 120 children, aged 12-15, to the USA to study (including the famous Rong Hong (1828-1912) [Rong, 2005]).
2. *Second generation* (1877): Nearly 100 navy students sent to Europe in early years of Qing Dynasty's Emperor Kuang-Xu.
3. *Third generation*: Students going to Japan in the early 20th century (including Zhou En-Lai [Hong Kong Museum of History, 2003]).
4. *Fourth generation*: Students going to the USA under the auspices of the Boxer Indemnity.
5. *Fifth generation*: Students going to France to study and work (including Yan Ji-Ci [Jin, 2000], Zhou En-Lai [Hong Kong Museum of History, 2003] and Deng Xiao-Ping).
6. *Sixth generation*: Students going to USSR during the 1920s.
7. *Seventh generation* (1927-1937): Students going abroad (including Ren Zhi-Gong, Wu Da-You [Qiu, 2001], Shi Ru-Wei [Zhao, 2005], Wu Chien-Shiung [Chiang, 2013] and Qian San-Qiang).
8. *Eighth generation* (1938-1948): Students going to Europe and USA (including Peng Huan-Wu and Huang Kun; China's first two Nobelists, Yang Chen-Ning [Chiang, 2002] and Lee Tsung-Dao [Zi, 2010]; and Wang Hao [He & Wen, 2006]).

Continuing with this counting, after 1949, we have three more generations of Chinese students going abroad, viz.:

9. *Ninth generation* (1949-now): A large number of students from Taiwan and Hong Kong, and a few from Macau, went to USA and Europe (including the four Nobelists: Ting Chao-Chung and Lee Yuan-Tseh from Taiwan, Tsui Chee and Kao Kuen from Hong Kong).
10. *Tenth generation* (1950s): Students going to USSR sent by the Chinese government (including Guan Wei-Yan, Pu Fu-Ke, Hao Bai-Lin and Yu Lu).
11. *Eleventh generation* (1978-now): Students from mainland China, going to USA, Europe, etc., sent officially or going privately [Cao, 2009].

The history of students *returning* to China to settle down and work is equally interesting. Before 1949, a large number of these students willingly returned to China (even during the resistance-against-Japan war years) and contributed to the modernization of their motherland. After 1949, there are three generations who go abroad privately *and* return to mainland China voluntarily, which are collectively and informally called *haigui* (sea turtles, meaning "returnees from overseas"):

1. *First generation* (1950s): Those coming back mainly from USA and Europe, soon after People's Republic of China was established (including Qian Xue-Sen [Chang, 1995], Li Yin-Yuan and Xie Yu-Zhang).
2. *Second generation* (mostly 1975-1985): Nearly 100 students of the 9th generation returned to China, mostly after the Cultural Revolution.
3. *Third generation* (after 1980): These are the 11th generation students returning when the reform-and-opening up process in China is picking up speed.

References

Baker, K. M. [1975] *Condorcet: From Natural Philosophy to Social Mathematics* (University of Chicago Press, Chicago).
Bishop, A. R. & Schneider, T. [1978] *Solitons and Condensed Matter Physics* (Springer, New York).
Cao Cong [2009] "'Brain drain,' 'brain gain,' and 'brain circulation' in China," *Science & Cultural Review* **6**(1), 13-32.
Chang, I. [1995] *Thread of the Silkworm* (BasicBooks, New York).
Chen Cheng-Jia & Yu Li-Sheng [2008] *Demeanor of a Great Scholar: Remembering Huang Kun* (Peking University Press, Beijing).
Chen Ruo-Xi [2008] *Perseverance, Regretless* (Chiuko Press, Taibei).
Chiang Tsai-Chien [2002] *Yang Chen-Ning: The Beauty of Gauge and Symmetry* (Bookzone, Taibei).
Chiang Tsai-Chien [2013] *Madame Wu Chien-Shiung: The First Lady of Physics Research*, trans. Wong, Tang-Fong (World Scientific, Singapore).
Committee on Physics Terms [1996] *Physics Terms* (Science Press, Beijing).
Dietrich, C. [1986] *People's China: A Brief History* (Oxford University Press, Oxford).

Hamann, D. R. & Isaacs, E. R. "Philip Moss Platzman," [2012] *Phys. Today* **65**(5), 64-65.
Hao Bai-Lin [2012] "My mentor Wang Zhu-Xi," *Physics (Wuli)* **41**, 455-357.
He Zhao-Wu & Wen Jing [2006] *Going to School* (SDX Joint Publishing, Beijing) pp 220-229.
Hong Kong Museum of History (ed.) [2003] *Boundless Learning: Foreign-Educated Students of Modern China* (Hong Kong Museum of History, Hong Kong).
Jin Tao [2000] *Yan Ji-Ci* (Liaoning Educational Press, Shenyang).
Kuo Chia-ling [1977] *Social and Political Change in New York's Chinatown: The Role of Voluntary Associations* (Praeger, New York).
Lai, H. M. [2010] *Chinese American Transnational Politics* (University of Illinois Press, Urbana).
Lam, L. [1971] "Facing squarely Hong Kong's political future: From supporting to burying the Legalization of Chinese Language Movement," *Zhongwen Yundong*, Issue No. 2, August, p. 1.
Lam, L. [1973] "Surfaces de Fermi, profil Compton et effets a N-corps," *Phys. Lett. A* **45**, 409-410.
Lam, L. [1992] "Solitons and field induced solitons in liquid crystals," in *Solitons in Liquid Crystals*, eds. Lam, L. & Prost, J. (Springer, New York) pp. 9-50.
Lam, L. [1994] "Bowlics," in *Liquid Crystalline and Mesomorphic Polymers*, eds. Shibaev, V. P. & Lam, L. (Springer, New York) pp. 324-353.
Lam, L. (ed.) [1997] *Introduction to Nonlinear Physics* (Springer, New York).
Lam, L. [2010] "The first 'non-government' visiting-scholar delegation in the United States of America from People's Republic of China, 1979-1981," *Science & Culture Review* **7**(2), 84-94.
Lam, L. [2014] "The founding of the International Liquid Crystal Society," in *All About Science: Philosophy, History, Sociology & Communication*, eds. Burguete, M. & Lam, L. (World Scientific, Singapore).
Lam, L. & Prost, J. (eds.) [1992] *Solitons in Liquid Crystals* (Springer, New York).
Li Ya-Ming (ed.) [2004] *Guan Wei-Yan's Memoir: An Oral History* (National Tsinghua University, Hsinchu).
Lin Lei (Lam, L) [1978a] "Microscopic theory of first-order phase transitions in liquid crystals," *Kexue Tongbao* **23**, 715-718.
Lin Lei [1978b] "Liquid crystals lead us in walking," *Guangming Daily*, Sept. 29.
Lin Lei [1979] "Nematic-isotropic transitions in liquid crystals," *Phys. Rev. Lett.* **43**, 1604-1607.
Lin Lei [1982a] "Solitons in condensed matter," in *Recent Developments in Statistical Mechanics and Condensed Matter Theory* (Huazhong Institute of Technology Press, Wuhan) pp.71-86.

Lin Lei [1982b] "Liquid crystal phases and the 'dimensionality' of molecules," *Wuli* (*Physics*) **11**, 171-178.
Lin Lei, Shu Changqing, Shen Juelian, Lam, P. M. & Huang Yun [1982] "Soliton propagation in liquid crystals," *Phys. Rev. Lett.* **49**, 1335-1338.
Lukes, S. & Urbinati, N. (eds.) [2012] *Condorcet: Political Writings* (Cambridge University Press, Cambridge).
Luo Ping-Han [2008] *Spring: Chinese Intellectuals in 1978* (People's Press, Beijing).
Madhusudana, N. V., Palierne, J. F., Martinot-Lagarde, Ph. & Gurand, G. [1992] "Charged twist walls in nematic liquid crystals," in *Solitons in Liquid Crystals*, eds. Lam, L. & Prost, J. (Springer, New York) pp. 253-263.
Peng Huan-Wu [2001] *Poems and Essays by Peng Huan-Wu* (Peking University Press, Beijing).
Qiu Hong-Yi [2001] *Wu Da-You: The Father of Chinese Physics* (Triumph, Taibei).
Rigden, J. S. [2000] *Rabi: Scientist & Citizen* (Harvard University Press, Cambridge, MA).
Rong Hong [2005] *My Life in China and America* (Unity Press, Beijing).
Shu Chang-Qing & Lin Lei [1984] "Theory of homologous liquid crystals. I. Phase diagrams and the even-odd effect," *Mol. Cryst. Liq. Cryst.* **112**, 213-231.
Sun Mu (ed.) [2008] *IPCAS 80th Anniversary* (Institute of Physics, Beijing).
Tang Tao, Yao Nan & Yang Lei [2010] "Feng Kang: The story of an outstanding mathematician," *Mathematical Culture* **1**(1), 24-38.
Tang Ying-Wu [1998] *Choices: The Road of Chinese Reform since 1978* (Economic Daily Press, Beijing).
The Seventies Monthly (ed.) [1971] *Truth Behind the Diaoyutai Incident* (The Seventies Monthly, Hong Kong).
Tsui, H. & Lam, L. [2011] "Making movies and making physics," in *Arts: A Science Matter*, eds. Burguete, M. & Lam, L. (World Scientific, Singapore) pp. 204-221.
Wang De-Lu & Liu Zhi-Guang [2012] "The home-bound journeys in the 1950s and later experience of some American-trained Chinese scientists," *Science & Cultural Review* **9**(1), 68-87.
Wang Shi-Ping & Yang Guo-Zhen [2012] "Eighty years of the Chinese Physical Society," *Physics* (*Wuli*) **41**, 506-512.
Wu Da-You [2005] "Reminiscences of early development of Chinese physics," *Physics* (*Wuli*) **34**(3), 165-170; **34**(4), 233-239; **34**(6), 399-404; **34**(8), 551-554.
Xiong Wei-Min [2012] "Resumption of China's foreign scientific exchange: An interview with Hu Ya-Dong," *Science & Culture Review* **9**(5), 106-115.
Zhao Jian-Gao [2005] "A pioneer in China's modern magnetism enterprise: Academician Shi Ru-Wei," *Physics* (*Wuli*) **34**(10), 758-764.

Zhao Jing-An, Tong Shou-Sheng & Ruan Liang [1980] "Research on the electro-optic effect due to cholesteric-nematic phase transition used in television displays," *Wuli* (*Physics*) **9**, 10-13.

Zhao Qing-Yuan [1998] *Cai Yuan-Pei* (Anhui People's Press, Hefei).

Zhao Yan, Chen Wei, Wang Yu-Peng & Sun Mu [2008] "Searching knowledge for eighty years, glorious physicists: Eighty years anniversary of Institute of Physics," *Physics* (*Wuli*) **37**(6), 363-371.

Zhu Bang-Fen (ed.) [2000] *Selected Papers of Kun Huang with Commentary* (World Scientific, Singapore).

Zhu Guo-Zhen [1982] "Experiments on director waves in nematic liquid crystals," *Phys. Rev. Lett.* **49**, 1332-1335.

Zi Cheng [2010] *Lee Tsung-Dao* (International Cultural Press, Beijing).

13

Scientific Culture in Contemporary China

Bing Liu and Mei-Fang Zhang

In recent years, the term "scientific culture" has been used frequently by scholars in mainland China. Although this term is very popular in many disciplines such as history, philosophy and sociology of science as well as in science policy and education studies, researchers in these fields have different study objectives, positions, approaches and styles. It is thus interesting to analyze this phenomenon, which may help readers to understand Chinese scholars' attitudes toward science. Wang Rong-Jiang has reviewed in 2011 the history of scientific culture studies in China during the past 20 years. He has systemically (though not comprehensively enough) introduced conferences, research books and articles on scientific culture, and briefly analyzed the differences between different approaches. Building on this review, scientific culture in contemporary China is presented from our own observations and perspectives here.

13.1 A Brief History

Although scientific culture hardly existed in ancient China [Wang, 2011] things changed in the early 20th century. [1] At that time, Chinese intellectuals started paying serious attention to science and its values. Due to defeat in the wars and corruption in the government of the Qing Dynasty, along with a deep sense of national crisis, intellectuals began to look for helpful ideological resources to save the nation, and they found science, among other things. Meanwhile many young intellectuals who deeply appreciated the important role science played in modern civilization while studying abroad, returned to their motherland. Once

---
[1] Section 13.1 is based on Wang Rong-Jiang's review [2011].

back to China, they eagerly spread Western modern science and technology. Especially, members of the Science Society of China worked very hard to advocate the scientific spirit, making it a movement to spread science and its values to the public.

Naturally, spread of the scientific worldview encountered resistance from the traditional intellectuals. The "Debate between Science and Metaphysics" that occurred in 1923 was an example. However, the status of science in China's ideology has already become very strong during this time. As Hu Shi [1923] said, "During the last three decades, one word has got the supreme dignity in our country. Whether you know it or not, whether you are a conservative or idealist, you should not speak it with contempt openly. This word is exactly 'Science'. People respect it but never discuss whether it is worthy of it".

Science was placed in a position of supremacy, and was thought to be capable of solving all social and cultural problems by the intellectuals and the public. This view of science has the trend of becoming scientism and, after 1949, due to the backward state of science and technology as well as the urgent need of country-building, this trend continued in the People's Republic of China.

During the Great Proletarian Cultural Revolution (1966-1976), cultural and academic developments have been greatly hampered, and discussion about science was broken off consequently. Things changed after 1976. In fact, the academic community in mainland China witnessed dramatic changes after the Cultural Revolution, and the Reform and Opening-up policy of 1978 led to a surge of "culture study fervor" at that time, resulting in the introduction into China of a considerable number of Western publications on the history, philosophy and culture of science. In a word, science attracted the attention of Chinese scholars once again! However, it was subtly different this time. After World War II, especially during the 1960s, scientism has encountered unprecedented criticism in the Western academia. A large number of articles on scientism written from the perspectives of history, sociology and philosophy have been published, and Chinese scholars have been influenced by these trends of thought inevitably. For this reason, many scholars in China take the 1980s as the startup phase of China's scientific-culture study. Indeed, the starting point and aims of it are obviously different from those in the science-advocating period in the 1920s and 1930s.

Initially, scholars began to emphasize the integration of scientific culture and humanistic culture. In fact, as a compound word made out of science and culture, "scientific culture" has been broadly used in Chinese academic literature and popular discourse especially after the 1950s. It always refers to the scientific and cultural literacy of the public, and the level of development of science and culture. However, since the end of the 1980s, influenced by a lot of Western works about scientific culture which have been translated in Chinese, this compound word began to be understood as "scientific culture" or "culture of science". Among these works was the first translated Chinese edition of C. P. Snow's *The Two Cultures* (trans. Chen Heng-Liu & Liu Bing, 1987), George Sarton's *The History of Science and the New Humanism* (Chen Heng-Liu, Liu Bing & Zhong Wei-Guang, 1989), and M. N. Richter's *Science as a Cultural Process* (Gu Xin & Zhang Xiao-Tian, 1989).

Meanwhile, a good number of Chinese authors' articles appeared in academic journals, especially in *Journal of Dialectics of Nature* and in *Studies in Dialectics of Nature,* two leading STS (Science, Technology and Society, or Science and Technology Studies) publications in China. For example, Zhao Lei-Jin [1987] introduced *A Guide to the Culture of Science, Technology, and Medicine* (The Free Press, New York, 1980) to Chinese scholars in 1987, and recommended it as a mentor of Chinese scientific culture study. Afterward, lots of scholars began to analyze and discuss the topic of scientific culture. For example, Ma Lai-Ping [1989] discussed the conflicts between scientific culture and China's traditional culture; Jia Bin-Xiang [1989] tried to explore the nature of scientific culture; Han Min-Qing [1990] analyzed the historical role of scientific culture and its limitations; He Ya-Ping [1992] highly affirmed the positive role of scientific culture and its universality.

In 1993, the Youth Committee of Chinese Dialectics of Nature Research Society, the Chinese Dialectics of Nature Research Society in Hei Longjiang Province and the Harbin Normal School jointly organized a conference titled "Scientific Culture and Development of the Modern World". The relationship between scientific culture, humanistic culture and traditional culture in China, the truth and value of scientific culture, and the scientific sprit and awareness were all hotly discussed by young scholars from the field of Dialectics of Nature [Liu, 1993]. This conference was a landmark event in the study of scientific culture during the first half of the 1990s.

Afterward, more and more scholars have paid attentions to scientific culture. The popularity of such studies culminated in a series of books published by Chinese STS scholars during the late 1990s to the early 2000s; e.g., Xiao Feng's *Scientific Spirit and Humanistic Spirit* (China Renmin University Press, 1994), Zhou Chang-Zong's *The Cultural Spirit of Western Modern Science* (Shanghai's People Press, 1995), Meng Jian-Wei's *On Humanistic Value of Science* (China Social Sciences Press, 2000), Li Xing-Min's *The Spirit and Value of Science* (Hebei Education Press, 2001) [Meng, 2009], Wang Da-Yan and Yu Guang-Yuan's *On Scientific Spirit* (Central Compilation & Translation Press, 2001), and Xiao Feng's *Integration of Science and Humanities in Modern Society* (Jiangxu People's Press, 2001). As mentioned above, the focus in this time period is mainly about scientific spirit, the humanistic value of science, and the relationship between science and the humanities.

Compared with the above works, which generally hold a positive view of science, some other scholars such as Jiang Xiao-Yuan, Liu Hua-Jie, Tian Song and Wu Guo-Sheng began to pay more attention to the negative influence of science, criticizing especially scientism popular in Western and modern Chinese history. There are two landmark events for scholars in this group: (1) They published an important journal named *Science Review* (Jiangxi Education Press) in 1999 (Fig. 13.1) and did the analysis and comments critically on a variety of scientific and cultural phenomena from the position of anti-scientism. (2) They organized the first national symposium on scientific culture study in Beijing, in 2002, and further clarified their stand on anti-scientism.

Meanwhile, China began to advocate and implement education reform, emphasizing the importance of so-called "quality-oriented education" or "education for all-round development". Especially in basic education (grades 1-12), the new curriculum standard was carried out. In this process, influenced by the *2061 Project* (Science for all Americans) and the National Educational Technology Standards (NETS), both from America, a large number of books and articles written by China's teachers and educational scholars have been published. Most of these works emphasized the importance of combining science and the humanities in the training of students. Up to now China's scientific culture studies have spread from STS to the education research field. A new prosperity is emerging at present time.

Fig. 13.1. The first issue of *Science Review* (1999), *Science & Culture Review* (2004) and *Our Scientific Culture* (2007) (left to right).

13.2 Definitions of Scientific Culture

So, what is scientific culture? As mentioned above, scholars from different fields/disciplines, even from the same field or discipline, have different understanding of the term of scientific culture. It is reflected vividly in the English translation of this Chinese word/term (kexue wenhua). Roughly speaking, there are three relatively familiar translations. They are "scientific culture", "science culture"/"culture of science", and "science and culture";[2] the first two translations are most used. Meanings of these English terms are subtly different, but Chinese

---

[2] Outside of China, for some people, "scientific culture" could have three meanings: culture of science, culture through science, and culture for science [Vogt, 2012]. Culture through science is summed up by Paul Caro [2008]. Margaret Jacob's position is that since science is a societal activity, scientific culture is part of society's culture and helps to push the development of technology and industry [1997].

scholars often mean to convey all of these three meanings when they use one of them in their books or articles.

Yet, because these different translations are in many cases dictated by the publishers of the Chinese articles or books, they often do not absolutely and accurately express the authors' real intentions. Consequently, the English translation of this special word in Chinese publications should not be taken seriously. Moreover, when the two Chinese words of "science" and "culture" were combined together by scholars to constitute the new composite word, some of them could in fact deliberately want to make it a fuzzy, complex concept.

Here, to convey the meaning of this word implied by various authors, four representative examples from Chinese scholars are presented.

1. Science-based culture

According to Sun De-Zhong [2005] from Wuhan University of Technology,

> Scientific culture is a specific kind of social culture originated from people's observing and studying natural phenomenon. It is a ramification of natural knowledge in society, with two meanings. First, it is a kind of scientific knowledge, scientific spirit refined or sublimated from this knowledge, and a part of humans' general ideology. In particular, scientists and scholars doing scientific culture studies interpret and transfer it from natural science to the humanities and social sciences, and even to public discourses and ideologies. Then it is considered and solidified by the public as a dominate concept, spirit and belief influencing their thinking and behavior. Second, it is also a kind of material "force" and an important component in humans' practical life. Science enters the material world via technology, and enables people to live in a world that is humanized, cultural, scientific and technological; as a result, science becomes the grand background of this multi-faceted world. Thus, scientific culture is not an epistemological concept of natural science but an action-based, theoretical concept of philosophy and culturology.

This is the most comprehensive definition which is very popular in China's academia.

2. Culture of science

Similarly, Meng Jian-Wei [2009], a scholar and activist in scientific and cultural philosophy from Graduate School of the Chinese Academy of Sciences (CAS), defines this concept as follows.

> Scientific culture should include four different parts, viz., value, system, behavior and achievement (in the form of scientific theory, technology and something materialized). Among them, scientific spirit, scientific idea and scientific value constitute the "soul" of scientific culture; the technological, empirical, mathematical and logical parts make up the "body" of it. The intertwining soul and body turn scientific culture into an organic whole. Accordingly, we scholars in scientific culture study need to do both: analyze its material content and observe its soul, especially investigating the relationship between the two.

3. Integrating science-related areas

While some scholars define scientific culture from the perspective of its research contents, other scholars emphasize the cooperation or separation between different research areas related to this concept. For example, Yuan Jiang-Yang [2007], historian of science from Institute for the History of Natural Sciences, CAS, writes,

> Scientific culture study is an interdisciplinary research area still filled with conflicting views and attitudes in today's China, and so we cannot give a clear definition of it. In this field, scholars with different academic background have developed several research approaches from different positions. Their attitudes toward science and its role in society, and their understanding of today's scientific culture are different too. So I suggest that scientists; historians, philosophers and sociologists of science; and decision makers in science and technology matters should seek more cooperation with each other when resolving the problems in science, technology, society and culture. And through this cooperation, the traditional boundaries between disciplines such as history of science, philosophy of science, and sociology of science will disappear gradually. Unfortunately, now in China there is still no clue how "scientific culture study" or "science and technology study" can

replace those fully institutionalized disciplines, or integrate themselves with other kinds of study on science and technology as a big discipline.

4. All science-related studies

Wu Guo-Sheng, philosopher from Peking University, even clearly indicates [2003] that

> The word "scientific culture" can be defined as the academic study in the history, philosophy and sociology of science; science policy and management; other areas of "Science and X", and even the popular writings in all these fields.

In this chapter, the term "scientific culture" embraces all the four meanings above, plus science and culture.

13.3 Scientific Culture Studies in Recent Years

The statistics of research articles published and (master and doctoral) theses on scientific culture from China, collected through keyword search from the China Knowledge Resource Integrated Database (CNKI), is presented in Table 13.1.

Table 13.1. Statistics of articles and (master and doctoral) theses on scientific culture from China (2001-2010).

Year	Number of articles	Number of theses
2001	106	0
2002	139	3
2003	122	6
2004	135	6
2005	159	10
2006	136	5
2007	134	8
2008	120	6
2009	117	5
2010	120	2

Table 13.1 shows that the number of research papers and theses (especially master theses) grew steadily after 2001; peaked in 2005 and declined slightly in recent years. And if articles published in newspapers are included, the numbers will be even higher. To a large extent, it reflects the sustained attention of Chinese scholars on this issue.

Scholars in scientific culture (SC) can be divided into two groups: (1) those mostly from STS (involving disciplines such as history, philosophy and sociology of science, and science policy and management), and (2) scientists from research institutions and instructors from high schools or universities. The questions are: Why are there so many scholars interested in this topic? And why have they paid attention to it for such a long time?

In fact, research in China is to a large degree constrained by the "discipline system" established jointly by the Academic Degrees Committee of the State Council and the State Education Commission. Only by fitting their research projects into this disciplinary structure (e.g., first-level, second-level and third-level disciplines) that the researchers are able to apply for research funding used to recruit students and cultivate professional activities. Unfortunately, SC cannot be classified neatly within any single discipline in this system, even though it continues to attract attention of many scholars and becomes an academic field "with Chinese characteristics". There are three reasons for this development.

Firstly, although it does not belong to any single discipline, SC can be taken as a specific research direction or research topic in many disciplines such as philosophy (especially philosophy of science), history (history of science), education (science education) and communication (science communication), even in political theoretical research. The students who have done SC studies counted in Table 13.1 actually obtained their degrees in philosophy, history, education, sociology, etc. In fact, SC research projects are often easier to get funding at the national level.

Secondly, the deeper reason for its popularity is that SC is a very "appropriate" subject of study for scholars. There are several reasons: (1) It is well known that Science (called Mr. S during the 1919 May Fourth Movement) has been seen as an important material, cultural and spiritual strength for national salvation by intellectuals and the public in modern China. (2) Chinese government has begun to implement the "Strategy of Invigorating China through Science and Education" since the mid-1990s. (3) Presently, science does play an important role in China's social, economic and political developments; academic study on science thus becomes very necessary and important. When the surge of "cultural study fervor" is set off in the academia, SC—as one of the most important components in culture—has become the "hottest spot". In fact, most SC scholars are those from the old school who still hold onto the logical, positivist view of science and stress the proactive roles of science and technology in the progress of social evolution.

Finally, since SC is a relatively vague concept it provides a lot of research space for scholars and instructors. (1) The first group of SC scholars (mostly from STS) can share comfortably the academic space provided by the "big box" of Scientific Culture Studies (SCS). This group includes those who

- support sincerely the positivist view of science;
- emphasize the positive impact of scientific development to culture and society;
- follow actively the government policies on scientific and technological education and popularization;
- focus on introducing into China the Western theories in science and technology studies;
- try to do case study on science in Chinese context;
- do research on the complex relationship between science and society from the perspective of anti-scientism, revealing the limitations of science.

Especially for those in anti-scientism, SCS is an admirable label for them to avoid being labeled as the "bad guys" who oppose the development of science. The reason is that in China, anti-scientism and anti-science are frequently confused, intentionally or unintentionally, by some people.

(2) For the second group (scientists and instructors), China's education reform prompts them to think about the goal and direction of education and provides them a good opportunity to participate in SCS. Frankly, most of SCS are still in the initial stage of introducing Western

thoughts and literature, advocating scientific spirit and value, and emphasizing the importance of combining science and the humanities, at least in theory; really influential research works are still very few in number. In fact, many articles were published by members of this group to cope with the official assessment of their schools or research institutes. Although there are few creative viewpoints or new theories proposed in these articles, the publication of which always meets the ideological needs of the time.

However, apart from "academic research", there are also a lot of noteworthy quasi-academic researches and popular-science writings on SC. In the opinion of the authors and Jiang Xiao-Yuan (historian from Shanghai Jiao Tong University), the term SCS should be used in its broad sense, to include both academic and quasi-academic researches, even though the latter means mainly the dissemination and popularization of SC. Moreover, these kinds of quasi-academic research and popular-science writing are really interesting and can influence the public more efficiently and strongly than academic papers do.

13.4 Academic Journals in Scientific Culture Studies

By their respective nature, the two groups of SC scholars publish their works in two different types of journals. The first type includes educational journals which publish a good number of articles discussing the relationship between science and the humanities. Journals of the second type are directed mainly to philosophy of science (usually called "dialectics of nature" in mainland China, the same name of Friedrich Engels' book *Dialectics of Nature*) that deal with the Marxist view of nature and science. For example, there are the

- *Journal of Dialectics of Nature*, published by CAS;
- *Studies in Dialectics of Nature*, by The Chinese Research Society of Dialectics of Nature;
- *Science Technology and Dialectics* (renamed *Studies in Philosophy of Science and Technology*), by Shanxi University;
- *Studies in Science of Science*, and *Science and Society* (formerly called *Impact of Science on Society*), by CAS.

Furthermore, since the government has paid increasing attention to the dissemination and popularization of science in recent years, some new journals such as *Science Popularization* (beginning in 2007) have emerged and played an important role in SC publishing. Additionally, some comprehensive publications (e.g., university journals) have also become the arena for SC articles.

Moreover, the journal *Science & Culture Review* and the book series *Our Scientific Culture* should be mentioned here (Fig. 13.1). As its name indicates, *Science & Culture Review*, initiated by Liu Dun and published by the Institute for the History of Natural Sciences, CAS, publishes mainly research articles on SC. In the Forward of the first issue, the editors [Liu & Cao, 2004] write

> We do not want to define the scope of scientific culture studies in this journal too clearly because the condition of a blur boundary often facilitates the diversity in academic research and all-round discussion on related subjects. We can interpret scientific culture either as the counterpart of humanistic culture in C. P. Snow's sense or regard science as a sort of subculture from the point of view of cultural anthropology, or think of it as a kind of philosophical, historical or sociological research on science, without excluding those scientific culture studies from the perspective of deconstructionism. In short, we hope to create a space for open dialogue in which scholars from different fields with different research interests can express their opinions from different perspectives. It would be beneficial to improving people's understanding of science and culture and eventually leads to a new culture.

This vision in the first-issue statement has basically been fulfilled in its subsequent issues. Nevertheless, the journal has maintained its focus on articles in external history of science.

In comparison, *Our Science Culture* edited by Jiang Xiao-Yuan and Liu Bing (Tsinghua University, Beijing) is a journal/magazine published as a book series, in lieu of a journal's ISSN number which is strictly controlled by the government. It has published one or two volumes per year since 2007. Its editors and most of its contributors have formed a small academic/mass media community called "Scientific Culture

Group". Most group members are scholars in history and philosophy of science who adhere to the position of anti-scientism. They exert their influence in both academic research and mass communication, and try to establish a School in SCS by publishing this journal regularly. In the Editors' Note [Jiang & Liu, 2007] of the first volume, it is written that

> Sometimes people would scruple about or try to avoid the term "collegial publication", but we do not have to avoid mentioning the color of "collegial publication" of *Our Scientific Culture* at all. This publication series aims to publish research results from members of our small group and from other scholars who hold positions on science and culture that are similar to ours. We try to make it inclusive while maintaining its own characteristics in the diverse academic environment existing in mainland China.

13.5 Conclusion

Although scientific culture studies have continued to be the research "hotspot" in China's academia since the end of 1980s, most of the articles and books published obviously lack originality. This incongruity between quantity and quality in academic publications is a common phenomenon shared by other research fields in mainland China.

However, everything has its pros and cons. Incidentally, the research fervor in scientific culture has brought about some positive effects. It has

1. facilitated researches in areas such as history, philosophy and sociology of science,
2. introduced some new ideas and theories to the areas mentioned above,
3. provided a bridge between scientific culture and humanistic culture,
4. introduced new academic views and achievements to the public;
5. it might benefit China's educational reform in the future.

From a practical point of view these positive effects on scholarship and education have, to some extent, compensated the lack of academic originality. But much remains to be done. We are optimistic that studies in scientific culture will continue to grow steadily in the future.

References

Caro, P. [2008] "Culture *through* science: A new world of images and stories," in *Science Matters: Humanities as Complex Systems*, eds. Burguete, M. & Lam, L. (World Scientific, Singapore) pp. 41-51.

Han Min-Qing [1990] "The historical role and its limitations of scientific culture," *Qilu Journal*, No. 6, 95-100, 116.

He Ya-Ping [1992] "Scientific culture and big science," *Studies in Dialectics of Nature* **8**(7), 1-8.

Hu, Shi [1923] "Preface to *Science and the View of Life*," in *Science and the View of Life* (Yadong Library, Shanghai) pp. 2-3.

Jacob, M. C. [1997] *Scientific Culture and the Making of the Industrial West* (Oxford University Press, Oxford).

Jia Bin-Xiang [1989] "Flourish of the view of scientific culture, *Journal of Dialectics of Nature* **11**(5), 9-16.

Liu Dun & Cao Xiao-Ye [2004] "Introducing *Science & Culture Review*," *Science & Culture Review* **1**(1), 1-5.

Liu Xiao-Ting [1993] "Introduction and summary of the first youth symposium on scientific culture and development of the modern world," *Studies in Dialectics of Nature* **9**(4), 72.

Ma Lai-Ping [1989] "The conflicts of scientific culture and Chinese traditional culture," *Shandong Social Science*, No. 2, 40-42.

Meng Jian-Wei [2009] "On scientific culture," *Bulletin of National Natural Science Foundation of China*, No. 2, 89-92.

Sun De-Zhong [2005] "On scientific culture and its contemporary value orientation," *Studies In Dialectics of Nature* **21**(3), 87-90.

Jiang Xiao-Yuan & Liu Bing [2007] "Editors' Note," *Our Scientific Culture* (East China Normal University Press, Shanghai) pp. 7-9.

Vogt, C. [2012] "The spiral scientific culture and cultural well-being: Brasil and Ibero-America," *Public Understanding of Science* **21**(1), 4-16.

Wang Hong-Sheng [2014] "The predicament of scientific culture in ancient China," in *All About Science: Philosophy, History, Sociology & Communication*, eds. Burguete, M. & Lam, L. (World Scientific, Singapore).

Wang Rong-Jiang [2011] "Review on indigenous studies of scientific culture in 20 Years (1990-2009)," *Journal of Dialectics of Nature* **33**(1), 81-88

Wu Guo-Sheng [2003] "Rethinking science communication and scientific culture," *China Reading Weekly*, Oct. 29.

Yuan Jiang-Yang [2007] "A brief discussion on scientific culture study," *The Chinese Journal for the History of Science and Technology* **28**(4), 480-490.

Zhao Lei-Jing [1987] "A mentor of Chinese science culture study: A book review of *A Guide to the Culture of Science, Technology, and Medicine*," *Journal of Dialectics of Nature* **9**(1), 80.

Part IV
Communication of Science

14

Science Communication: A History and Review

Peter Broks

At first it looks simple. Communication is the transfer of information from A to B, from one person to another. Therefore "science communication" (scicomm) must be the transfer of scientific information from A to B. This common sense idea has been the dominant view of scicomm for many years and has helped to shape debates about "scientific literacy" and the "public understanding of science". However, it is a view that is dependent upon a specific history of science and it is an idea that has been challenged in recent years. This chapter will show: Firstly, how present day ideas about scicomm were created in the 19th century; secondly, how two historical moments from the 20th century helped shape current understanding of scicomm; and finally, how that understanding has recently been challenged. The main aim of the chapter is to argue that scicomm should not be seen simply as the transfer of information but rather that it is better understood in terms of how meanings about the natural world are created and negotiated. Material for this chapter is drawn principally from the UK and the United States. However, similar developments can be found in other countries not least because they often copied what was happening in the UK and the US.

14.1 Nineteenth-Century Origins of "Popular Science"

Much of what we now think of as "science communication" (scicomm) and "popular science" (popsci) has its origins in the 19th century. The period sees a shift from an inclusive and participatory form of popsci to one where there was a clear separation of experts from lay public. It is a separation with which we are now familiar. Equally familiar is the idea that the gap is to be bridged by scicomm.

14.1.1 Early-Nineteenth Century: Republic of Science

In the early-19th century popsci was much more inclusive and participatory than what we understand it to be now. The science that appeared in popular publications was aimed at encouraging scientific activities amongst their readers. The raw materials for cutting-edge science such as rocks and plants were easily accessible. The scientific equipment to study these was relatively inexpensive (geologist's hammer, magnifying glass, book to identify plants).

There was no easy separation of science from public. Studies of the period have found a range of non-elite scientific activities or what Susan Sheets-Pyenson has called "low science". In her examination of popsci periodicals Sheets-Pyenson found that the editors used their publications to help build a "low scientific culture". They emphasized the universal accessibility of science and encouraged their readers to take up scientific activities. The dominant rhetoric was of a *Republic of Science* which all could join. Science was to be open to all. Membership of the Republic was not based on specialist knowledge but only on an eagerness to collect facts and make simple observations [Sheets-Pyenson, 1985].

A good example of such low science can be found in the artisan botanists studied by Anne Secord. According to Secord these botanists were at the intersection of elite and popular culture. Socially they were part of a network that included gentleman botanists, medical botanists, herbalists, gardeners, nurserymen and plant dealers. Culturally their interest in botany straddled a class divide. On the one hand, their botanical activity was a perfect example of the "rational recreation" which middle-class reformers tried to foster. On the other hand, their societies usually met in public houses which were the locations for the kind of immorality and drunkenness that the rational recreation was supposed to replace. Similarly their collective practice based on a fair exchange of plants and information was at odds with the more individualistic practices of gentleman botanists, collectors and plant dealers. These would be more concerned with profit, ownership of specimens and the control of botanical knowledge through the control of print media [Secord, 1994].

The result was that in the first half of the 19th century popsci was contested territory:

> For many, including some in the scientific elite, the contest revolved around the issue of whether science should be popular, in the sense of being open to a wide range of participants (including women and artisans) who could contribute to and benefit from the production of knowledge. [Secord, 1994, p. 297]

In short, the artisan botanists were claiming their right to participate in science and claiming this right, not simply as interested members of the public but through their activities *as* botanists .

Also contested were the meanings associated with these activities. In the early part of the 19th century the pursuit of science meant different things to different people. For example, within the culture of radical politics science was seen as a weapon with which to attack the established church. In contrast, the established church frequently looked to science as a way to paper over the cracks of doctrinal differences. Natural Theology or the study of God's design in nature provided common ground for those who might otherwise find themselves in disagreements over church ritual or the various aspects of Christian belief [Broks, 2006].

Alongside these views was a more utilitarian belief that the pursuit of science could be justified simply on the grounds that it was useful. The usefulness of science has a history that can be traced back at least to the early-16th century and the writings of Francis Bacon. However, in the early-19th century we can see it most clearly expressed in the utilitarianism of Jeremy Bentham. In the Benthamite philosophy the principle of utility ("the greatest happiness of the greatest number") was best served by the spreading of "useful knowledge" as a way to help foster social stability. The Society for the Diffusion of Useful Knowledge (established in 1826) was one of a number of Benthamite initiatives driven on by social reformers.

Of course, not all knowledge was seen as "useful". Indeed some knowledge was seen as downright dangerous. For example, in the Mechanics' Institutes that proliferated in the 1820s and 1830s the middle-class founders of the institutes were keen that discussion of politics and religion should be avoided as far as possible. Instead, the working mechanics who were supposed to attend the institutes were to be given lectures on science and the study of nature. The institutes were conceived of as an instrument of social control and science was to play a key part. From their study of these institutes Shapin and Barnes found that their creators believed "a regimen of scientific education for certain members of the working class would render them, and their class as a whole, more docile, less troublesome and more accepting of the emerging structure of industrial society" [Shapin and Barnes, 1977, p. 32].

14.1.2 Late-Nineteenth Century: The Rise of the Expert

In the second half of the 19th century the idea of science as useful came to be pre-eminent. Other meanings—such as the association of science with radical politics or Christian belief—survived but were increasingly pushed to the margins. Also marginalized was the low scientific culture of artisans as a clear division was established between science and the public. The rhetoric of science as useful was clearly associated with the increasing professionalization of science, so too was the perception that the public were to be merely passive recipients of scientific knowledge.

The professional sense of useful science can be found in the early decades of the Royal Institution (founded in 1799). The establishment of the Royal Institution marks a shift from science as an aristocratic form of polite knowledge as seen in the Royal Society to science as an instrument for the maintenance of a stable social machine.

> England, on this view, did not have social problems. Rather it had "technical difficulties", and thus the various facets of the condition-of-England question and the modernization of the nation were scientific matters, amenable to scientific treatment. Thus was born, as a political force, the technological fix. [Berman, 1978, p. 109]

At the Royal Institution the scientific management of social problems came to be expressed through a concern for improved sanitation and street lighting as well as testimony to courts and government inquiries.

Conveniently, the disinterested, objective solutions that were needed to tackle society's problems could be provided by an emerging class whose profession was disinterested objectivity. State intervention into many aspects of public and private life created new demands for scientific expertise in food standards, pollution, contagious diseases, and factory inspections. These new demands on science in turn created a stronger desire in the scientific community to protect its independence, status and integrity.

Beginning with the British Medical Association in 1856 a new and much larger wave of professional institutions came into existence for doctors, dentists, mechanical, mining and electrical engineers, naval architects, accountants, surveyors, chemists, teachers and others, expressly to exclude quacks and charlatans who put profit before professional service [Perkin, 1969, p. 429].

However, the professionalization of science not only excluded quacks and charlatans but also the public more generally. From around the 1860s onwards professionalization is accompanied by a rhetorical shift in popsci periodicals, a shift from active participation to passive support. The earlier rhetoric of a Republic of Science open to all was dropped. Instead popsci periodicals now sort to rally their readers into supporting professional "high science". Where once there had been a low scientific culture that was active, participatory and often in opposition to the high scientific establishment, by the end of the century popsci periodicals had become "accomplices of the high scientific community" and the public were asked merely to cheer from the sidelines [Sheets-Pyenson, 1985, p. 563].

14.1.3 *Popular Science Redefined*

In 1833 William Whewell coined the term "scientist" in an attempt to bring a defining unity at a time when science was increasingly characterized by a bewildering array of specialized languages, journals, institutions and equipment. It was not a term that found much favor amongst those to whom it might apply, but by the end of the century those same specialized languages and institutions helped provide a sense of self-identity in the consolidation of a professional scientific community. An appeal to specialized knowledge and practices enabled scientists to draw the boundary between what was science and what was not. In this way, excluding the public became a defining feature of what it meant to be scientific.

More particularly, by helping to draw the boundary between science and public the professional consolidation of science necessitated a redefining of what was meant by "popular science".

> Increasingly, however, the term "popular science" was used by the dominant culture to signify bodies of literature and scientific activity that had little or no interaction with elite science.....More than any other area of knowledge production, scientific practice became increasingly associated with specific sites from which "the people" were excluded. [Secord, 1994, p. 297]

With the public excluded from scientific enterprise "popular science" was no longer the science done *by* the public but rather it became the science that was given *to* the public. By the end of the 19th century popular science was the science that had been *popularized*.

This is how we now commonly think of popsci and the role of scicomm. It is the transfer of knowledge from the professional scientist to the lay public. It is, as one scholar puts it, a form of alms giving, charity bestowed upon the public by a benevolent scientific community [Bucchi, 1998]. It is the product of a particular set of historical circumstances, but has become the common sense view. Until recent years it would also be the dominant view in studies of scicomm. This can be seen in the two key moments of the 20th century which are the subject of the next section.

14.2 Two Moments

In the second half of the 20th century two key moments helped to crystallize the dominant view of scicomm. In the 1950s the launch of the first artificial satellite triggered fears about scientific literacy, and in the 1980s concern for the "public understanding of science" set out scicomm as a new duty for scientists.

14.2.1 *Sputnik and Fears about Science Literacy*

In 1957 a high-pitched beep-beep-beep frightened the US. The radio signal from Sputnik was a constant reminder that the Soviet Union had the technology to place a satellite into orbit. If the Soviets could launch a satellite over the United States and send down radio signals, then clearly they would have (or would soon have) the capability to launch missiles over the United States and send down bombs. In the Cold War the red menace had seized the really high ground. The threat to national security was also accompanied by a blow to national pride, and self-esteem was not restored by the US attempt to launch its own satellite. Brought forward several months so that it could be an immediate response, the rocket exploded on take-off providing the Soviet press to make even more political capital with gleeful headlines such as "Kaputnik" and "Stayputnik" [Watson, 2000, p. 484]. If this really was the start of the space race then so far there appeared to be only one runner.

Even though the technological response was far from successful, the government was quick to take action. Within 12 months of the launch of Sputnik the National Defense Education Act was passed to help remedy what was seen as the shortcomings of US education when compared with that of the Soviet Union. More specifically, the launch of Sputnik highlighted the lack of science education. Attitudes to science were generally positive but a survey by the National Association of Science Writers found that levels of factual knowledge were low [Gregory and Miller, 1998, p. 4]. Science and technology were now firmly placed on the national agenda and with this new public prominence came a growing concern for what would later be known as "science literacy".

The key figure in the attempt to disseminate science to a wider public was Warren Weaver. Trained as a mathematician Weaver acted as a "Catalyst" for many of the public science activities and initiatives involving the American Association for the Advancement of Science, the National Science Foundation, the Rockefeller Foundation, the Sloan Foundation and the Council for the Advancement of Science Writing [Lewenstein, 1992]. However, in the history of scicomm what is more important is the work he did with Claude Shannon, an engineer at the Bell Telephone Laboratories. In a two-part article published in 1948 Shannon developed a mathematical model for how information is transmitted in physical systems such as two telephones connected by a wire. Weaver extended the model in an attempt to understand all systems of interpersonal communication. The result was a co-authored book published in 1949, *The Mathematical Theory of Communication*.

The Shannon and Weaver model of communication quickly established itself as a scientific way to understand what happens when people try to communicate. It reduces communication to the simple form of Sender-Message-Receiver. Information is transmitted as a message from one person to another. In the process the message can become distorted or corrupted by "noise". We can easily understand this when we think about the crackle we hear on a telephone line, but as a general theory of communication the idea of a crackly line is applied to all forms of communication

With Weaver's central role in fostering public science and his work with Claude Shannon we should not be surprised to find that this model of communication being applied to the problem of science literacy. The model of Sender-Message-Receiver would easily translate as Scientist-Popularization-Public. Christopher Dornan has characterized the approach as:

> ... a camp of inquiry that operates with a rigidly linear model of the communication process. Scientists are the sources of information, the media are the conduit, and the public is the ultimate destination. The goal is to minimize media interference so as to transmit as much information as possible with the maximum fidelity. [Dornan, 1989, p. 102]

The Shannon and Weaver model also gave scientific legitimacy to the common sense idea of scicomm that had been established at the end of the 19th century. This combination of common sense and mathematical description would help ensure that the model dominated thinking about scicomm for the rest of the century. By 1990 Stephen Hilgartner could refer to it simply as the "dominant view" [Hilgartner 1990].

The culturally-dominant view of the popularization of science is rooted in the idealized notion of pure, genuine scientific knowledge against which popularized knowledge is contrasted. A two-stage model is assumed: First, scientists develop genuine scientific knowledge; subsequently popularizers disseminate simplified accounts to the public. Moreover, the dominant view holds that any differences between genuine and popularized science must be caused by "distortion" or "degradation" of the original truths. Thus popularization is, at best, "appropriate simplification"—a necessary (albeit low status) educational activity of simplifying science for non-specialists. At worst, popularization is "pollution", the "distortion" of science by such outsiders as journalists, and by a public that misunderstands much of what it reads [Hilgartner, 1990, p. 519].

14.2.2 Bodmer and the Public Understanding of Science

The publication of Thomas Kuhn's book *The Structure of Scientific Revolutions* in 1962 helped to instigate critical sociological studies into science. Throughout the 1960s and 1970s science was increasingly examined as a social activity rather than as a simple narrative of progress. Similarly, in the 1960s and 1970s a more critical approach was taken to the study of media and communication. As Stuart Hall has explained there was a major shift in mass communications research away from the mainstream approaches of American behavioural science to a more European critical or "ideological" perspective. Within this new perspective concern was not so much with the transmission and impact of particular media messages (as would be found with the Shannon-Weaver model), but more with the politics of signification, the struggle over meaning, and the production of social consent [Hall, 1982].

Nevertheless, despite two decades of critical inquiry into science and two decades of critical inquiry into communication, by the 1980s there was still a notable absence of any similar examination of scicomm. This can clearly be seen in the approach to be found in a report produced by the Royal Society and the initiatives that it fostered.

The Royal Society had set up an ad hoc committee in 1983 with the remit:

> to review the nature and extent of public understanding of science and technology in the UK and its adequacy for an advanced industrialised democracy, to review the mechanisms for effecting the public understanding of science and technology and its role in society, [and] to consider the constraints upon the processes of communication and how they might be overcome. [Bodmer, 1985, p. 9]

The committee's report published in 1985 was both clear and influential. Usually referred to as the Bodmer Report (after the committee's chairman Walter Bodmer, Director of Research at the Imperial Cancer Research fund), it set out the importance of scicomm for an industrialized democracy:

> A basic thesis of this report is that the better public understanding of science can be a major element in promoting national prosperity, in raising the quality of public and private decision-making and in enriching the life of the individual. . . Improving the public understanding of science is an investment in the future, not a luxury to be indulged in if and when resources allow. [Bodmer, 1985, p. 9]

Moreover, it now became a duty for scientists to communicate to the public. Whereas previously popularizing science had often been left to those on the fringes of the scientific community, now it was seen as an essential part of the work carried in science departments if only because it was increasingly a required element in application for public funding.

The immediate impact of the Bodmer Report was a flurry of activities and initiatives many of which were promoted by the newly formed Committee on the Public Understanding of Science (COPUS)— a joint venture of the Royal Society, the Royal Institution, and the British Association for the Advancement of Science. Workshops were set up to provide media training for scientists, a prize was awarded each year for the best popsci book, and small grants were handed out for community-based projects. The report also provided a spur to the establishment of two new chairs in the public understanding of science (at Oxford University and Imperial College London) as well as the setting up of a new journal *Public Understanding of Science* which provided an outlet for academic research.

However, it is difficult to avoid the sense of panic about all this "PUS" activity (as the public understanding of science was soon to be called). Indeed, Steve Fuller has described it as "our latest moral panic" [1997]. According to Brian Wynne it was a panic of the scientists' own making:

> In short, the re-emergence of the public understanding of science issue in the mid-1980s can be seen as part of the scientific establishment's anxious response to a legitimation vacuum which threatened the well-being and social standing of science. It is arguable that this legitimation vacuum, which manifests itself as the widely publicized and lamented public ignorance of science, is the direct result of the way that science in the past legitimated its harvest of public funds by *distancing* itself from the ordinary public. ... [S]cience now finds itself hoist with its very own petard, namely the cultural alienation whose establishment it actively, if innocently, promoted. [Wynne, 1992, p. 38, original emphasis]

Scientists would appear to be victims of their own success. Worried about possible declining public support for science it could be argued that PUS was not so much an attempt to legitimize popularisation but rather that popularisation was an attempt to legitimize science. To put it more crudely, it was very easy to see PUS as little more than a massive public relations exercise for science. It was never considered what would happen if the public did understand science and *because* of that understanding decided not to support it.

14.3 New Challenges and New Models

Although "PUS" originated in the UK it quickly became an international phenomenon—as can be seen in the pages of the journal *Public Understanding of Science*. In some cases it helped to foster new initiatives, in others it simply provided a convenient label for exiting activities and interests. In either case the model for communication was the same: the common sense model of transmitting information from A to B. However, in the past 20 years this model has been dismissed as inadequate. What had long been known in Media Studies was finally being applied to scicomm.

14.3.1 PUS: Problems and Politics

The essence of PUS thinking can be found in the first issue of the journal *Public Understanding of Science*. In the very first article the demand from Bodmer and Wilkins was simple: "We need to know the most effective models to use to get messages across to a wide variety of target audiences" [Bodmer & Wilkins, 1992, p. 7]. Here we can clearly see the dominant, common sense understanding of communication with its language of transmitting messages, being effective and "targeting" audiences. In media studies it is an idea that has been labelled as the "effects model" or the "hypodermic model". The public is simply a target to be hit or injected with information. The effectiveness of communication is then measured in number of hits or how much information has been injected. For example, the effectiveness of an advertising campaign would be measured in numbers of sales before and after, or a political campaign measured in changes in voting intentions. Within scicomm it quickly became labelled as the "deficit model". Here the public were seen as having a deficit (in science information) that had to be remedied, in the same way that empty vessels needed to be filled. Levels of scientific literacy were to be the measure of how "full" the public were.

As an idea "The Public Understanding of Science" and its associated "deficit model" was fraught with difficulties. Years of work in social studies of science and in media studies left it open to questions on its unproblematic portrayal of science as an uncontested body of knowledge, its linear model of communication, its framing of literacy as a measure of "understanding", and its simplistic conceptualization of the public (even the idea that there is such a thing as "the" public). Criticisms of the deficit model were becoming more widespread through the 1990s. Nevertheless, such criticisms "barely impacted on scientists' discussions of public understanding of science" [Gregory and Miller, 1998, p. 17]. Not only were the criticisms ignored but there seemed to be a more general reluctance to give any consideration at all as to how to think about scicomm:

> In discussions of science, popularization and the public, talk is frequently of activity—of what one should do in order to achieve better or more public understanding of science. Rather less attention has been devoted to articulating the philosophies and models that inform and drive popularizing activities. [Gregory and Miller, 1998, p. 81]

There are a number of reasons why this might be the case. As mentioned earlier, the dominant idea of scicomm seemed to have the winning combination of common sense and mathematical modeling. To reject it would be to go against common sense (even though much of science might be equally counter-intuitive) and to dismiss an approach that could make use of quantification (i.e., apparently be "scientific").

However, it may also be that the lack of critical attention given to thinking about scicomm may also be because *the commonsense model is an oversimplification that works in the political interests of scientists*. In its relationship with the public, science needs to perform a delicate balancing act. On the one hand it needs to maintain its authority by setting itself apart from the general public; on the other, it needs public support to maintain its legitimacy. The dilemma for scientists is that being set apart increases their alienation, but being more "popular" undermines their authority. As Simon Locke has argued,

> One means of doing this is to accentuate the impoverished condition of public knowledge and the public's dependence on professional expertise. Hence the periodic revival of concern with public understanding, and the characteristic rhetoric aimed at "assessing" the extent of and "improving" the public's knowledge deficiency. The "discovery" of such deficiency then justifies the experts' belief in their expertise and the call for support to raise the public standard. ...

... The deficit model is readily identifiable as the rhetoric of professional ideology, masking inequities of power in the language of democratization; citizenship through science comes at the price of expressing knowledge in ways acceptable to professional (natural) scientists—it is our way or not at all. Hence the presence of competing knowledge claims and knowledge bases are rejected as simply "anti-science"... [Locke, 1999, p. 78]

This political dimension is also highlighted by Christopher Dornan. By seeing the public as the "problem" and improved communication as the "solution", the commonsense model diverts attention away from scientists. It is a sleight of hand that results in accepting science on its own terms: "It holds that science is indeed an inherently objective, rational and heroic endeavour, and that the project of science communication should be to affirm just this view in the popular understanding" [Dornan, 1989, p. 102]. Similarly, Stephen Hilgartner argues that the commonsense model "remains a useful political tool for scientific experts" because it

> sets aside genuine scientific knowledge as belonging to a realm that cannot be accessed by the public, but is the exclusive preserve of scientists. It thus buttresses the epistemic authority of scientists against challenges by outsiders. [Hilgartner, 1990, p. 530]

Thus we can see the need to examine not only scicomm but also how we think about scicomm.

14.3.2 *From PUS to PEST*

The Bodmer Report ushered in an era of frenetic activity around the public understanding of science. The activities would continue but the PUS way of thinking was clearly in decline by the end of the 1990s. In 2000 the Select Committee on Science and Technology from the UK's House of Lords reported that "despite all this activity and commitment, we have been told from several quarters that the expression 'public understanding of science' may not be the most appropriate label." The committee had been told that it was a "rather backward looking vision", even that it was "outmoded and potentially dangerous".

> It is argued that the words imply a condescending assumption that any difficulties in the relationship between science and society are due entirely to ignorance and misunderstanding on the part of the public; and that, with enough public-understanding activity, the public can be brought to greater knowledge, whereupon all will be well. [Select Committee on Science and Technology, 2000, 3.9]

Instead there was a "new mood for dialogue" (a phrase repeated several times in their lordships' report). The new mood brought with it a new acronym. Efforts too improve PUS would now be directed towards PEST (Public Engagement with Science and Technology).

Criticisms of the deficit model became commonplace as scholars of scicomm turned to other ways of understanding the relationship of science to public. Stephen Hilgartner was an early critic and had already suggested the analogy of a stream to describe the various degrees of communication. Rather than a simple binary categorization of science and public there was instead, he suggested, a whole spectrum of knowledges and contexts—upstream towards scientists at work and downstream towards the public in everyday life. Upstream is the unpublished shoptalk in labs, meetings and technical seminars. Downstream are TV programs and the tabloid press. "The point is simply that 'popularisation' is a matter of degree. The boundary between real science and popularized science can be drawn at various points depending on what criteria one adopts, and these ambiguities leave some flexibility about what to label 'popularization'" [p. 528].

In science policy circles moving public debates upstream were seen as a way of heading off potential criticisms of new technology before they became too serious. In the UK, for example, the government was anxious that the development of nanotechnology would not face the same opposition as did genetically modified (GM) food and crops. However, for Wilsdon and Willis moving debates upstream is not just a process of removing obstacles but is also something that can provide an opportunity:

Above all, we need to move away from the idea that new technologies are developed by scientists and then presented to the public as a done deal. Science and society must work together to shape the direction of a technology—with research processes opened up to scrutiny and debate, and assumptions challenged from the start. It is only through these new forms of "upstream" public engagement that we will develop technologies that people want. [Wilsdon and Willis, 2004a]

This more dialogic or PEST approach can be seen in the development of consensus conferences, citizens' juries and science shops as well as the involvement of patients interest groups and activist groups in the policy process. In consensus conferences and citizen's juries, for example, selected members of the public act as a lay panel cross-examining expert witnesses on a specific issue related to science and technology. It was after just this kind of process that the Danish government decided in 1987 not to fund projects on animal gene technology.

However, as Alan Irwin has pointed out, the new PEST agenda led to frustration amongst policy makers and practitioners, "that one way communication may not be enough but that two-way communication is much more challenging than it might at first appear" [Irwin, 2009, p. 4]. This can be seen in the *GM Nation?* debate carried out in the UK in 2003. Tackling the issue of the commercialization of GM crops this was a major exercise in public dialogue with hundreds of local, regional and "top-tier" meetings, thousands of feedback forms and nearly three-million hits on the debate's website. Nevertheless,

> What we can especially identify in this case is a tendency for "public communication and dialogue" to be seen as a discrete phase within the policy process: an activity to be fed into decision-making at the appropriate time, alongside other forms of evidence, before business as usual can return. [Irwin, 2009, p. 11]

In this respect upstream "dialogue" becomes little more than an updated version of the earlier public relations exercises of winning over the public to new technologies but now carried out in an earlier phase of development. In short, "not dialogue at all but simply a more sophisticated form of the old (and zombie-like) deficit model". [Irwin 2009, p. 13]

The problem is not only one of context (i.e., dialogue within the policy process) but also one that arises from the nature of the dialogue. This is evident in two ways: Firstly, in who sets the agenda for the dialogue; and secondly, in the unequal weighting given to the two sides. For example, public engagement is most often called on to debate whether a technology is safe rather than whether it is necessary. "Possible risks are endlessly debated, while deeper questions about the values, visions and vested interests that motivate scientific endeavour often remain unasked or unanswered" [Wilsdon & Willis, 2004b, p. 18]. More fundamental questions are at stake in any new technology, "Who owns it? Who benefits from it? To what ends will it be directed?" [p. 23].

> Questions such as those about whose interests are served by different kinds of science and scientific representation, and about the basis of trust and social accountability of different institutional forms of control and ownership of science, are effectively deleted. [Wynne, 1992, p. 38]

The public may display low levels of scientific literacy but can still be acutely sensitive to the political contexts of science. In the BSE (bovine spongiform encephalopathy, or "mad cow disease") crisis of the early 1990s the British public may have distrusted government science advisers because they lied about the safety of eating beef, but "what is frequently ignored is the public's (correct) perception that the lies originated in the influence of the beef industry over the Ministry of Agriculture, Food and Fisheries" [Dickson, 2000, p. 918].

Thus, by setting the agenda for dialogue PEST is more an invitation for the public to engage with science rather than for science to engage with the public. Furthermore, there is an additional asymmetry in the dialogue—science has "facts", the public has "opinions". The devaluing of opinion and linking it to public ignorance is a comparatively recent development (see [Bensuade-Vincent, 2001]) and masks the role played by opinion within the practice of science. Science is driven by dissent, by scientists taking different positions and it is especially important to remember this in the context of scicomm because *it is in those areas where science is still undecided that there is likely to be most public interest and debate.*

It can be expected that moving those debates "upstream" will, in turn, expose more differences of opinion as the public gets to see science as a work in progress. Nevertheless, as Steve Miller says, "Controversy and uncertainty are still regarded as things that should be kept within the scientific community. ... 'Not in front of the children' is still the attitude. But this, too, has to change" [Miller, 2001, p. 118].

14.3.3 *Meanings and Trust*

While the shift from PUS to PEST is to be welcomed it still considers scicomm as a transfer of information albeit now in a two-way exchange rather than a one-way flow. Public sensitivities to the political context for scicomm suggest that even this more nuanced account misses the point, viz., that *information transfer is only part of communication and not always the most important part.* Science is how we make sense of the world, and scicomm is one way in which the public make sense of science. If we are able to make sense of the world it is because we are able to make it make sense to us within our own sense-making environment or culture. The information a statement contains might remain the same, but how we make sense of it changes with context, medium and the relationship between the people involved. Our concern should not just be with asking whether information is accurate or not, but also asking what does it mean?

Meaning is not something that is transferred and passively received but something that is actively constructed. In turn those active constructions of meaning will differ not only with context and medium but also with culture, with gender, ethnicity, age, class, etc. For example, take the simple statement "Eat more fruit". The meaning will change from circumstance to circumstance. A doctor telling their patient "eat more fruit" is different to a mother telling their daughter. The meaning will be different again if we were to see it on an advertising poster in a supermarket. Similarly, the meaning changes with the medium. Having a face to face meeting to be told you have been made redundant is different to being told by text message or finding out via the TV news. A flipchart or PowerPoint presentation may work well enough at a conference or in a lecture but imagine one in a barroom discussion—not only would it be inappropriate but even the inappropriateness would have a meaning (signifying a lack of social skills to put it politely).

Science communication is not immune to these nuances. Many of the pronouncements from science in public discourse take the form "I am a scientist. I know what I am talking about. Listen to me." Now consider how the meaning might change—regardless of specific subject content—if this is a scientist employed by the military or a large pharmaceutical company, or if they are part of a press conference organized by state officials. Imagine how the meaning might change if it is an elderly white man talking to disaffected black teenagers. Imagine again how it might be if it is a national official talking to a local community or in a TV debate to a national audience.

What emerges from these considerations are a whole new set of questions. What counts as being a scientist? Where do we draw the boundaries between science and non-science, between scientists and non scientists? Similarly what counts as expertise? Who are the experts? What about lay experts? What about someone who is a scientist but is talking outside their own discipline or realm of competence? More fundamentally, why listen? Engagement presupposes particular social and political relationships which in turn raises questions about authority and democracy.

In short, we expose a fresh concern when we try to understand scicomm as a process of making meanings rather than a process of transferring information. *The central problem is no longer a deficit of knowledge but a deficit of trust.*

14.4 Conclusion: Scientific Literacy

The commonsense idea of scicomm sees it as a process of transferring information from scientist to the public. Successful communication is judged as effective communication, i.e., that which has a greater effect or impact. The desired impact is to reduce the public deficit in science, a deficit that is measured in levels of scientific literacy. It is this commonsense view that has informed science policy at least since the World War II and most particularly over the past 30 years. It is also an idea that is fundamentally flawed.

Although there may be common agreement about the desirability of increasing levels of *scientific literacy*, it is not at all clear what scientific literacy is, nor how it should be measured. The history and philosophy of science is littered with failed attempts at defining what science is. How are we expected to measure just how much the public understands if we don't know what it is that should be communicated? One solution might be to accept that *science is both a body of knowledge and a set of practices*, but even this serves only to highlight the diversity of practices across the various sciences (and note the need for "sciences" to be plural). Are we to expect someone to be "literate" in practices as diverse as high energy physics, palaeontology and genetics?

Trying to define scientific literacy cannot be separated from the reasons for wanting to improve scientific literacy. Surveys regularly show that in the US and the UK a round 30% of the population think that the Sun goes round the Earth. Should we be shocked by this? It does seem to suggest a high level of public ignorance, but *does it matter*? Adding up the number of correct answers might be easy to measure, but which chippings from the body of scientific knowledge is it essential for the public to know? *Less easy to measure is the public understanding of science as process.* Surveys might ask questions about scientific method (e.g., about double-blind experiments and placebos), but here again the question to ask is which methods and, more importantly, why?

In recent years attention has become focussed on the need for evidence-based policy (often in contrast to policy-based evidence). This does suggest a way forward in thinking about scientific literacy, but only at the expense of the commonsense model of communication. If there is a danger that our concept of scientific literacy is either too simplistic (counting correct answers) or too problematic (what is science anyway?), then we could aim for a literacy that at least appreciates the nature, strength and importance of evidence. This may be equally difficult to measure and therefore not be helpful if we wish to cling to the idea of a public "deficit" in science, but it does bring to the fore the importance of understanding science in context or what is sometimes called *civic scientific literacy*. For the public to be literate in this sense is for them to be able to relate their knowledge of science (as product and as process) to the contexts of their own lives and the world around them. In this respect evidence is not just a collection of facts but is the basis for social, political and economic action. Similarly, our ideal science literate would not only understand the need for evidence but would also be able to weigh one kind of evidence against another.

However, the more that we *contextualize* science the more difficult it becomes to subject "literacy" to measurement and quantification. Contextualization takes us away from simple metrics to the more complex realm of meanings. We make sense of evidence by seeing it in context and how we make sense will change from person to person, culture to culture, context to context. Meanings are not given; they are constructed and contested. Consequently, it may be easy to count how many people think the Sun goes round the Earth but a different approach is needed if we wish to understand how people make sense of the evidence concerning such things as evolution, climate change, or genetic modification.

There is also another way we should see the commonsense model as inadequate—the world of information and communication is changing. Television, press, and radio may differ in how they communicate science, but access to the internet changes the very nature of scicomm. Not only are the boundaries between scientist, science communicator and public becoming increasingly blurred by the growth of science blogging, but the spread of user-generated content opens up a new world of *user-generated science*. This is a challenge that scientists may be unwilling to face. So long as scicomm was seen as information transfer the public could be identified as the problem, but as science is opened up to the public so it becomes easier to see how scientists lose control over the meanings that the public construct. The situation becomes even more problematic for scientists as user-generated content increases the public's capacity to produce its own "science" in the first place. It may be that over the next few years we will see a return to that "Republic of Science" open to all which was a feature of the early-19th century. If we do, then our understanding of scicomm will have to address the nature of expertise. If we can all generate our own evidence then who are the experts, where does their authority come from and should we trust them?

References

Bensaude-Vincent, B. [2001] "A genealogy of the increasing gap between science and the public," *Public Understanding of Science* **10**, 99-113.

Berman, M. [1978] *Social Change and Scientific Organisation: The Royal Institution, 1799-1844* (Heineman Educational, London).

Bodmer, W. [1985] *The Public Understanding of Science* (Royal Society, London).

Bodmer, W. & Wilkins, J. [1992] "Research to improve public understanding programmes," *Public Understanding of Science* **1**, 7-10.

Broks, P. [2006] *Understanding Popular Science* (Open University Press, Maidenhead).

Bucchi, M. [1998] *Science and the Media: Alternative Routes in Scientific Communication* (Routledge, London).

Dickson, D. [2000] "Science and its public: the need for a 'Third Way'", *Social Studies of Science* **30**, 917-23.

Dornan, C. [1989] "Science and scientism in the media," *Science as Culture* **7**, 101-121.

Fuller, S. [2000] "Science studies through the looking glass: an intellectual itinerary," in *Beyond the Science Wars: The Missing Discourse about Science and Society*, ed. U. Segerstråle, U. (State University of New York Press, Albany, NY).

Gregory, J. and Miller, S. [1998] *Science in Public: Communication, Culture, and Credibility* (Plenum, New York).

Hall, S. [1982] "The rediscovery of 'Ideology': return of the repressed in media studies," in *Culture Society and the Media*, eds. Gurevitch, M, Bennett, T, J. Curran, J. & Woollacott, J. (Methuen, London).

Hilgartner, S. [1990] "The dominant view of popularization: conceptual problems, political uses," *Social Studies of Science* **20**, 519-539.

Lewenstein, B. V. [1992] "The meaning of 'public understanding of science' in the United States after World War II," *Public Understanding of Science* **1**, 45-68.

Locke, S. [1999] "Golem science and the public understanding of science: from deficit to dilemma," *Public Understanding of Science* **8**, 75-92.

Miller, S. [2001] "Public Understanding of Science at the crossroads," *Public Understanding of Science* **10**, 115-120.

Perkin, H. [1969] *The Origins of Modern English Society, 1780-1880* (Routledge, London).
Secord, A. [1994] "Science in the pub: Artisan botanists in early nineteenth century London," *History of Science* **32**, 269-315.
Select Committee on Science and Technology (House of Lords) [2000] *Science and Society* (HMSO, London).
Shapin, S. and Barnes, B. [1977] "Science, Nature and Control: Interpreting mechanics' institutes," *Social Studies of Science* **7**, 31-74.
Sheets-Pyenson, S. [1985] "Popular science periodicals in Paris and London: The emergence of a low scientific culture, 1820-1875," *Annals of Science* **42**, 549-572.
Watson, P. [2000] *A Terrible Beauty: A History of the People and Ideas that Shaped the Modern Mind* (Weidenfeld and Nicolson, London).
Wilsdon, J. & Willis R. [2004a] "Techno probe," *Guardian*, September 1.
Wilsdon, J. and Willis R. [2004b] *See-through Science: why public engagement needs to move upstream.* [Demos, London]
Wynne, B. [1992] "Public Understanding of Science research: new horizons or hall of mirrors?" *Public Understanding of Science* 1, 37-43.
Irwin, A. [2009] "Moving forwards or in circles? Science communication and scientific governance in an age of innovation," in *Investigating Science Communication in the Information Age: Implications for Public Engagement and Popular Media*, eds. Holliman, R., Whitelegg, W., Scanlon, E., Smidt, S. & Thomas, J. (Oxford University Press, Oxford).

15

Popular-Science Writings in Early Modern China

Lin Yin

Early science communication in modern China lasts from the 1840s till the beginning of the 20th century. For about 60 years, popular-science articles and books was one of the main approaches to bring scientific and technological knowledge to the public and to arm them with scientific thinking. This chapter deals with how popular-science writing has sprouted in the modern Chinese society and gives a historical introduction to the popular-science works at different stages.

15.1 Introduction

There used to be splendid cultural and scientific achievements in ancient China. Our Chinese ancestors contributed greatly to human civilization in the fields of agriculture, medical science, mathematics, astronomy, geo-science, engineering and so on. Due to the cruel reign of dictatorial system and severe restraint of feudal thinking, Chinese traditional science and its thoughts declined gradually from the middle of Ming Dynasty (1368-1644) around the beginning of the 15th century. And why modern science did not arise in China becomes quite a confusing and interesting issue for researchers in history of science and technology (S&T) [Lam, 2008, pp. 31-32]. When China was forced to face the issue of S&T later, about 300 years had passed. In fact, it was not until the late Qing Dynasty (1644-1911)—the beginning of China's modern period when the foreigners massively brought in the scientific culture[1] and Western religion to China along with "invincible fleets and guns"—that scientific thought started booming in this rich land [Wang et al, 2007, pp. 202-215].

During the process of enlightenment with S&T, popular-science (popsci) writings played an important role in science communication (scicomm). Here, popsci writings refer to the articles and books written with the aim to deliver S&T knowledge to the public at large, irrespective of the readers' educational background or knowledge of the subject. In this chapter the evolution of these popsci writings is described, along with the development of S&T in Chinese literature during the same period.[2]

15.2 Science Writings before Modern Time

Before the 1840s there were a lot of works regarding science. However, due to their profoundness in contents and the fact that the laypeople at that time were not encouraged to be literate, their circulation were rather restrained. Actually, they can hardly be seen as popsci works. But they are worth mentioning for their precious value in the history of S&T and scicomm. Basically there were two kinds of science writings in which we could find the rudiments of what we call popsci writings later. One is some Chinese classics concerning science; the other is translations of foreign science works.

---

[1] In the history of China, there have been two periods of scientific culture spreading from the West into China. The first period begins from late Ming Dynasty and ends at the beginning of Qing Dynasty when Western technology was accepted rather than the scientific thoughts. The second period spans over 50 years starting from late Qing Dynasty. In this period, the Western natural sciences and social sciences were extensively introduced and exerted great influence on the social development of China. These two periods are called by historians as "Westernization Movements". This chapter concerns mainly the second period.

[2] A brief history of popsci writing in the West is provided by Jon Turney [2013]. In particular, book-length accounts of the Victorian era and of the early 20th century Britain are given in [Lightman, 2007] and [Bowler, 2009], respectively. Popsci publishing in contemporary China is summarized in [Wu & Qiu, 2012].

15.2.1 Chinese Classics Concerning Science

Plenty of materials of great scientific value are recorded in Chinese ancient classics such as *Shanhaiching* (*Classic of Mountains and Seas*),[3] *Chuang-tse* [4] and *Mohist*.[5] For example, in the poem *Tian Wen* (*Ask the Heaven*) by Qu Yuan[6] we find the following sentences.

> At the far reach of great heaven, what settles and possesses the heaven? So much stars at corners of the sky, who knows how many are they?
>
> East, West, South and North, which is farther than others in the directions? If the South and North is the major axis of an ellipse, how much longer is it than the short axis?

These sentences involve early hypotheses about the sky and the universe, which are described so poetically by the litterateur. Other examples are *Shui Jing Zhu* (*Commentary on Waterways Classic*) and *Meng Xi Bi Tan* (*The Dream Creek Essays*). The former is written by Li Dao-Yuan in the sixth century which appears to be a comprehensive geographical treatise recording the river water system, the physical geography and economic geography of the regions along the system. The latter, in the form of written notes, is completed by Shen Kuo from the Song Dynasty (960-1279) in the 11th century. The book contains the life observations and insights of the author. It documents the outstanding contributions in S&T by the Chinese working people and the author's own research results, reflecting the brilliant achievements in natural science of ancient China, especially in the Northern Song Dynasty (960-1127). Western scholars

---

[3] *Shanhaiching* is one of the important classics in Pre-Qin period (?-221 BC), which can be regarded as the oldest geographic work with legends and tales. It covers the contents of geography, botany, zoology, religion, medicine, etc.

[4] Chuang-tse was an influential Chinese philosopher who lived around the 4th century BC during the Warring States Period, a period corresponding to the philosophical summit of Chinese thought—the Hundred Schools of Thought. He is credited with writing (in part or in whole) a work known by his name, the *Chuang-tse*.

[5] *Mohist* is the collection of speeches by Mo-tse who was a Chinese philosopher during the Hundred Schools of Thought period (early Warring States period). *Mohist* records Mo-tse's political ideology, ethics, philosophy, logical thinking and military thinking.

[6] Qu Yuan (340 BC-278 BC) is one of the earliest romantic poets in China.

regard this work as the ancient Chinese encyclopedia and there are a variety of foreign language translations. Besides, *The Travel Diary of Xu Xia-Ke* by Xu Hong-Zu[7] and *Tian Gong Kai Wu* (*Exploration of the Works of Nature*) by Song Ying-Xing[8] are also worthy of attention for their high value in both Chinese literature and scientific history. These works are simultaneously collections of beautiful prose and carriers of science.

15.2.2 Translation of Foreign Scientific Literature

At the end of the 16th century (late Ming and early Qing Dynasty), the first wave of "science communication" from the West to China began. This first movement is the beginning of introduction, communication and transplantation of Western S&T to China. It mainly took the form of translating S&T works into Chinese by Chinese intellectuals and Western missionaries. During these 200 years, Matteo Ricci,[9] Johann Adam Schall von Bell,[10] Ferdinand Verbiest,[11] and so on, not only

---

[7] Xu Hong-Zu (1587-1641) was known by the name Xu Xia-Ke. Xu traveled throughout the provinces of China, often on foot, to write his enormous geographical and topographical treatise, documenting various details of his travels, such as the locations of small gorges or mineral beds. *The Travel Diary of Xu Xia-Ke* is largely systematic, providing accurate details of measurements.

[8] The book by Song Ying-Xing (1587-1666), a scientist of Ming Dynasty, is a comprehensive scientific writing about agriculture and handicraft in ancient China. It is also known as an encyclopedic book of the Ming Dynasty.

[9] Matteo Ricci (1552-1610), an Italian Jesuitical missionary and scholar, came to live in China in the 1580s. He was the fist Western scholar who was able to read, study and learn Chinese literature. He not only devoted himself to spread Catholicism but also tried to make friends with Chinese government officials and celebrities in order to diffuse among them knowledge of astronomy, mathematics, geography, etc.. He contributed greatly to Sino-Western cultural exchange.

[10] Johann Adam Schall von Bell (1592-1666), a German Jesuit and astronomer, spent most of his life as a missionary in China and became an adviser to the Shunzhi Emperor of the Qing Dynasty.

[11] Ferdinand Verbiest (1623-1688) was a Flemish Jesuit missionary in China during the Qing Dynasty. He was an accomplished mathematician and astronomer and proved to the court of Kangxi Emperor that European astronomy was more accurate than Chinese astronomy. He then corrected the Chinese calendar and was later asked to rebuild and re-equip the Beijing Ancient Observatory, being given the role of Head of the Mathematical Board and Director of the Observatory. He worked as a diplomat and cartographer, and

preached in China, but also brought in a large amount of knowledge about Western science and civilization. Their methods of missionary work with knowledge spreading had gained the support of enlightened scholar-bureaucrat and even the emperor. Collaborating with some well-educated Chinese officials, they translated and introduced a large number of Western classics. According to statistics, during the 200 years the missionaries of the Society of Jesus in China totally translated 437 Western books, including 251 purely religious books, which is 57 % of the total; 55 humanistic science books, a 13 %; 131 natural science books, a 30 % [Qiu, 2004, p. 11]. In particular, Matteo Ricci and Xu Guang-Qi[12] translated Euclidean's *Geometry*, which not only brought China much advanced scientific knowledge and philosophical thoughts, but also introduced to China many fresh scientific words, such as point, line, plane, curve, surface, right angle, acute angle, obtuse angle, vertical, parallel, diagonal, triangle, quadrilateral, polygon, circle, center of a circle, circumference, geometry, etc. During this movement, called "West Learning Spreading to the East", a lot of books on astronomy and calendar mathematics have been introduced into China and greatly furthered development of those studies.

The translation of science literature in late Ming and early Qing Dynasty often involved many individuals and took a very long time. Quite difficult the task was, but the outcomes were not that many, because the books covered broadly in natural science and plenty of interdisciplinary professional terms which are hard to translate.

In fact, the scientific literature before modern times can hardly be regarded as popsci writings. First, they are not "popular" because they were written for the feudal scholar-bureaucrats and the literati but not for the masses. Second, apart from the translated scientific books, quite a few works are literary pieces with few scientific contents. Anyway, one thing is certain: Science writings in China's modern period began with the translation of Western scientific books and followed with original

---

also as a translator, because he spoke Latin, German, Dutch, Spanish, Hebrew and Italian. He wrote more than 30 books.

[12] Xu Guang-Qi (1562-1633), scholar in late Ming Dynasty, was good at mathematics, astronomy, engineering, agriculture and military science. He is also a famous politician, thinker and one of the communicators dedicated to Sino-Western cultural exchange.

writings modeling after the contents, structures and composition styles of these translated works.

15.3 Science Writings at the Beginning of Modern Time (1840–1860)

After the two Opium War finally ended in 1860, Western countries, with the signing of unequal treaties, gained the right to spread religion, to found schools and hospitals, and other privileges in treaty ports. The missionaries (mostly Protestants) moved their activity base from Southeast Asia to the southeast coastal cities in China. While they preached, the missionaries also passed on to the Chinese people modern Western scientific knowledge on Western medicine, machinery products and other new findings in S&T. In the very early part of modern times, the missionaries were instrumental in spreading S&T to China. The publishing houses affiliated to churches had basically occupied the field of translation and publication of the Western books for the first 20 years in late Qing Dynasty.

15.3.1 *Science Writing Activities of the Western Missionaries*

The Western missionaries did great contributions in communicating science in China by publishing magazines and compiling books. They opened windows for the Chinese to look at an unseen world that shocked their perception of the old one.

In Hong Kong the missionaries published a monthly magazine *Xia Er Guan Zhen* (English name is *Chinese Serial*) in 1853, which was the first Chinese magazine in Hong Kong after the Opium War. It gave considerable space to introduce S&T, astronomy, medical knowledge, etc. We can see this through the articles' titles such as "Brief theory of fire marine engine manufacturing", "Comet", "Brief exposition of geology", "Summary on biology", "Brief exposition of body", "Brain dominates the human body", etc. This magazine was exported to the mainland provinces, and was extremely popular among the readers from the viceroys and governors to the businessmen, literati and officialdom, and the common people [Xiong, 1994, p. 146].

In Guangzhou, the British missionary Benjamin Hobson wrote *On Astronomy* (1849), *New Thesis of Human Body* (1851) and *New Edition of Natural History* (1855). *On Astronomy* comprehensively and timely introduced Western astronomy achievements before 1840, including Copernicus and Galileo's theory. *New Thesis of Human Body* was the first modern Chinese book which systematically introduced Western anatomy. *New Edition of Natural History* involved astronomy, geography, physics, chemistry, optics, electricity, biology and other topics. The contents were quite rich, covering a lot of knowledge introduced to China for the first time ever since the "West Learning Spreading to the East" movement, such as the conservation law of matter, the law of universal gravitation, and knowledge of chemical elements. Many Chinese scientists later had read the book carefully and made experiments according to the instructions.

Shanghai was the center of communicating Western culture by the missionaries. From 1843 to 1860 Mohai Publishing House did the biggest contribution to this cause by book translation and publication and it was the very first publisher adopting letterpress printing. The business was managed by the British missionaries Walter Henry Medhurst and Alexander Wylie, and involved Chinese scholars such as Wang Tao and Li Shan-Lan who were good at mathematics, astronomy, botany and mechanics. The Chinese and foreign scholars cooperated to translate and publish a large number of scientific works, such as *Mathematics Enlightenment* (1853), *Brief Exposition of Gravity Mechanism* (1858), *Talking about the Heaven* (1859), *Botany* (1859), *On Western Medicine* (1857), *Q&A for Making a Thorough Inquiry* (1851), *Science Manual* (1856), etc. Mohai Publishing House occupied an important position in the spreading of western learning in the early modern Shanghai and even the whole country [Qiu, 2004, pp. 11-14].

15.3.2 *Representative Science Writings by Chinese Authors*

During this period the representative of science writing was Wei Yuan who was a famous thinker and historian in the late Qing Dynasty. His master work is *Hai Guo Tu Zhi* (*Illustrated Treatise on the Maritime Kingdoms*) (Fig. 15.1). Wei's book is based on the manuscript *Si Zhou*

Zhi[13] (*China Chronicle*) by Lin Ze-Xu.[14] In fact, Lin entrusted his manuscript to Wei, instructing him to gather further information on foreign countries to enrich the manuscript and complete the book. In 1843 Wei published *Hai Gu Tu Zhi* in 50 volumes based on what Lin had compiled, documents he himself had collected and his own writings. Later, the 1847 edition was expanded to 60 volumes while the 1852 edition had 100 volumes. *Hai Guo Tu Zhi* is a classic which systematically introduces the world's history, geography, politics and culture, advocating the learning of Western S&T achievements. It is the first elaborate work on world history written by a Chinese. Wei wanted to enlighten his fellow countrymen and enlarge their vision with natural science knowledge, different cultures, social institutions and Western customs. It woke up the awareness of Chinese people to take advantages of modern S&T. In 1854, the work has been reprinted in Japan which greatly influenced the Meiji politicians and thinkers.

15.4 Popular-Science Writings during the Westernization Movement (1861–1895)

After the two Opium War, the rulers of Qing Dynasty was divided into two groups, the radicals (called "Yang Wu" by historians, meaning "foreign affairs") and the conservatives ("Wan Gu", meaning "stubborn"), who held different views on how to solve a series of domestic trouble and how to deal with foreign invasion. The Yang Wu Clique advocated the use of Western advanced military and civil technologies to make the country rich, to cast off the country's predicament and to develop industry and commerce in Chinese society by means of national capitalism, for the sake of maintaining the feudal reign of Qing Dynasty. From the 1860s to 1890s, the Clique carried out a national reform called "Yang Wu Movement" (the second

---

[13] *Si Zhou Zhi* is a translated work briefly introducing the geography, history and political system of over 30 countries in Asia, Europe, Africa and America. It was translated and compiled under the guidance of Lin Ze-Xu.

[14] Lin Ze-Xu (1785-1850) is a Chinese scholar and official of Qing Dynasty, who is famous for his fight against the forced importation of opium by the Western powers.

Westernization Movement) to learn from the advanced technologies in the West in order to resist the invasion of the Western powers.

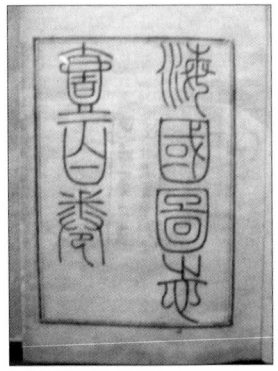

Fig. 15.1. *Hai Guo Tu Zhi* (*Illustrated Treatise on the Maritime Kingdoms*). *Left*: title page; *right*: sketches of steam ships.

In this period, a few insightful intellectuals and officials came to realize that it was science but not the advanced technology that can really save China from being oppressed by and lagging behind the Western powers. Educational organizations for teaching modern science have been gradually established, attracting students from all over the country. Science writing, as one of the powerful means of scicomm, consequently stepped into a stage of rapid growth either in terms of the scale and speed, or the depth and diversity of the content. Let us take a look at some of the typical newspapers, magazines and books to see how science writing develops from the previous stage.

15.4.1 Newspapers and Magazines

The most influential newspapers and magazine in scicomm in late Westernization Movement are *The Church News* and *Ge Zhi Hui Bian* (*The Chinese Scientific Magazine*). *The Church News* was founded in 1868, widely introducing a large number of modern S&T products, from daily necessities to the means of production. In 1874 its name changed into *The Universal Gazette,* changing its coverage from main focus on religion to more secular and scientific contents. It became a comprehensive magazine with the main contents of daily news and S&T. *Ge Zhi Hui Bian* (Fig. 15.2) was the first modern Chinese specialized scicomm magazine from 1876 to 1892, which was initiated and managed by the British missionary John Fryer. The aim of the magazine was: "On the one hand, promoting the spirit of inquiry; on the other hand, communicating popular and practical knowledge of science in the Qing Empire. It will introduce the translated scientific publications, publish short commentary of science courses and scientific presentations, and serve as intermediary for consultation and acquirement of interested science information by the local educated people..." [Wang, 1996, p. 36-47]. Its main contents can be divided into four categories: knowledge of natural sciences, technical knowledge, Chinese and foreign scientists' life stories, and answers to what the readers asked. There are two columns in each periodical, "Discourses" and "Inquiries", in which modern Western scientific knowledge were translated and compiled [Cai & Liu, 2007, pp. 59-65].

Fig. 15.2. *Ge Zhi Hui Bian* (*The Chinese Scientific Magazine*), established in 1876.

Before its publication, the translation of modern Western science books in China was mostly too recondite to be understood by people who had no specified knowledge of science. The translation of scientific articles in *Ge Zhi Hui Bian* was concise and popular that gained much welcome from the beginners. Still more, the amount of print and the breadth of distribution of the magazine were incomparable with other contemporary publications. The magazine had high prestige in the intellectual circles; for quite a long time, many people regarded it as the guidebook for science learning and it was included in the required reading list by some schools at that time.

15.4.2 Translation and Publication of Science and Technology Books

Publication of translated science books still played a very important role in scicomm. According to the statistics, during the second half of the 19th century, various types of organizations around China translated totally 500 Western books, in which the vast majority was on themes of natural sciences and its related applications, while social-science works also counted for a quite large proportion. Differing from the beginning of modern times, in this period both church publishing institutions and Chinese official publishing organizations were getting involved in this cause.

The representatives of the church publishing institutions are Yizhi Book Association, Guang Xue Society, Mei Hua Library, etc. Yizhi Book Association was founded in 1877 where textbooks were edited and published by missionaries in late Qing Dynasty. During the decades of 1870s and 1880s they had edited two sets of complete system of textbooks for the junior and the senior learners. Each set covered arithmetic, geometry, algebra, survey, natural history, astronomy, geography, chemistry and other disciplines. Until 1890 they had published more than 98 kinds of textbooks, which was 20 years earlier than the Chinese-edited new textbooks, and influenced very deeply the Chinese common school textbooks and curriculum afterwards.

Among the Chinese official publishing organizations the most important one was the Translation House affiliated to Jiangnan Manufacturing Bureau which was the most famous factory manufacturing military arms in the late Qing Dynasty. The Translation House was founded in 1868 and for about 40 years thereafter it had translated 160 kinds of books regarding applied science and engineering technology, basic theory of natural sciences, social science and so on. The translation of Western books generally took the form of cooperation between foreigners and local scholars which meant that the foreigners interpreted the texts into oral Chinese first and then the local scholars put them down into grammatically and rhetorically correct Chinese. The quality of the works by the Translation House was better than that of others and exerted great impact on the acceptance of scientific culture by Chinese intellectuals.

15.5 Popular-Science Writings from After the Westernization Movement to the Beginning of the Twentieth Century

Along with the modern science gradually being rooted in the traditional culture, there emerged a strong trend of scicomm which mainly demonstrated as the founding of scientific associations and the print and publication of the scientific newspapers as well as magazines. Just from 1895 to 1898, the reported associations in China rose to 76, which dispersed in 31 different cities of 10 provinces [Dong, 1995, p. 366]. The number had arisen to over 600 before the Qing Dynasty ended, which included political groups, social learning societies and academic associations. These associations became the important organizations for communication of new knowledge, new ideas and new values. Their activities not only aroused the enthusiasm of the mass population to learn S&T and how to use them, but also improved the position of S&T in Chinese society. In this period, more and more popsci publications came out and we see the coalition of science with literature.

15.5.1 *Springing Up of Local Newspapers and Magazines*

Along with the first climax of societies and associations, the amount of newspapers also arose. According to incomplete statistics, from 1895 to 1898, the newly published newspapers in Chinese increased to about 120 kinds, of which 80% were founded by Chinese locals [Qiu, 2004, p. 61]. They are divided into two categories: one was more professional S&T compilations, such as *Journal of Agronomy*, *Journal of Arithmetic*, *Journal of New Knowledge*, etc., and the other was those for political news, current affairs and public opinions. They dealt mainly with political and social contents, but also sometimes had the communication of scientific knowledge. In fact, most of the newspapers at that time took as their duty to disseminate new knowledge of S&T for the purpose of enlightening the wisdom and vision of the public.

For the magazines, it is easy to understand its increase if one relates this to the fast development of science associations. During the two decades from 1900 to 1919, there were more than 100 kinds of domestic magazines of S&T. Some were managed by the government while most were private, many of which were run by scientific communities. The professional scientific magazines had relatively fewer readers in comparison with the comprehensive ones; the latter played a more helpful role in spreading scientific thoughts and culture.

Some of the more influential comprehensive magazines are:

- *Yaquan Magazine,* founded in 1900 in Shanghai and edited by Du Ya-Quan,[15] aimed to "discover and record science of Physics, Mathematics, Chemistry, Agronomy, Business, and Technology". It was the earliest natural science magazine managed by Chinese, which marked the birth of modern S&T magazines in China.

- *Scientific World,* founded in 1903 in Shanghai, was hosted by Shanghai Scientific Instrument Museum. It was the earliest natural science magazine in China and was named after the Chinese word "science" was created (which was transplanted from Japan). Its purpose was to "trigger science-based industry, make our people's knowledge and skills grow day by day". The magazine also published many science history and scientists' biographical articles.

At the beginning of the 20th century the magazines spreading technical knowledge were many, especially those communicating agricultural technology. This was partly because the Chinese were deeply impressed by the power of technology when they first came to know the Western culture and because China's economy was based on its agriculture. Magazines communicating scientific theories and knowledge

---

[15] Du Ya-Quan (1873-1933) was a famous publisher and translator in popsci in modern China.

were relatively few, and they basically focused their contents on the introduction of foreign S&T achievements.

15.5.2 Two New Forms of Popular-Science Writing

Apart from the popsci articles, at the beginning of the 20th century two new forms of popsci writing came out which brought more glamour to science. The same as modern science, it was also an "imported good".

1. Science fiction

Science fiction, as the name implies, combines both scientific and literary contents. *Eighty Days around the Earth* (1900) is known as the first science fiction introduced to China so far. The author of this book is Jules Verne, the "father of science fiction". His other two fictions *Travel on the Moon* and *Underground Travel* (Fig. 15.3) were translated and introduced to china by Lu Xun[16] in 1903. Early Chinese science fictions are almost all translations of Jules Verne's works, such as *Fifteen Little Heroes* (1903), *Nameless Hero* (1905), *A Journey to the Center of the Earth* (1906) and so on. They were very welcome by the literate people; e.g., *Eighty Days around the Earth* had four Chinese versions at that time.

In 1905, the Novel Forest Press published *New Legend of Mr. Ridiculous*, authored by Xu Nian-Ci,[17] which was the birth of Chinese science-fiction writing [Ye, 2004]. It tells the story about one Mr. Ridiculous who travels to the Moon, Mercury and Venus, and finally comes back to Earth trying to save his country from invasion with his acquired "superpower". Some of the science fictions, although coming from the West, were translated into classical Chinese in the form of Chinese novel with each chapter headed by a couplet giving the gist of its content, which can be regarded as the beginning of science-fiction writing in China.

---

[16] Lu Xun (1881-1936), the pen name of Zhou Shu-Ren, was one of the major Chinese writers of the 20th century. Lu was a shor-story writer, editor, translator, critic, essayist and poet.
[17] Xu Nian-Ci (1875-1908) is known as a famous translator.

2. Fiction with illustrations

Also in those years, Novel with Illustration, a sort of novel with illustrated fine-lined portraits of main characters in serial magazine was published in Shanghai. It tried to use literary form to fight superstition and explain science. The novel *Sao Mi Zhou*, in the manner of debate between two persons, denounced superstition as "great harm that hinders China evolution" and pointed out that "reforming feudal customs must be done first to save China". The novel *Legend of Blind Swindler* exposed the experience of a blind fortuneteller trying every trick to mislead the public. There were other works like the *Word Analysis Teller*, *Fortune Teller*, etc. and they all tried to uncover the harm of superstition and blind faith which had been deeply rooted in the mind of old Chinese. The literary value of these works may not be high, but what they convey are like sharp weapons fighting against the blindness and ignorance of the readers. They exposed the hypocrisy of religious idols and put forward ideas to cast aside superstition, and thus help to change the cultural customs.

Fig. 15.3. Two Chinese books (1903) translated from Jules Verne's *From the Earth to the Moon* (left) and *A Journey to the Center of the Earth* (right).

15.6 Conclusion

In the span of more than one and a half century of early modern China, popsci works gradually become a specific writing style and category. In contrast, the writings in scientists' scientific articles or in the science textbooks in the early period cannot be classified neatly as science writing or popsci writing. The reason is that these two styles of writing are merged together due to the Chinese scientists' lack of experience in writing technical articles in the fashion of modern science.

Like modern S&T originating from the West, popsci writing either began with the translations or was modeled after Western popsci works. In this period, the science content was not popular enough and the writing styles were not rich enough to attract more readers. However, they still played an important role in enlightening people, improving their livelihood and stimulating their thoughts.

References

Bowler, P. [2009] *Science for All: The Popularization of Science in Early Twentieth-Century Britain* (University of Chicago Press, Chicago).

Cai Wen-Ting & Liu Shu-Yong [2007] "Science communication in the late Qing Dynasty: A case study on *Ge Zhi Hui Bian*," *Science Popularization* **6**(1), 59-66.

Dong Guang-Bi [1995] *History of Modern Science and Technology in China* (Hunan Education Press, Changsha).

Lam L. [2008] "Science Matters: A unified perspective," in *Science Matters: Humanities as Complex Systems*, eds. Burguete M. & Lam L. (World Scientific, Singapore) pp. 1-38.

Lightman, B. [2007] *Victorian Popularizers of Science: Designing Nature for New Audiences* (University of Chicago Press, Chicago).

Qiu Ruo-Hong [2004] *Communication and Enlightenment: Research on Scientific Culture in Modern China* (Hunan People's Publishing House, Changsha).

Turney, J. [2013] "Commentary for *Public Understanding of Science* special issue," *Public Understanding of Science* **22**(5), 570-574.

Wang Lun-Xin, Chen Hong-Jie, Tang Ying & Wang Chun-Qiu. [2007] "Popular-science writings in modern China," in *History of Science Popularization in Modern China* (Popular Science Press, Beijing).

Wang Yang-Zong [1996] "*Ge Zhi Hui Bian* and dissemination of Western knowledge of modern science and technology in the late Qing Dynasty," *China Historical Materials of Science and Technology* **17**(1) 36-47.

Wu, G. & Qiu, H. [2012] "Popular science publishing in contemporary China," *Public Understanding of Science* **22**(5), 521-529.

Xiong Yue-Zhi [1994] *Western Learning Introduced into China and the Society of Late Qing Dynasty* (Shanghai People's Press, Shanghai).

Ye Yong-Lie [2004] "Retrospect on one-hundred-years development of science fiction in China," *China Reading Weekly*, September 17.

Part V
Other Science Matters

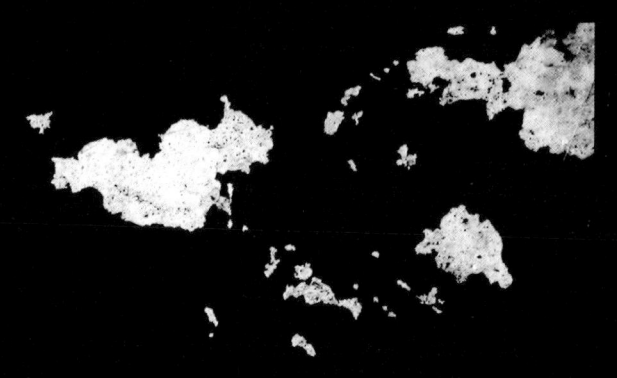

16

Understanding Art through Science: From Socrates to the Contextual Brain

Kajsa Berg

Scholars from Socrates and Leonardo to Charles Le Brun, Friedrich Wölfflin and recently David Freedberg have explored how viewers empathize with art works. Together with many other thinkers, they have suggested that empathy is best understood through studying human nature. Freedberg applies neuroscientific data on mirror neurons (brain-cells that provide the basis for empathic reactions) to examine viewer engagement with a variety of art works. His argument privileges neural processes and challenges the importance of cultural explanations of viewer engagement (for example social, political and religious). The *contextual brain* is a concept that seeks to reconcile neuroscientific arguments with context based explanations. It is based on an understanding of mirror neurons and neural plasticity (how the human brain changes as a result of experience or training) and is applied to the particular case of Caravaggio's paintings in 17th century Rome.

16.1 Introduction

Empathy and the viewers' emotional reactions were already important to Socrates and to find answers he posed a series of questions to the sculptor Cleiton [Xenophon, 1926, p. 235]. In Renaissance Italy, the viewers' emotional responses became crucial to art theorists and artists, who found inspiration in the writings of Horace, for example. This focus was intensified in the late 16th century and early 17th century. This intensification is important here as this coincides with the period when

Caravaggio painted most of his works.[1] Following this focus on emotional response, the European academicians began to classify emotional reactions in the 17th century. When the topic was again updated in the 19th century it was attached to aesthetics. With increasingly cultural explanations of art in the 20th century, the emotional responses of the viewer came to be understood wholly in cultural terms. This emphasis is now being questioned by neuroarthistorians such as David Freedberg.

The *contextual brain* is based on two strands of neuroarthistory. Freedberg, who works with the neuroscientist Vittorio Gallese, applies mirror neurons to argue that humans react emotionally and empathetically to art because of the depiction or suggestion of movement. Mirror neurons are brain cells that respond when a person makes a particular movement or facial expression, or when she sees someone else make that same movement. In particular, they are activated by goal directed movements such as tearing or grasping. Similar neurons are activated when making and seeing facial expressions, such as smiling or looking disgusted. There are also neurons that react to being touched and seeing touch, and being in pain as well as seeing someone else in pain. Many scientists are suggesting that these types of neurons are the basis for empathy [Gallese, 2001]. He is thus following on from all of the scholars since Socrates who have been examining empathic reactions to art by looking closely at human nature. Through focusing on universal emotional responses, he criticizes the predominance of cultural and cognitive explanations [Freedberg & Gallese, 2007].

John Onians, the founder of neuroarthistory, uses the data on mirror neurons to explain the emergence of representational art in the Chauvet caves [Onians, 2007b]. He also applies research on neural plasticity,

---

[1] Michelangelo Merisi da Caravaggio (1571-1610) was one of the most influential artists in 17th century Italy. He was trained in Milan but moved to Rome in the 1590s where he became known for his portraits, still life paintings and genre scenes. His first large scale religious commission (*Calling of St Matthew* and *Martyrdom of St Matthew*) for S. Luigi dei Francesi launched him as a successful painter of religious subjects. He became famous for his realist methods and his bold chiaroscuro as well as a tumultuous lifestyle. In 1606 he fled Rome after killing Ranuccio Tommasoni, travelling to Naples, Malta and Sicily before dying on his way back to Rome at Porto Ercole in 1610.

which is how the brain changes as a result of learning, training and experience [Onians, 2002; 2007a]. This is important as he focuses on explaining why art looks different at different times and different places. Neural plasticity allows the art historian to take context into consideration while still being able to utilize the many insights neuroscience has in how humans interact with visual material.

The *contextual brain* is also an extension of the art-historical concept *period eye* which was coined by Michael Baxandall to explain how differences in people's skills and experiences impact on their visual perception. For example, he argues that a 15th century, Florentine, educated viewer of Masaccio's *Expulsion from Paradise* will understand the meaning of the piece because of his understanding of gestural signs used by the Benedictine order. With this precise skill the viewer will understand the gestures of Adam covering his face as *shame* and Eve pressing her hand against her chest as *grief* [Baxandall, 1972, p. 61]. This type of explanation relies very heavily on the cultural context and the cognitive rather than the emotional abilities of the viewer. Similarly, scholars have sought contextual explanations to argue that viewers empathized with Caravaggio's paintings. For instance, Pamela Jones points out the connection between emotionally based devotional practices and the viewing experience of Caravaggio's *Madonna di Loretto* [Jones, 2008, pp. 149-158]. However, the empathic connection is not explained in any great detail.

As a concept the *contextual brain* differs from Baxandall's *period eye* in two respects. Firstly, through replacing *period* with *contextual* it acknowledges spatial differences as well as temporal ones. This is particularly important as art history as a discipline encompasses studies of art across the world. Secondly, as the *brain* is substituted for the *eye*, it is possible for art historians to understand visual perception in neural terms, and benefit from the advances made in neuroscience. The concept is building on 2000 years of explorations of human nature to understand art and the emotional responses to art. With more complex knowledge of how the human brain works, it is now possible to explain what types of changes occur in the human brain as a result of training and experience, thus deepening the understanding of cultural products.

16.2 Empathic Responses to Art from Socrates to Gombrich

The data on mirror neurons suggest that humans empathize because of the commonality between seeing movements and facial expressions and actually making those movements and facial expressions. The brain's involvement in the process is not referred to directly until the 17th century with Charles Le Brun's lecture on the matter. However, the connection between movement and expressions of the one hand and emotional engagement on the other can be traced back to Socrates.

16.2.1 Ancient Greece and Rome

Xenophon's (c. 435-354 BC) *Memorabilia* includes the possibly earliest account of emotional engagement to art. In this Socrates (469-399 BC) asks the sculptor Cleiton a number of questions and establishes that the illusion of life in art is a result of "accurately representing the different parts of the body as they are affected by the pose—the flesh wrinkled or tense, the limbs compressed or outstretched, the muscles taut or loose" [Xenophon, 1926, p. 235]. This "exact imitation of the feelings that affect bodies in action also produce[s] a sense of satisfaction in the spectator" [p. 235]. Aristotle (384-322 BC) suggested that the basis for the arts, including the visual arts, is imitation as "imitation comes naturally to human beings from childhood" [Aristotle, 1997, p. 17]. Even at this early stage here there is a notion of empathetic engagement being innate.

References to empathic engagement were also important in other contexts. Cicero (106-43 BC) notes that for effective delivery of a speech the orator necessarily had to be skilled in expressing emotion with his body, his hands and his face. Cicero argued that an orator should express the emotions he wished to arouse in the audience [Cicero, 1942, p.169]. The use of emotion and emotional expression so commonly used to stir the audience in theatre was also useful to the orator. The most widely quoted statement of empathy through emotion from ancient Rome concerns poetry; Horace states "As men's faces smile on those who smile, so they respond to those who weep. If you would have me weep, you must first feel grief yourself: then.... will your misfortunes hurt me" [Horace, 1926, p. 459].

16.2.2 The Renaissance

In the 15th century some of the most eminent thinkers in art theory extrapolated on the ideas of these writers for the practical use of the renaissance artist. Leon Battista Alberti (1404-1472) paraphrased Horace's statement about poetry in the first European art theory treatise, *De Pictura* (1435), stating that "we mourn with the mourners, laugh with those who laugh and grieve with the grief-stricken" [Alberti, 1972, p. 81]. He suggested that a good painter should paint movements and expressions in a lifelike manner to engage the spectator emotionally. Furthermore, the correct and appropriate depiction of movements and facial expressions would also impact on the viewers' ability to understand the narrative as they would internally mimic the characters.

Leonardo da Vinci (1452-1519) followed and developed Alberti's notions in his writings on anatomy and motion. He emphasized that in order to paint well, an artist has to study closely the movements of the body as these betray the motions of the mind [Leonardo, 1989, pp. 130-153]. His drawings of anatomy show the scientific aspirations behind his theories and his experience as a practicing painter makes his claims particularly persuasive. He suggested further that it is the experience, or judgment, of the artist's own body that enables him to perceive, or judge, other bodies, anticipating modern neuroscience and the data on mirror neurons.

While these two writers are very well known, the most emphatic statement of emotional engagement comes from a lesser known source. Giovanni Paolo Lomazzo (1538-1600) was an artist and art theorist working in Milan in the second half of the 16th century. His ideas were spread and discussed in the city's artistic circles. Significantly, it is likely that Caravaggio, who was trained in Milan, knew of Lomazzo's theories. Lomazzo's description of the physical and emotional engagement of the viewer goes beyond his predecessors:

Therefore, just as it naturally happens that someone who laughs or cries or makes some other expression moves others who see him to have the same emotion of happiness or of grief, as he [Horace] said "if you want to see me weep you first have to suffer pain yourself so then your misfortune harms me"; thus and not differently a picture composed with gestures taken from life as I said above without doubt will cause [the viewer] to laugh with he who laughs, to think with he who thinks, to grieve with he who cries, rejoice with he who rejoices and furthermore to marvel with he who marvels, to desire a beautiful girl for a wife when seeing a nude, to suffer with he who is afflicted and to feel hungry when he sees someone eating precious and delicate food, to fall asleep at the sight of someone sleeping sweetly, to feel moved and almost become infuriated with those who fight in a spirited way in battle represented with their own appropriate and fitting movements, to be moved with contempt and revulsion at the sight of those doing disgusting and shameful deeds and an infinite number of similar emotions.[2] [Lomazzo, (1584) 1968, p. 105]

Lomazzo proposed that looking at a correctly painted character can make the viewer feel sleepy, hungry, amorous and suffer with someone in pain (which he then substantiated by the quote from Horace). This is important as not only did he mention emotional states as bound to empathetic responses, but also purely bodily functions such as fatigue, hunger and desire. Crucially Lomazzo saw a common base for all of these types of engagement. To make the viewer engage viscerally, emotionally and morally the correct depiction of movement is essential. According to Lomazzo, sight is vital in both emotional and empathetic engagement of the human mind/brain and body.

16.2.3 Charles Le Brun

The first mention of the human brain in relation to empathic engagement comes a century after Lomazzo's statement, in 1688, when Charles Le Brun (1619-1690) gave a lecture on the depiction of expression. This

---

[2] I have received help from Matthew Sillence, John Onians, and Silvia Evangelisti in the translation of this text.

occasion was important and Le Brun's talk was subsequently published in several different editions. It was held at the Academie Royale in Paris with the intent of classifying and codifying a way to correctly depict all of the different emotions. He agreed with his predecessors that a picture cannot be perfect without expression, since the representation will not appear *real* without it [Le Brun, (1688) 1994, p. 126]. The lecture focused on detailed descriptions of how the emotions are expressed in the face, through the movement of muscles and nerves, and Le Brun used his own drawings to demonstrate his argument. Emphasizing the scientific component of his claims, he promised the audience to come back and address the value of physiognomy (unfortunately he failed to deliver). He also stated that "it is my opinion that the soul receives the impressions of the passions in the brain, and that it feels the effect of them in the heart" [p. 126]. He continues by referring to the variety of expressions and how these are felt; "JOY is an agreeable emotion of the soul which consists in the enjoyment of a good which the impressions of the brain represents as its own" [p. 127]. By referring to the brain in these terms his statement anticipates the types of claims made by neuroscientists today. Similar systems of classification continued to be published and used throughout the 18th century culminating with Charles Darwin's (1809-1882) *The Expression of the Emotions in Man and Animals* which was published in 1872 [Darwin, (1872) 1998] . However, the role of the brain in empathetic reactions is left unexplored.

16.2.4 Nineteenth-Century Empathy Theory

The 19th century saw the immense developments of Camillo Golgi (1843-1926) and Santiago Ramón y Cajal (1852-1934). Golgi developed a particular type of staining process that made it possible to see individual neurons and Cajal thereby managed to depict three layers of retinal neurons [Glickstein, 2004]. However, even though the complexity of the brain was examined in more detail than even before, it was neglected in studies of empathy.

In the 19th century there was an increasing emphasis on empathy in aesthetics. Empathy had been discussed in 1866 by Friedrich Theodor Vischer (1807-1887) who believed that humans intuitively project their emotions on the rest of the world: "Thus we say, for example, that this place, these skies and the colour of the whole, *is* cheerful, *is* melancholy, and so forth" [Vischer, (1866) 1998, pp. 687-688]. His son, Robert Vischer (1847-1933), expanded on this idea and applied it to the viewer's experience of an object in 1873. He suggested that humans empathetically transpose and transform themselves into the objects they look at, be it a *proud* fir tree, an *angry* cloud or a *prickly stubborn* cactus [Vischer, (1873) 2004, pp. 690-693]. Art historians immediately realized the importance of the Vischers' arguments. Heinrich Wölfflin (1864-1945), whose contribution is more widely recognized, was influenced by Robert Vischer in his doctoral thesis *Prologomena to a Psychology of Architecture* (1886). While Vischer seems to have been content with the aesthetic experience engaging the eyes and taking place in the human imagination, Wölfflin emphasised instead that empathy involves the whole body and while looking at classical columns it is "as if we ourselves were the supporting columns" [Wölfflin, (1886) 1998, p. 714].

16.2.5 Ekman, Gombrich and Bryson

Empathy and emotional engagement were sidelined in many of the important developments in art history in the 20th century. Culturally based explanations focusing on cognitive skill, economic factors, social aspects and political frameworks provided new answers to questions concerning viewers' engagement with objects. Treating art as a system that is similar to a language suggested a variety of different ways to explore art works' meanings. With this focus on context and meaning, applying theories based on emphasizing a universal *human nature* became problematic. Emotions came to be considered by many as culturally specific. However, human emotions and expressions were explored in great detail in other disciplines. Most notably, the psychologist Paul Ekman made huge leaps in understanding innate facial expressions, demonstrating that some facial expressions were indeed understandable across the globe. This led him to revise his expectation that emotions were context specific and conclude that some emotional states were indeed universal.

Even the perhaps most famous art historian of the 20th century, Ernst Gombrich was ambivalent about the use of biological science in discussions on art, even though he often based his approaches on psychology and collaborated with psychologists. In *Physiognomic Perception* (1963) he acknowledges that recognizing facial expressions is a natural process for human beings and argues that when looking at someone's face "we see its cheerfulness or gloom, its kindliness or harshness, without being aware of reading *signs*", thus criticizing language based theory [Gombrich, 1963, p. 47]. In 1982 he finally suggests that the 19th-century ideas about empathy may be partially right as he notes "Unless introspection deceives me, I believe that when I visit a zoo my muscular response changes as I move from the hippopotamus house to the cage of weasels" [Gombrich, 1982, p. 128]. This section follows an explanation of why humans project themselves onto animals, as happens in caricature for example, and he believes that this response is instinctual, automatic and involuntary.

These references to human biology and human experience are important as the dominant theoretical approaches for discussing art were focusing entirely on the historical context of the art production and reception [Bryson et al, 1991, p. 1]. In 1991 Norman Bryson, Michael Ann Holly and Keith Moxey suggest that approaches based on studies of perception, psychology and phenomenology cannot be useful, because they would not be adequately historically specific. Bryson particularly criticizes Gombrich's approach in *Art and Illusion* and summarizes by stating that "Perceptualism, the doctrine whose most eloquent spokesman is undoubtedly Gombrich, describes image-making entirely in terms of these secret and private events, perceptions and sensations occurring in invisible recesses of the painter's and the viewer's mind" [Bryson et al, 1991, p. 65].

16.2.6 *Baxandall and Bryson (Revised)*

Interestingly, Michael Baxandall begins his chapter on the *period eye* in *Painting and Experience in Fifteenth-Century Italy*, with several references to human biology and its importance in perception. This may seem surprising as the subtitle of his book is *A Primer in the Social*

History of Pictorial Style. After a brief introduction to the way the human eye functions, he mentions the brain as the point at which perception is no longer homogeneous [Baxandall, 1988, p. 29]. He continues by exploring what impact experience can have on differences in perception:

> In practice these differences are quite small, since most experience is common to us all: we all recognise our own species and its limbs, judge distance and elevation, infer and assess movement, and many other things. Yet in some circumstances the otherwise marginal differences between one man and another can take on a curious prominence. [Baxandall, 1988, p. 29]

Baxandall privileges these marginal differences. He identifies a few crucial elements of what he terms the cognitive style of primarily male patrons in 15th-century Italy which relates to a very particular set of skills (for example religious or mathematical) that are applied to the viewing of painting [pp. 29-108]. He argues that these skills were particularly important in 15th-century Italy because a painter's skill became more important as an economic commodity (compared, for example, to the cost of labor-time or the amount of costly paints used) [pp. 1-27].

There are two principal areas in which this approach can be updated. Firstly, the focus on the cognitive style leads to a neglect of other components of perception, such as emotional engagement. Secondly, the narrow range of skills discussed can only really be expected of a small number of educated men. Both problems become evident as he discusses the use of religious images.

Baxandall refers to original sources from the 15th century (similar statements recur into the 17th). These ask for three necessary components of religious imagery. The first is clarity; the narrative needs to be easily understood by all different types of people, including women and illiterate people. The second criterion is that the image needs to be memorable. The third is that it should be emotionally engaging, so that the viewer is more likely to follow a pious example and be put off sin [pp. 40-43]. It is clear that focusing on the cognitive skills of the educated male patron (as Baxandall indeed does) neglects a large proportion of the audience as well as the emotional experiences of these viewers.

Baxandall admits to having a narrow focus and when he focuses on gestures he chooses those that are most likely to need very particular types of skill. For example, the different narrative episodes of the Annunciation require familiarity with a variety of gestures that signify different parts of that narrative. The Annunciation was traditionally divided into five episodes (Disquiet, Reflection, Inquiry, Submission and Merit), depicted with slight gestural differences. In this particular case, only an initiated viewer would have full access to the meaning of the imagery and indeed a patron would most likely specify which part of the annunciation he would like to have depicted. However, this type of explanation and what Baxandall recognizes as the prevalent argument about religious imagery of the time do not correspond very neatly.

There is then a tension between the necessity for simplicity, in order for the illiterate to understand the imagery, and the complexity of gestural signification, required for different narrative episodes of the Annunciation. Baxandall's focus on a small number of viewers means that he does not need to address this tension. Similarly, his focus on a small set of learned skills means that he does not need to explain how these stock poses from a painter's vocabulary would be used to move the viewer. Since Baxandall is focusing on the marginal differences in 15th-century patrons' cognitive skills, these issues were not a priority. He is at the same time justified in claiming that the differences between people's experiences and skill impacts on their engagement with the work.

A statement from that advocate of textually based approaches, Norman Bryson, suggests that the key to understanding particularly emotional responses better lies in neuroscience. Acknowledging the limitations of Wittgensteinian philosophy, Deconstructionism and Psychoanalysis, because they are necessarily focused on the textual, the symbolic and thus the cognitive he writes instead that

> The radicalism of neuroscience consists in its bracketing out the signifier as the force that binds the world together: what makes an apple is not the signifier 'apple'...., but rather the simultaneous firing of axons and neurons within cellular and organic life. [Bryson, 2003, p. 14]

He further suggests that neuroscience can facilitate discussions of "feeling, emotion, intuition, sensation—the creatural life of the body and of the embodied experience" [Bryson, 2003, p. 14].

16.3 Mirror Neurons

The discovery of mirror neurons has substantiated the claims of Socrates, Leonardo, Le Brun and Wölfflin and is now being used as an integral tool in neuroarthistory. The initial findings appeared in 1988 when Giacomo Rizzolatti and his team in Parma first found motor neurons in the macaque monkey brain that respond both to making goal-oriented hand and mouth movements as well as to seeing those types of movements [Rizzolatti et al, 1988]. In 1996 "mirror neuron clusters" were found in human brains [Rizzolatti et al, 1996] and finally, in 2010, single mirror neurons were recorded in humans [Mukamel et al, 2010].

Mirror neurons in a macaque's or human being's brain respond to the movements of the individual's own body as well as to seeing the movements of an external body. Every time the individual sees an action performed, the brain responds in the same way as if that individual were in fact moving. This provides a basic link, not only between human beings but also between viewers and painted characters. Vittorio Gallese brings the data on the mirror neurons to bear on philosophical issues of consciousness and argues that beyond action recognition and learning from looking, these neurons constitute the basis for empathetic engagement [Gallese, 2001]. Similar types of neurons have now been traced in various areas of the human brain.

A team of neuroscientists demonstrated how both the actual touch of a leg (not seen by the person examined) and the seeing of someone else's leg being touched activated neurons in the secondary somatosensory cortex [Keysers et al, 2004]. Subsequent experiments showed that touch considered more widely, including the observation of inanimate objects touching and humans being touched by objects (rather than hands), had the same effect on the brain. Tactility as a phenomenon is thus treated mainly in one area of the brain that links any seen touch to the experience of touch. The researchers connected this tactility link to empathetic responses. The team calls the neuron function *touching sight* as the data shows how sight can be a vehicle for understanding touch through this empathetic link.

In 2005, Philip Jackson, Andrew Meltzoff and Jean Decety found similar types of neurons in the pain areas of the brain [Jackson et al, 2005]. As a consequence of understanding how humans react to seeing others in pain they are also able to investigate the evolutionary advantages of empathy. Wicker and others have made similar discoveries in studies on facial expressions particularly focusing on disgust [Wicker et al., 2003].

There is also evidence for changes in the types of responses to movement in the premotor cortex where the mirror neurons were first established. Calvo-Merino and his collaborators have found neural plasticity in the case of dance movements [Calvo-Merino et al, 2005]. In the experiment, ballet dancers and non-experts watched ballet. The ballet dancers' brain responded more than the others. The response was particularly strong in the premotor cortex where the mirror neurons are located. Because they were familiar with the movements through constant training, their brains could process the seen material more easily. The evidence suggests that human beings understand movement through simulation. The non-experts did not have these particular movements in their own repertoire. The plasticity of mirroring has also been suggested in autism studies where scientists have hypothesised that thinning of the grey matter (lack of neural connections), in areas that would include mirror neuron systems leads to decreased empathetic ability [Hadjikhani et al, 2006].

16.4 Neuroarthistory

These developments in neuroscience have encouraged interdisciplinary approaches to art and aesthetics. The neurobiologist Jean Pierre Changeux was the first to have referred to Rizzolatti's discoveries of "mirror neurons" in macaque monkeys and state that these are important in gesture recognition [Changeux, 1994].

One of the first art historians to connect mirror neurons to viewer experience is David Freedberg. Together with Vittorio Gallese he presents a framework for understanding embodied aesthetic responses.

They explain the bodily reactions that can arise from looking at images and use viscerally engaging pictures. Michelangelo's sculptures of *Prisoners* show figures embedded in stone and as they seemingly struggle out of the material; their straining bodies are understood and almost felt by the viewer. In Goya's illustrations in *Desastres de la Guerra* the viewer is faced with a scene of a man's body being mutilated. In Caravaggio's *Doubting Thomas* Thomas pushes a finger into the wound of Christ. Freedberg argues that viewers react to these images as a result of the mirror neurons connecting the viewer to the painted characters. He also stresses that this phenomenon is extensive and includes many different types of movement as well as a variety of emotions.

Indeed, Freedberg emphasizes emotion, as opposed to cognition, as a critical aesthetic component in the viewing of imagery. By presenting the foundations of emotional responses to art in this way, Freedberg hopes to "challenge the primacy of cognition in responses to art" [Freedberg & Gallese, 2007, p. 197]. The neuroscientific data enables him to suggest that to investigate human behavior should now necessitate an understanding of both the brain's anatomy and chemistry [Freedberg, 2007]. He further notes the embodied response that can occur when looking at, for example, a drip painting by Jackson Pollock (1912-1956) where the movement of the painter is automatically implied through the trace, the applied paint. He also gives the example of Lucio Fontana (1899-1968) [Freedberg & Gallese, 2007, p. 197]. Fontana's slit canvases presents a different sort of maker's mark and Freedberg argues that a bodily understanding of the making of that mark becomes embedded in the viewer's brain and body because of the mirror neurons.

John Onians focuses on the mirror neurons as he investigates the first marks made by humans in the Chauvet caves [Onians, 2007b]. These are placed directly above the marks left by bears clawing at the rock face. This suggests that mental and then actual imitation was the basis for these images. Onians further uses neural plasticity to account for the representation of subject matter. He demonstrates how the attentive looking at animals crossing over the river near the cave lead to humans creating some of the first representational art in Europe. Context is particularly important in his argument. A cave with bear marks, near a major crossing for animal migration combined with human hunters whose visual perception was honed to be particularly acute in looking at animals resulted in extraordinary imagery [Onians, 2011]. In suggesting that visual attention and empathic mirroring may be at the base for this specific art making behavior Onians is criticizing earlier theories that assume a cultural context of ritual use where there is very little evidence to support such claims.

Eric Kandel [2012] offers a neuroscientific angle on this debate, particularly focusing on the art of Gustav Klimt, Oskar Kokoschka and Egon Schiele in Vienna. His work is similar to that of Freedberg in that he does not explain historical specificity in neural terms. However his approach includes more detailed descriptions of the neurochemistry involved in viewer engagement, something that will enhance the understanding of empathic responses.

16.5 The Contextual Brain, Empathy and Caravaggio

While Freedberg focuses on the universal responses of empathy to Caravaggio's religious images, many art historians have suggested contextually specific explanations to such engagement. By combining the two approaches, and acknowledging that the mirror neuron connections may be strengthened by training, it may be possible to satisfy both the need to explain emotional engagement and to consider the context. Using the *contextual brain* in this way can also address the problems in Baxandall's approach. In the early 17th century, the specifications for religious art were equivalent to those in the 15th. Religious images needed to be clear (accessible to a wide range of viewers), memorable and emotionally stimulating.

One of the prerequisites of empathic engagement is the depiction of movement. This is emphasized in Freedberg's example of Caravaggio's *Doubting Thomas* and movement is an important factor in most of Caravaggio's works. Freedberg could thus be right in suggesting that empathic engagement with his work is universal. Interestingly, Caravaggio's most successful commission, *The Entombment* of Christ (1602-1604), is particularly focused on movement. It presents St John and Joseph of Arimathea lowering Christ over a stone ledge. In the cramped composition, the Virgin mimics the crucifixion with her outstretched arms just behind the main characters. Beside her the two Marys are visibly grieving; one with her arms stretched towards heaven and the other with her head bowed. The movement is emphasized by Mary's and the Virgin's outstretched arms, the lowering of the lifeless body and Christ's hand which together with the shroud hangs over the stone slab.

The context for this painting is Chiesa Nuova, an Oratorian church where the decoration was planned by Philippo Neri (1515-1595); one of the most important spiritual leaders in Rome in the late 16th century. He expressed the wish that all altarpieces in the Chiesa Nuova should be used in meditational practices and was known to spend hours in contemplation in front of the church's altarpiece by Federico Barocci [Langdon, 1999, pp. 241-245]. This instruction would still have been enforced by the time Caravaggio painted *The Entombment*.

One of the most common contextual arguments connects Caravaggo's paintings with this devotional practices [Chorpenning, 1987; Jones, 2008, p. 109]. These were most notably developed by St. Ignatius of Loyola (1491 or 1495-1556) in the *Spiritual Practices* proper. However, as Chorpenning has observed, the actual practice has its roots in medieval traditions and was commonplace in Rome. Spiritual exercises are meditational practices, in which the practitioner uses the imagination to make the Christian mysteries tangible and real. This is done by focusing on the actual space where a religious narrative takes place and by using all the senses to engage with and become part of a spiritual narrative.

The writings of St. Ignatius were a particularly successful example of a wider tradition. Two versions of St. Ignatius's *Spiritual Exercises* were approved by the pope, Paul III, in 1548, while the most common version used today was compiled in 1593 and widely disseminated in Rome by 1615. Similar practices were also encouraged in other treatises, both in Latin and Italian, an example being the *Spiritual Combat* by the Theatine Lorenzo Scupoli (1530-1610), which was published in over thirty Italian editions between 1589 and 1610. All of the orders which commissioned works from Caravaggio, including the Oratorians, not only practiced some form of the exercises but disseminated them to the public through preaching. Thus, the practice of placing oneself in the narrative of a saint's, the Virgin's or even Christ's life was popularized. Through these exercises, empathy was a skill exercised by most Romans on a regular basis.

The basic structure of St. Ignatius's exercises is fairly simple. The exercises are divided into four weeks and the practitioner is required to engage in some exercises every day. Ignatius suggests that if possible they should be adapted to the capacity of the individual, making the exercises accessible to all age-groups and classes. Ignatius particularly encouraged spreading the exercises to the illiterate [Ignatius of Loyola, 1606, p. 16]. The exercises begin with the exerciser imagining a particular setting for the meditation, such as hell, the place of the Nativity or the Crucifixion [Ignatius of Loyola, 1955, p. 66]. The second task is for the subject to ask for the appropriate emotion. Notably, the language used in these tasks is very similar to that used by the art theorists, Alberti, Leonardo and Lomazzo. In the case of hell, the exerciser is to "ask for an interior sense of the pain which the lost suffer" [p. 66] or in the case of the Resurrection the participant is to ask for "joy with Christ in His joy" [p. 54]. The senses are then activated one by one to aid the exercise and make it more real for the participant. In imagining hell, the exerciser is to see the fires and souls burning, to hear the screams and groans of those in the flames, to smell the smoke, the brimstone and the corruption, to taste the bitterness of tears and sadness and finally to feel the touch of the fire. The entire process is both highly sensual and emotive. The exerciser is further encouraged to revisit his exercises in his daily routine. For example, in week three the participant is required to imagine the Last Supper (among other scenes). He is to rethink the scene as he takes his own food: "let him do so as if he saw Christ our Lord eating with his disciples, and consider how he drinks, and looks, and speaks; and let him endeavour to imitate Him" [p. 146].

The continuous training in empathy involved in these spiritual practices necessarily brought about changes in the practitioner's neural connections. Indeed, as in Calvo-Merino's study of ballet-dancers mentioned above [Calvo-Merino et al, 2005], various neuron systems are very likely to have been involved in this process, especially since the exercises were supposed to engage the practitioner emotionally. As suggested above, mirror neurons show plasticity and the repeated practice of mentally placing oneself in someone else's shoes would necessarily have impacted on the ability to empathize. The data further suggests that the 17th century viewers in Chiesa Nuova were more enabled to engage with his images because of their continuous training in spiritual practices [Onians et al, 2012, p. 621].

Caravaggio's painting was most likely used as a tool in these types of exercises. In extension the composition of *The Entombment* gives the appearance of Christ's body is being lowered onto the altar, out into the viewer's space and forms a crucial role in the Eucharist where Christ's body is offered to the viewer in the form of the wafer. As the figure of Christ seems to be lowered into the viewers' space, it becomes comparable to the Host which would have been held up in front of the picture and then offered to the audience. Here the movement implied in the imagery makes the representation a part of the ritual. The responses of viewers in the churches of Rome were expected to contain an element of emotional engagement. Training in spiritual exercises would have made these images more accessible to contemporary viewers, who were used to imagining themselves as taking part in the religious narratives. The training would most likely have increased their capability to respond to imagery, encouraging them to engage with it emotionally and empathetically.

16.6 Conclusion

Empathy is an important form of viewer engagement with art. Understanding this emotional and physical engagement as a type of internal imitation of representations and suggestion of movement and expressions is in itself nothing new. Scholars throughout history have explored and utilized this connection in their work. Artists such as Cleiton and Leonardo copied gestures and expressions from life, in order to have an emotional impact on their viewers. Theorists like Lomazzo and Le Brun tried to further the arts by systematizing the representations of movement. Art Historians like Wölfflin and Gombrich tried to account for how viewers respond to different types of art.

The discovery of mirror neurons has enabled art historians to investigate this phenomenon in more detail. Freedberg and Onians have complementary suggestions for how this data can be used within art history. Freedberg presents a coherent, albeit general, framework for understanding embodied aesthetic responses while Onians refers to neural plasticity as a tool for understanding context specific phenomena. Kandel's main contribution lies in the specificity of neuro-chemical responses.

A neuroarthistorical approach based on the *contextual brain* necessarily considers both the context and the human brain. Freedberg's argument that it is necessary to include the brain and the emotional life of humans in order to understand viewer responses cannot be ignored. However, it is also crucial that the need for context specific explanations be met. Baxandall's *period eye* and Freedberg's argument can be reconciled. Using data on both mirror neurons and neural plasticity suggest how the human brain is flexible even though the basic structure is universal. In the specific case of Caravaggio's *Entombment* it is possible to combine data on the mirror neuron function and its plasticity with a contextual factor like the particular training supplied by the spiritual exercises to understand how people in 17th century Rome in particular would have empathized with the imagery.

When Bryson commented on the value of a neurologically based approach, he was particularly struck by the applicability of the tools provided by neuroscience. This is an important point. Neuroarthistory can be applied to areas, like emotional response, that have resisted systematic analysis by available approaches. It is indeed necessary for art historians to keep up to date with neuroscientific material, or risk making unfounded statements about human nature. Art historians need to understand how the brain works in order to make claims about features such as movement in works of art. Tempering the focus on cognition and cultural contexts with neuroscientific material does not diminish their importance. Instead such an approach offers a more detailed and flexible way of examining how humans engage with art.

References

Alberti, L. B. [1972] *On Painting and On Sculpture*, trans. Grayson, C. (ed.) (Phaidon, London).
Aristotle [1997] *Poetics*, trans. Heath, M. (Penguin, London).
Baxandall, M. [1972] *Painting and Experience in Fifteenth Century Italy*, 2nd ed. (Clarendon Press, Oxford).
Bryson, N. [2003] "Introduction: The neural interface," in *Blow-up: Photography, Cinema and the Brain*, ed. Neidich, W. (Distributed Art Publishers, New York) pp. 11-19.
Bryson, N., Holly, M. A. & Moxey, K. (eds.) [1991] *Visual Theory* (Polity, Cambridge, UK).
Calvo-Merino, B., Glaser, D. E, Grèzes, J., Passingham, R. E. & Haggard, P. [2005] "Action observation and acquired motor skills: An fMRI study with expert dancers," *Cerebral Cortex* **15**, 1243-1249.
Changeux J.-P. [1994] "Art and Neuroscience," *Leonardo* **27**(3), 189-201.
Cicero [1942] *De Oratore*, 2 Vols., trans. Sutton, E. W & Rackham, H. (The Loeb Classical Library, William Heinemann, London).
Chorpenning, J. [1987] "Another look at Caravaggio and religion," *Artibus et Historiae* **8**(16), 149-158.
Darwin, C. [(1872) 1998] *The Expression of the Emotions in Man and Animal* (HarperCollins, London).
Freedberg, D. & Gallese V. [2007] "Motion, emotion and empathy in esthetic experience," *Trends in Cognitive Sciences* **11**(5), 197-203.
Freedberg, D. [2007] "Empathy, motion and emotion," *Wie sich Gefühle Ausdruck verschaffen: Emotionen in Nahsicht*, eds. Herding K. & Krause Wahl A. (Driesen, Berlin) pp. 17-51.
Gallese, V. [2001] "The 'shared manifold' hypothesis; from mirror neurons to empathy," *Journal of Consciousness Studies* **8**(5-7), 33-50.
Glickstein, M. [2004] "Vision structure and function: The early history," in *Visual Neurosciences*, Vol. 1, eds. Chalupa L. & Werner J. (MIT Press, Cambridge, MA) pp. 3-13.
Gombrich, E. [1963] Meditations on a Hobby Horse and other Essays on the Theory of Art (Phaidon, London).
Gombrich, E. [1982] The Image and the Eye; Further Studies in the Psychology of Pictorial Representation (Phaidon, Oxford).
Hadjikhani, N., Joseph, R. R., Snyder, J. & Tager-Flusberg, H. [2006] "Anatomical differences in the mirror neuron system and social cognition network of autism," *Cerebral Cortex* **16**, 1276-1282.
Horace [1926] *Ars Poetica*, trans. Fairclough, H. R. (The Loeb Classical Library, William Heinemann, London).
Ignatius of Loyola [1955] *The Spiritual Exercises*, 5th ed., trans. Longbridge, H. W. (Mowbray, London).

Jackson P., Meltzoff, A. & Decety, J. [2005] "How do we perceive the pain of others? A window into the neural processes involved in empathy," *Neuroimage* **24**(3), 771-779.
Jones, P. [2008] *Altarpieces and Their Viewers in the Churches of Rome from Caravaggio and Guido Reni* (Ashgate, London).
Kandel, E. R. [2012] *The Age of Insight* (Random House, New York).
Keysers, C., Wicker, B., Gazzola, V., Anton, J. L., Fogassi, L., Gallese, V. [2004] "A touching sight: SII/PV activation during the observation and experience of touch," *Neuron* **42**, 335-346.
Langdon, H. [1999] *Caravaggio: A Life* (Pimlico, London).
Le Brun, C. [(1688) 1994] "Lecture on expression," in *The Expression of the Passions: The Origin and Influence of Charles Le Brun's Conférence sur l'Expression Générale et Particulière*, ed. Montague, J. (Yale University Press, New Haven) pp. 126-140.
Leonardo Da Vinci [1989] *On Painting: An Anthology of Writing*, trans. Kemp, M. (ed.) & Walker, M., (Yale University Press, New Haven).
Lomazzo, G. P. [(1584) 1968] *Trattato dell' Arte de la Pittura* (Georg Olms Verlagsbuchhandlung, Hildesheim).
Mukamel, R., Ekstrom, A.D., Kaplan, J., Iacoboni, M. & Fried, I. [2010] "Single-neuron responses in humans during execution and observation of actions," *Current Biology* **20**, 750-756.
Onians, J. [2002] "The Greek temple and the Greek brain," in *Body and Building: Essays on the Changing Relation of Body and Architecture*, eds. Dodds, G. & Travernor, R. (MIT Press, Cambridge, MA) pp. 44-63.
Onians, J. [2007a] *Neuroarthistory. From Aristotle and Pliny to Baxandall and Zeki* (Yale University Press, New Haven).
Onians, J. [2007b] "Neuroarchaeology and the origins of representation in the Grotte de Chauvet," in *Image and Imagination: A Global History of Figurative Representation*, eds. Renfrew, C. & Morley, I. (McDonald Institute for Archaeological Research, Cambridge, MA) pp. 307-320.
Onians, J. [2011] "Neuroarthistory: Reuniting ancient traditions in a new scientific approach in the understanding of art," in *Arts: A Science Matter*, eds. Burguete, M. & Lam, L. (World Scientific, Singapore) pp. 78-98.
Onians, J., Anderson, H. & Berg, K. [2012] "Neuroscience and the nature of visual culture," in *The Handbook of Visual Culture*, eds. Heywood, I. & Sandywell, B. (Berg, London) pp. 606-627.
Rizzolatti, G., Camarda, R., Fogassi, L., Gentilucci, M., Luppino, G. & Matelli, M. [1988] "Functional organization of inferior area 6 in the macaque monkey: II area F5 and the control of distal movements," *Experimental Brain Research* **71**(3), 491-507.
Rizzolatti, G., Fadiga, L., Gallese, V. & Fogassi, L. [1996] "Premotor cortex and the recognition of motor actions," *Cognitive Brain Research* **3**, 131-141.

Vischer, F. T. [(1866) 1998] "Critique of my Aesthetics," in *Art in Theory: 1815-1900*, eds. Harrison, C., Wood, P. & Gaiger, J. (Blackwell, Oxford) pp. 687-688.
Vischer, R. [(1873) 1998] "The aesthetic act and pure form," in *Art in Theory: 1815-1900*, eds. Harrison, C., Wood, P. & Gaiger, J. (Blackwell, Oxford) pp. 690-693.
Wicker, B., Keysers, C., Plailly, J., Royet, J. P., Gallese, V. & Rizzolatti, G. [2003] "Both of us disgusted in *my* insula, the common neural basis of seeing and feeling disgust," *Neuron* **40**(3), 655-664.
Wölfflin, H. [(1886) 1998] "Prologomena to a psychology of architecture," in *Art in Theory: 1815-1900*, eds. Harrison, C., Wood, P. & Gaiger, J. (Blackwell, Oxford) pp. 711-717.
Xenophon [1926] *Memorabilia*, Book 3, trans. Marchant, E. C. (The Loeb Classical Library, William Heinemann, London).

17

Spy Video Games after 9/11: Narrative and Pleasure

Ting-Ting Wang

Role-playing games (RPG) is the most important type of video games, for both PC and online games. In the last decade or so, video games with a spy theme were pretty common and have become an important genre in games. This chapter discusses, from the narrative and cultural perspectives, the identity and pleasure of the RPG player, and analyses the text construction in the multinarration of spy games. The discussion is divided into three parts: (1) How does a spy game tell a story? We will do a narrative analyze and find out how the narrative scheme causes pleasure in the player. (2) What stories are told by spy games? Putting these stories among the various post-9/11 texts, we can see what kind of "tacit writing" is provided by the spy games. (3) Starting from these game texts we discuss what special role is played by spy games in the video game industry, in the global political and economic system, and what effects result from it.

17.1 Introduction

Video games as a new type of cultural industry are attracting a lot of attention and are becoming a multibillion dollars enterprise [Donovan, 2010]. Role-playing games (RPG) in which the player can interact and change the development of the game, is the most important type of video games, for both PC (personal computer) and online games. Starting from the simple word games, the development of RPG has gone through many generations of changes. By including more and more elements into the game, RPG always enjoy a large number of players and is undeniably the king of video games [Barton, 2008]. (Games or video games in this chapter refer to RPG.) On the one hand, differing from a novel's reader and a movie's audience, the game's player is simultaneously the narrative's reader and object. This provides a new problem in the theory of narratives in literature. On the other hand, as a new scene in the "spectacles society", the production and consumer mechanisms of video games are a worthy topic for investigation.

In the last decade or so, video games with a spy theme were pretty common and have become an important genre in games. This includes the PC games (e.g., *Splinter Cell*, *Death to Spies* and *Alias*), online games (*The Agency*) and games derived from movies (e.g., *The Bourne Conspiracy*, *Spyhunter* and *Metal Gear Solid*). We could say that these spy games became mature and popular after the 9/11 event of 2001, which thus must have some intertextuality with other post-9/11 texts.

This chapter discusses, from the narrative and cultural standpoint, the identity and pleasure of the video game player, and analyses the text construction of the multi-narration of spy games. The discussion is divided into three parts: (1) How does a spy game tell a story? We will do a narrative analyze and find out how the narrative scheme causes pleasure in the player. (2) What stories are told by spy games? Putting these stories among (along with) the various post-9/11 texts in literature, we can see what kind of "tacit writing" is provided by the spy games. (3) Starting from these game texts we discuss what special role is played by spy games in the video game industry, in the global political and economic system, and what effects result from it.

17.2 Narrative and Pleasure

Alfred Hitchcock (1899-1980), in his conversation with François Truffaut (1932-1984), differentiates "surprise" from "suspense". If the audience sees a person sitting in front of a desk and then there is an explosion, then this is a surprise and not suspense. But, if the audience sees someone puts a bomb under the desk before this person comes in, then the audience will worry about this person's fate and we have suspense. Here Hitchcock linked up the question of focus with the question of pleasure: When the narrator knows more than the person in the film does, there will be suspense. (Of course, this is only one particular example; not all narratives are like this.) And this is exactly the methodology we will follow in the discussion of this section, viz., we will try to establish the relationship between the narrative in spy games and the experience received by the player.

As the identity of the movie audience was constructed through a century of movie history,[1] the identity of video game players does not exist a priorily. If Hollywood movies induce pleasure by letting the audience think what they see is real then the video game player's pleasure comes from the illusion that she herself is the narrator. In fact, the player is unlike an audience in a movie; she, through choices in game playing to accomplish various missions, thinks she is the narrator of the story. And so the pleasure provided by RPG to the player is to let her being a narrator even though she is really just an audience and not the real narrator of the game's story.

The player constructed by RPG is a subject with a double identity: She is the audience and at the same time, a participant. While she reads the game's text and edits the codes (*signs*), she also acts in the game and interacts with the text. It is important to point out that as a player acts out the character, she is *not* the character herself. Thus, in a game, we are talking about the interacting relationships between the three roles of a player, viz., character/participant/audience. The challenge to narrative theory presented by this new form of a game's narrative is exactly this: In the text of a novel or a movie, there is no participant in the character/reader or the character/audience relationship, respectively.

Among the player's three roles, only the character/audience relationship could be said to be the same as that in a movie. Viewing a "passing-through video"[2] is such a case. The experience a person receives by viewing a passing-through video is the same as that by

---

[1] Movies construct two audiences: one is the audience sitting in the movie house; another is the person who holds the camera in making the movie. But a Hollywood drama movie through its narrative form removes the existence of the camera, and let the person sitting in the cinema think that what she sees is real and has nothing to do with what the camera man wants the audience to see. Yet, avant garde movies show the existence of the camera man and let the audience know what they see is not real but is only a narrative. However, Hollywood movies can bring more pleasure to the audience than avande garde movies do because the former looks more real.

[2] A video made by a successful game player recording every step he went through in the game, which is shared online.

viewing a movie: The character in a game is controlled completely by someone else, which has nothing to do with the viewer.

In the character/participant relationship, what the player experiences is pleasure coming from action. Except for possessing a basic identity, the character in a game is totally empty, without thoughts and without personality. Only with the help of the participant that the character can act and finish an assignment. Similarly, a game's text is in fact constituted by a narrative with limited scope; the character's behavior and performance decide which narrative the text ends up with. Thus, the character and narrative are both in a growing state; the narrative can only become complete through the effort of the participant. Consequently, aiming for a perfect narrative motivates the player's action. And, in a game, every action basically will lead to certain consequences which will help to attain the perfect narrative and hence, the motivation of the player will never be exhausted. Then, the questions are: Is this a form of writing? Is the participant also a narrator? My answers are "Yes". This point is more pronounced in online games. Online games are more open than PC games because the text of a online game is produced by all the players.

Exactly because of this, online games are replacing the PC games since the player can interact not only with the game system but also with other players. Moreover, a perfect narrative can never be attained and thus the game will never end.

Participant/audience is the player's double roles. The relationship between these two roles is more like that of an object and its mirror image. The participant, the mirror image, is the idealized image of the audience's self towards which the audience projects her desires. What the player in the audience role sees in the player in the participant role is a perfect self.

Spy games were developed from the "first-person shooter" (FPS) games but were different from the pure FPS games represented by *Counter-Strike* (USA, four parts, 2001, 2004, 2004, 2012). In a FPS game the screen shows what the player—the first person shooter—sees, and the aim is to eliminate the enemies on the screen while the player's image does not show up on screen. In contrast, the aim in spy games is no longer eliminating the enemy but "penetration" or "infiltration", i.e., avoiding the enemy's sighting, passing through a few scenes, and finally reaching a destination. This change is to coordinate with the spy story's plot but also to answer the society's continuous condemnation of bloody violence permeating video games. Since the game's aim is penetration, the player's action is no longer shooting down the enemy. More precisely, there are two methods the participant uses to avoid the enemy: The player can choose either to eliminate the enemies (such as guards) or slip through them (see more details below). If the player chooses elimination the game will give lower point to the player because the game encourages penetration but not killing. Sometimes, selective killing—a relatively "benign" approach reflecting America's national value—is allowed. An example is *Medal of Honor: Rising Sun* (USA, 2003) which shows that the main tactic in antiterrorism is selective killing (see Section 17.3).

The two methods of *avoiding* the enemy are:

1. In spy games, there are always shadows on the screen. One way for the spy (the player) to avoid being detected by the enemy is to hide in the shadows. Out of the shadow means the possibility of being detected by the enemy and the spy puts himself in great danger and could lose his life. This is a metaphor that shadow represents safety and brightness represents danger. Light and shadow is an important topic in Western philosophy; light represents freedom and rationality while darkness represents ignorance and bondage. In movie language, light is usually used to convey a safe situation and a happy mood while shadow implies danger and anxiety.[3] But in spy games, the opposite is true. In these games, the spy going through light and shadow hints at his anxiety over identity.

2. The other method of avoiding the enemy is more indirect and requires the participant to think more. For example, in *Death to Spies* (USA, 2007) the participant could knock out an enemy and the character puts on his dress. In so doing, even bumping into an

---

[3] However, Michel Foucault (1926-1984) reversely rewrites this proposition. He sees gaze as a kind of monitoring, a form of power. Light, gaze and body are simultaneously subjected to an exercise of power in the open. In video games, if the character is not sighted he is free to move around. This set up has a more postmodern meaning than what is available in pure FPS games.

enemy face to face, the character would not be recognized. In addition, if the participant kills an enemy she must hide the body; otherwise, the patrolling enemy would sound an alarm if the body discovers. Moreover, before entering a room the character first uses a mirror, taken from the car before she leaves the car, to view inside the room. These tricks are borrowed from various spy movies and TV series. The player seems to be shuttling between two intertwined texts constructed from a spy movie and a game, respectively. The two texts interact and reinforce each other, making the spy world more realistic to the player.

17.3 Narratives in Post-9/11 Spy Games

Following the dissolve of USSR and the unification of Germany, the post-Cold War period unfolded in the early 1990s. The polarity of the two opposing camps of socialism and capitalism weakened over time. Politics, economics and culture began their globalization. But at the same time the workings of ideology became more complicated because the dualistic structure was no longer effective. It can be said that the 9/11 was one of the most important events that occurred during the post-Cold War period. After 9/11, terrorists replaced USSR as America's most important, helped by mass culture that followed suit. While the United States began her determined, global antiterrorism operation, most of other countries either supported or joined this operation. A new set of dualistic relationships seems to emerge: terrorists/international society gradually replaced the socialism/capitalism and became a new form of ideology. In mass culture, terrorists were being constructed into a new enemy. Iraq war scenes appeared more or less in Hollywood movies such as *Transformer* and *Iron Man*. At the same time, popular games like *Close Combat*, *Special Forces* and *Full Spectrum Warrior* use Middle-East war zones, partially or wholly, as their backdrop (background). These games create an antiterrorist soldier—a courageous and effective spy in fact. His mission is to liberate Arabian cities one after one from the hands of the terrorists and transform them into democratic cities.

Yet, as Marc Quellette points out in his article "I hope you never see another day like this: Pedagogy and allegory in 'post 9/11' video games", these texts often involve a double-narrative: On the one hand it creates such an antiterrorist warrior; on the other hand the soldier is being questioned. In Quellette's opinion, *Medal of Honor* is such a double sword. In this game, called a 9/11 fable by some, the player in his role as an American soldier defeats the Japanese through his efforts and wins the Pearl Harbor war. However, the player can choose how to win the war. He can achieve it by killing innocent people or by using some strategies. In this way the game, while positively endorsing the correctness of antiterrorist wars, hints at the existence of multiple routes in carrying out the fight such that violence is not the only possible solution.

Here, we discuss the multi-narrative in spy games. Several popular games are chosen as examples in our text analysis; they are *Splinter Cell* (France, five parts, 2004, 2005, 2007, 2008 and 2010), *Alias* (USA, 2004) and *Metal Gear Solid* (Japan, five parts, 1998, 2001, 2003, 2008 and 2010).

Sun Bai points out that a kind of identity anxiety shows up in spy novels before the Cold War.[4] The character tries to search and confirm her identity. But in the Cold War period, this identity crisis no longer exists. Yet after the Cold War, this crisis resurfaces. Sun thinks that the Cold War's either-we-or-them logic provides people with a valid and clear identity, which vanished after the war and thus the reemergence of the issue of identity crisis. Accordingly, when a new kind of duality—international society/terrorists—appeared after 9/11, any identity anxiety should fade away gradually since any character with such an anxiety can be constructed as the main image of an antiterrorist. However, at least in some spy games, this is not the case. In fact, as described above, such an image does appear in many spy games. But as soon as such an identity is being constructed, the identity is being questioned and deconstructed, until its final and complete destruction. Thus, the new antiterrorist ideology does not solve the identity crisis problem, and so the spies in post-9/11 spy games all live with an identity crisis.

The main characters, Sam, Snake and Sydney in the three games mentioned above respectively, are all spies employed by the US government. But after witnessing the organization's dark side and going

---

[4] Private communication, Peking University, 2010.

through a series of betrayal, treachery and parting/death, they all choose to quit their spy job.

- In *Alias*, Sydney, the major character, a CIA's SD-6 spy, is proud of her job until one day, her fiancé got killed and she finds out Arvin Sloane, the SD-6 boss, has turned the unit into a terrorist organization and kills her fiancé. She asks for help at the real CIA and becomes a double agent working for CIA. Finally, after the problem is solved, she resigns from CIA and marries her new boyfriend and raise children. The reason: She is disillusioned by the betrayal of Sloane that she trusted very much before.

- In *Splinter Cell*, Sam is originally a veteran spy of National Security Agency (NSA). His boss and close friend Lambert, in order to prevent him from being threatened by the enemy, lied to Sam that his daughter has died in a traffic accident. Sam's NSA colleague Grim rescues his daughter and uses her as a threaten to force Sam to work for his project. Even though Grim is presented as a positive character in the drama, the tactic he uses is no different from that of the terrorists. In the end, Sam finally declines Grim's invitation to stay in NSA and leaves with his daughter to lead a peaceful life.

- In *Metal Gear Solid*, Snake is a member of the special-force's antiterrorist team Foxhound but is cheated by his boss. In the future epoch in the Game's Part 4, war is no longer fought between countries but big corporations. He thus builds his own military base MSF, which does not belong to any country or military power. The base is completely independent and its aim is to do whatever anyone pays him to do. In other words, the base is a terrorist camp. The player's job in the game is to run the camp.

The character's despair (after finding out that the government organization she belongs to can no longer be trusted) is the basic reason that these three spies all quit their job at the end. Even though the individual should make sacrifices for the country's benefit according to the ethics of America's middle class, not to mention the ethics of government agents whose job is to protect the country, does this mean that the country can use this as an excuse and hurts at will the individual's feeling? This questioning is without doubt humanistic in nature. But America's condemnation of antiterrorists is also based mostly on humanistic considerations. For example, they condemn the killing of innocent citizens. The humanism concept that an individual's value or rights has priority over the overall aims of a nation of a bigger whole, constructed during the Renaissance, seems to have a kind of transcendental value. Any action—coming from either side of a dualistic conflict—that contradicts it is unacceptable. When this conviction in humanism is lost, even an antiterrorist warrior can no longer continue to carry out his mission. Of course, the games also show clearly that, after all, there is a difference between the American government and terrorist organization. The latter kills innocent people while the former commits only deceit, threatening and corruption. Yet, Snake's final action of establishing the base hints that the transition from an antiterrorist to a terrorist is so entirely possible and not unreasonable; the difference between these two personas is much smaller than we think. Hence, a new enemy, the terrorists, established after 9/11 is not that effective in eliminating the dualistic ideology and thus the identity crisis. Here, we return to what is mentioned at the beginning of this Section, viz., a spy game's text is a multi-narrative which does not simply try to construct a new dualism but a new "other" image.

17.4 Games in Reality and Reality in Games

In 2009, two big events aiming at online games happened in China:

1. The *Warcraft* (USA, 1994) game with five millions players was off for 52 days long. In April, Netease replaced Blizzard-Nine Cities as mainland China's agent for the *Warcraft* game. Immediately after that, the China's State Administration of Press, Publication, Radio, Film and Television declared that *Warcraft* was an electronic "publication" and needed the government's approval. Thus, Netease has to turn off temporarily the game's server on June 7 and suffered a daily loss of seven-million Yuan. Interestingly, a game that has existed for four years suddenly needed to be officially approved by

the government and this happened right after the game's agent was replaced.

2. Yang Yong-Xin, an asylum doctor from Linyi, Shandong Province, used electroshock therapy to treat web-addition (which is mostly online-game addiction) patients. This became a widespread controversy and was reported by several mass media including the CCTV.

Based on these two events the netizen Kannimei produced a "video movie"[5] titled *Web Addiction War*. The genius of this movie lies in its description of a game player trapped in a complex political-economical-ethical net. The movie was called "The Uncrowned King of 2009 Chinese Movies" by Tianya BBS' netizens. In this movie, the character Kannimei (same name as the producer's), an ordinary game player, encounters the *Warcraft* game's stoppage. During this period he is kidnapped by Professor Yang to go through the electroshock therapy. In Yang's prison, he meets a female player of the online game. The two fall in love and escape from the prison together.

Video games as a cultural industry are always immersed in a discourse net formed from politics, economics and ethics. And the game player will be presented by various texts in such a net. On the one hand, video games are often blamed drugs full of bloody violence that poison teenagers; the game player is thus presented as a morally negative figure. On the other hand, in the players' own presentation, they are a group of figures who mutually support and love each other and fight together, in the games and in reality. In their view, the game is just a place; how they are presented or represented are solely up to the players themselves.

According to Johan Höglund [2008], since the first FPS game was born, this kind of games has always been linked intensely to the real world's military complex. The concepts of "military-industrial-media-entertainment network" [Der Derian, 2001] and "military-entertainment complex" [Lenoir & Lowood, 2013] mean the same thing. That is, the military industry, private enterprise and mass culture are merging

---

[5] A "video movie" is made completely by editing segments taken from one or more video games.

together, and the American government can do nothing about it. In 1999, the American military recruits reached its historical low in number. To free her from such a despair situation, the military started to invest in video games and used the games as a means to attract new recruits. In 2002 the American army developed the FPS game *America's Army* and provided free download for the public. Simultaneously, the American army invested in private game companies to develop software that helps in soldier training. *Full Spectrum Warrior* (2004) and *DARWARS Ambush* (2008) are two such examples. These games simulate realistically military training and war zones, most of which use Iraq as the background. In short, spy games after 9/11 are influenced by antiterrorist wars, vice versa. The merging of military industry, private enterprise and mass culture is exemplified in spy games.

Talmadge Wright et al [2002] point out that the player in an online game not just copies societal rules but also challenges them. As we analyzed above, in comparison to PC games, online games are a more open system. The rules obeyed by the player are often societal rules, not the game's rules, e.g., when she is trying to raise her rank by practicing her skills with senior players. And there are, of course, those who challenge societal rules, such as the PK devotees. Games are presented in society, and society is presented in games. Thus, the opposition of games and society are broken. Games are no longer games, and reality is no longer reality. Both are just a system of signs.

17.5 Conclusion

Spy games are an important genre of role-playing games which, in turn, are the most important type of video games. The RPG player's pleasure comes from the illusion that she herself is the narrator. Since a perfect narrative can never be attained, the game will never end. And, uniquely in RPG, there are interacting relationships between the three roles of a player, viz., character/participant/audience.

The 9/11 event of 2001 was one of the most important events that occurred during the post-Cold War period. The post-9/11 spy games involve a double-narrative: The identity of the antiterrorist soldier is being questioned at the same time when it is being constructed and so the spies in these games all live with an identity crisis. In other words, a new enemy, the terrorists, established after 9/11 is not that effective in eliminating the dualistic ideology and thus the identity crisis. A spy game's text is a multi-narrative which does not simply try to construct a new dualism but a new "other" image.

Using the events surrounding the popular game *Warcraft* in 2009 China as an example, we show that there is reality in games and there are games in reality. Video games as a cultural industry are always immersed in a discourse net formed from politics, economics and ethics. And the game player will be presented by various texts in such a net. However, the player in an online game not just copies societal rules but also challenges them. The opposition of games and society are broken. Games and reality are just a system of signs.

References

Barton, M. [2008] *Dungeons and Desktops: The History of Computer Role-Playing Games* (A. K. Peters, Wellesley, MA).

Der Derian, J. [2001] *Virtuous War: Mapping the Military-Industrial-Media-Entertainment Network* (Westview Press, Boulder, CO).

Donovan, T. [2010] *Replay: The History of Video Games* (Yellow Ant, East Sussex, UK).

Höglund, J. [2008] "Electronic empire: Orientalism revisited in the military shooter," *Game Studies* **8**(1), September.

Lenoir, T. & Lowood, H. [2013] "Theaters of war: The military-entertainment complex" (www.stanford.edu/class/sts145/Library/Lenoir-Lowood_Theater Of War.pdf, February 1, 2013).

Wright, T., Boria, E. & Breidenbach, P. [2002] "Creative player actions in FPS online video games," *Game Studies* **2**(2), December.

18

Statistical Physics for Humanists: A Tutorial

Dietrich Stauffer

The image of physics is connected with simple "mechanical" deterministic events: an apple always falls down; force equals mass times acceleration. Indeed, applications of such concept to social or historical problems go back two centuries (population growth and stabilization, by Malthus and by Verhulst). However, since even today's computers cannot follow the motion of all air molecules within one cubic centimeter, the probabilistic approach has become fashionable since Ludwig Boltzmann invented Statistical Physics in the 19th century. Computer simulations in Statistical Physics deal with single particles, a method called agent-based modeling in fields which adopted it later. Particularly simple are binary models where each particle has only two choices, called spin up and spin down by physicists, bit zero and bit one by computer scientists, and voters for the Republicans or for the Democrats in American politics (where one human is simulated as one particle). Neighboring particles may influence each other, and the Ising model of 1925 is the best-studied example of such models. This chapter will explain to the reader how to program the Ising model on a square lattice (in Fortran language); starting from there the readers can build their own computer programs. Some applications of Statistical Physics outside the natural sciences will be listed.

18.1 Introduction

Learning by Doing is the intention of this tutorial: Readers should learn how to construct their own models and to program them, not learn about the great works of the author [Stauffer et al, 2006] and the lesser works of his competitors [Billari et al, 2006].

Already Empedokles is reported to have 25 centuries ago compared humans to fluids: Some are easy to mix, like wine and water; and some, like oil and water, refuse to mix. Newspapers can give you recent human examples, and physicists and others have taken up the challenge to study selected problems of history and other humanities with methods similar to physics [Castellano et al, 2009]. In the opposite direction, Prados [2009] uses the physics dream of "unified field theory" to describe his history of the Vietnam War. Biological applications of statistical physics are more widespread [de Vladar & Barton, 2011]. The use of "differential equations" is recently reviewed by Vitanov and Ausloos [2012].

The next section will shortly deal with power-law plots of Zipf type and with random walks; the third one recommends ways on how to construct models, followed by one on how to program a simple Ising model on a square lattice. And a concluding section will list some applications. An appendix will introduce the Fortran programming language and another one gives some formulas for the Boltzmann distribution and Ising model. Finally, the third appendix will list some history titles that the author, and hopefully the reader, finds interesting.

18.2 Zipf Plots and Random Walks

Here we explain one standard analysis tool and one way to introduce randomness.

18.2.1 *Zipf Plots*

If one counts how often a word appears in written language, a power law was found by Zipf. Other power laws appear for the wealth or income of people, the volume of cubes and spheres, the area of circles and squares, and more generally of "fractals." Let us assume the most common word appears n_1 times, the second-most common word n_2 times, and more generally, the rth most common word n_r times. Then the frequency n_r as a function of the rank r may, at least for intermediate r, obey a power law proportional to $1/r^a$. Since the variable is the rank r and not the exponent a, this law is not an exponential law of Boltzmann type. To check if and for what region of r a power law is approximately valid, one may plot n_r versus r double-logarithmically, that means the logarithm of n_r is plotted versus the logarithm of r. On such logarithmic scales, the numbers 1, 10, 100, 1000, etc. are equidistant, just like 0, 1, 2, 3, etc. are equidistant on linear scales. Since $1000 = 10^3$ ($= 10 \times 10 \times 10$), we call 3 the decadic logarithm of 1000. Graphics software of computers (I prefer simple gnuplot) allows you to chose logarithmic axes and lets the computer do the needed calculations. If the data decay on a double-logarithmic plot with a straight line, the negative slope of that line is the exponent a of the power law $n_r = \text{const}/r^a$.

Usually at the end of such a plot, like for large r and small n_r, the data scatter a lot. Then binning may be useful; you collect all ranks between 1 and 10 in one bin, between 10 and 100 in the next bin, between 100 and 1000 in the third bin, etc. If the range of r is not very large, it may be better to bin in powers of two instead of powers of 10: 2 and 3, from 4 to 7, from 8 to 15, from 16 to 31, etc. Plotting the sums of n_r within one bin double-logarithmically against the average r of the bin, one may get a better plot since now the slope is $1 - a$ and thus larger than $-a$ for the un-binned data.

18.2.2 Random Walks

Throwing coins is a standard method to produce randomness. On a square lattice, one may start in the middle, and then move to one of the four nearest neighbors, selected randomly. This random choice is repeated again and again until the lattice boundary is reached. For example, with one coin throw one determines whether to move horizontally or vertically, and then with another coin throw one selects in which of the two remaining directions one moves. This is called a random walk. On an infinite lattice the average of the squared distance traveled since the origin is proportional to the number of coin throws. Appendix 18.1 contains, after the Fortran manual, a short computer program. Applications include the spread of innovations in the Stone Age by random walks from one settlement to a neighboring one [Sumour et al, 2011b].

18.3 Model Building

Here we give some general advice about how to define a model which then is investigated.

18.3.1 *What Is a Model?*

"Models" in physics and in this tutorial usually deal with the single elements of a system and how their interactions produce the behavior of the whole system. Outside of physics, a model may just be any mathematical law approximating reality. Thus the statement that human adult mortality increases exponentially with age is often called the Gompertz model by demographers but the Gompertz rule by physicists; the latter ones use the Penna model of individuals undergoing genetic mutations and Darwinian selection to simulate a large population perhaps obeying Gompertz [Stauffer et al, 2006].

18.3.2 *Binary versus More Complicated Models*

If no previous work on a general model is known to you, I recommend to start with binary variables where each particle has only two choices, called spin up and spin down by physicists, bit zero and bit one by computer scientists, occupied or empty for percolation, and voters for the Republicans or for the Democrats in American politics (where one human is simulated as one particle). Of course reality is more complicated, but we want to understand reality: Does it agree with the simplest possible model? (Simulations for pilot training, etc. are different [Bridson & Batty, 2010].) Thus we follow the opinion of Albert Einstein that a model should be as simple as possible, but not simpler. If you want to simulate traffic jams [Chowdhury et al, 2000] in cities, the color of the cars is quite irrelevant, but for visibility in the dark the color matters.

Once the binary case has been studied, one can go to more than two choices for each particle. If three groups are fighting each other, obviously three choices are needed in a model [Lim et al, 2007]. Even in a two-party political system like the USA, other candidates were important in Florida for the US presidential elections of 2000. The opinions of people [Malarz et al, 2011] are in reality continuous and can be modeled by one or several real numbers between zero and unity, or between minus infinity and plus infinity. Nevertheless it is standard practice in opinion polls to allow only a few choices like full agreement, partial agreement, neutrality, partial disagreement, full disagreement. And in elections one can only vote among the discrete number of candidates or parties which are on the ballot. In the Kosovo opinion of the International Court of Justice (July 22, 2010) one judge criticized the binary tradition of either legal or illegal, stating that tolerable is in between; nevertheless the court majority stayed with the binary logic of not illegal.

Physicists like to call the binary variables "spins" but readers from outside physics should refrain from studying the quite complicated spin concept in quantum mechanics. Spins are simply up and down (1 or 0; 1 or -1). Similarly, do not be deterred if physicists talk about a Hamiltonian; in most cases this is just the energy from high school.

18.3.3 *Humans Are Neither Spins Nor Atoms*

Of course, that is true, but it does not exclude that humans are modeled like spins. Modern medicine blurred the boundary between life and death; nevertheless we usually talk about people having been born in a certain year, and having died in another year, as if they were binary up-down variables. Reasons for death are complicated and I do not even know mine yet; nevertheless demographers like Gompertz estimated probabilities for dying at some age. Such probabilities are rather useless for predicting the death of one individual, but averaged over many people they may give quite accurate results. When I throw one coin I do not know how it will fall; when I throw thousand coins, usually about half of them fall on one side and the others on the other side (law of large numbers). If I throw 1000 coins and all show "head," most likely I cheated. Thus to simulate one person's opinion and decisions on a computer does not seem to be realistic; to do the same for millions of people may give good average properties, like the number of deaths at the age of 80 to 81 years, or the fraction of voters selecting the parties in an upcoming election. The whole insurance industry is based on this law of large numbers. Humans are not spins but many humans together might be studied well by spin models.

18.3.4 Deterministic or Statistical?

Non-physicists often believe that physics deals with deterministic rules: The apple falls down from the tree and not up; force equals mass times acceleration; etc. (Or they have heard of quantum-mechanical probability and apply that to large systems where such quantum effects should be negligibly small.) In this sense the cause of World War I was seen as a consequence of the arms race modeled by deterministic differential equations for averages [Richardson, 1935]. And the decay of empires was described [Geiss, 2008] as starting at the geographical periphery, since the influence from the center decreases towards zero with increasing distance, just as the gravitational force between the sun and its planets; see also [Diamond, 1997, Epilog].

Because of its historical importance let us look into Richardson's papers of 1935: Two opposing (groups of) nations change their preparedness x for war because of three reasons: (1) the war preparedness of the other side; (2) fatigue and expense; (3) dissatisfaction with existing peace treaties. Reasons 1 and 3 increase and reason 2 decreases the war preparedness; reason 1 is proportional to the x of the opponent, reason 2 to the own x, and reason 3 independent of x_1 and x_2. Even complete disarmament ($x_1 = x_2 = 0$) at one moment does not help if there is dissatisfaction with the existing peace. And if reason 1 is stronger than reason 2 for both sides, both x increase exponentially with time towards infinity. His second paper details the mathematical solutions for these linear coupled inhomogeneous differential equations.

More recently, the widespread use of computers shifted the emphasis to more realistic probabilistic models, using random numbers to simulate the throwing of coins or other statistical methods. This "Statistical Physics" and a simple example are the subject of the next section.

18.4 Statistical Physics and the Ising Model

The Ising model of 1925 leads to one of the simplest and versatile application of the much earlier Boltzmann probability.

18.4.1 *Boltzmann Distribution*

No present computer can simulate the motion of all air molecules in a cubic centimeter. Fortunately, Ludwig Boltzmann about 150 years ago invented a simple rule. The molecules move at temperature T with a velocity which can change all the time but follows a statistical distribution: The probability for a velocity v is proportional to $\exp(-E/T)$, where E ($= \frac{1}{2} mv^2$) is the kinetic energy and m is mass. The same principle applied to a binary choice; see Appendix 18.2. The temperature T is measured neither in Celsius (centigrade) nor Fahrenheit but $T = 0$ at the absolute zero temperature (about -273 Celsius below the freezing temperature of water) and moreover is measured in energy units. (When T is measured in Kelvin, the corresponding energy is $k_B T$ where k_B is the Boltzmann constant and set to unity in the present tutorial.) The function $\exp(x)$ is the exponential function, also written as e^x, which for integer x means the product of x factors e (i.e., $e \times e \times \ldots \times e$, with x numbers of e in the multiplication); $e \approx 2.71828$; $e^x = 2^{x/0.69315} = 10^{0.4343x}$. Exponential growth is sometimes also called geometric growth; for example the 1973 US Watergate crisis was compared to a cancer compounding itself geometrically each day.

These assumptions can be regarded as axioms on which Statistical Physics is built, like the Parallel Axiom of Euclidean Geometry, but in some cases they can be derived from other principles. Humanists are allowed to use these ideas of Boltzmann since history and sociology institutes were named after him.

18.4.2 *Ising Model*

In 1925, Ernst Ising (born in the heart of the city this author lives in) finished his doctoral dissertation on a model for ferromagnetism, which became famous two decades later and was shown to apply to liquid-vapor transitions half a century later. We assume that each site of a lattice (e.g., a square lattice where each site i has four neighbors: clockwise up, right, down, left) carries a spin $S_i = \pm 1$ (up or down). Neighboring spins "want" to be parallel, i.e., they have an energy $-J$ if they are in the same state and an energy $+J$ if they are in the two different states. Moreover, a "magnetic" field H between minus infinity and plus infinity (also called B) tries to orient the spins in its own direction. The total energy formula is given in Appendix 18.2.

As discussed before in the Boltzmann subsection (Section 18.4.1) the higher the energy E is the lower is the probability to observe this spin configuration; at infinitely high temperatures T all configurations are equally probable; at $T = 0$ all spins must be parallel to each other and to the field H in equilibrium. The "magnetization" M is the number of up spins minus the number of down spins. Computer simulations of this Ising model will be described in Appendix 18.1.

Applied to human beings, this Ising model could represent two possible opinions in a population; everybody tries to convince the neighbors of the own opinion (J), and in addition the government H may try to convince the whole population of its own opinion. The temperature then gives the tendency of the individuals not to think like the majority of their neighbors and the government. Zero temperature thus means complete conformity, and infinite temperature completely random opinions.

Theories with paper and pencil in two dimensions as well as computer simulations give $M(T, H)$ where M = number of up spins minus number of down spins. In particular, for $H = 0$ and in more than one dimension, the equilibrium magnetization $M = \pm M_0$ is a non-zero spontaneous magnetization for $T < T_c$ and is zero for $T \geq T_c$ where T_c is the critical or Curie temperature (named after Pierre Curie, not his more famous wife Marie Curie). On the square lattice, $T_c/J \approx 2.27$ is known exactly, in three dimensions only numerically; in one dimension there is no transition to a spontaneous magnetization, as Ernst Ising had shown: $T_c = 0$.

The above model obeys Isaac Newton's law *actio* = −*reactio*: The sun attracts the earth with the same force as the earth attracts the sun, only in opposite direction. Human relations can also be unsymmetrical:

He loves her but she does not love him. Then the bond between sites i and j may be directed instead of the usual case of an undirected bond. For example, if i influences j but j does not influence i, flipping the spin at j from parallel ($S_j = +S_i$) to antiparallel ($S_j = -S_i$) to the spin at i can cost an energy $2J$ while flipping S_i at constant S_j costs nothing. In this case no unique energy $E(S_i, S_i)$ is defined for this spin pair and thus such models have been much less studied in the physics literature. (In this example, one could gain a lot of energy from nothing by the cyclic process of flipping j from antiparallel to parallel, gaining energy $2J$, then flipping i from parallel to antiparallel, costing nothing, then flipping j again and so on. Such a perpetuum mobile does not exist in physics.)

Outside physics, of course, one can commit such crimes against energy conservation and forget all probabilities proportional to $\exp(-E/T)$. Instead one can assume arbitrary probabilities, as long as their sum equals one. For example, if an element has three possible states A, B and C, then one may assume that with probability p an A becomes B, a B becomes C, and a C becomes A; with probability $1 - p$ the element does not change, and there is no backward process from C to B to A to C. Then one has a circular perpetuum mobile, which may be realistic for some social processes. Usually to find a stationary or static equilibrium, one has to wait for many non-equilibrium iterations in the simulation (in a static situation, nothing moves anymore; in a stationary simulation the averages are nearly constant since changes in one direction are mostly cancelled by changes of other elements in the opposite direction.) There are many choices, one needs not mathematics, but mathematical thinking: precise and step-by-step. Outside of physics such simulations are often called "agent based" [Billari et al, 2006; Bonabeau, 2002]; presumably the first one was the Metropolis algorithm published in 1953 by the group of Edward Teller, who is historically known from the US hydrogen bomb and Strategic Defense Initiative (Star Wars, SDI).

18.5 Applications

Among various applications, the Schelling model is emphasized here since sociologists seem to ignore important corrections and improvement made to this model.

18.5.1 Schelling Model for Social Segregation

Economics Nobel laureate Thomas C. Schelling advised the US government on war and peace in the 1960s and published a methodologically crucial paper (cited nearly 700 times four decades later) [Schelling, 1971; Fossett, 2011; Henry et al, 2011] which introduced methods of statistical physics to sociology. Each Ising spin corresponds to one of two ethnic groups (black and white in big US cities) both of which prefer not to be surrounded by the other group. The above Ising model then shows that for $T < T_c$ segregation emerges without any outside control: The simulated region becomes mostly white or mostly black. Unfortunately, Schelling simulated a more complicated version of the Ising lattice at $T = 0$ which did not give large "ghettos" like Harlem in New York City, but only small clusters of predominantly white or black residences. Only by additional randomness [Jones, 1985] (cited only eight times, mostly by physicists) can "infinitely" large ghettos appear. Instead, it is easier to just simulate the above Ising model which also allows people of one group to move into another city (or away from the simulated region) to be replaced by residents of the other group [Sumour et al, 2008; 2011a]. Sumour et al also cite earlier Schelling-type simulations by physicists; see also [Rogers & McKane, 2011]. For nearly three decades physicists ignored the Schelling model; now the sociologists ignore the physics simulations of the last decade and the much earlier Jones [1985] paper.

As examples we give two Ising-model figures from Müller et al [2008] where people also increase their amount T of tolerance (= social temperature) if they see that their whole neighborhood belongs to the same group as they themselves. And afterwards they slowly forget and reduce their tolerance. With changing forgetting rate one may either observe lots of small clusters (Fig. 18.1) or one big ghetto (Fig. 18.2).

About simultaneously with this Schelling paper, physicist Weidlich [2000] started his sociodynamics approach to apply the style of physics to social questions; see also [Galam, 2008].

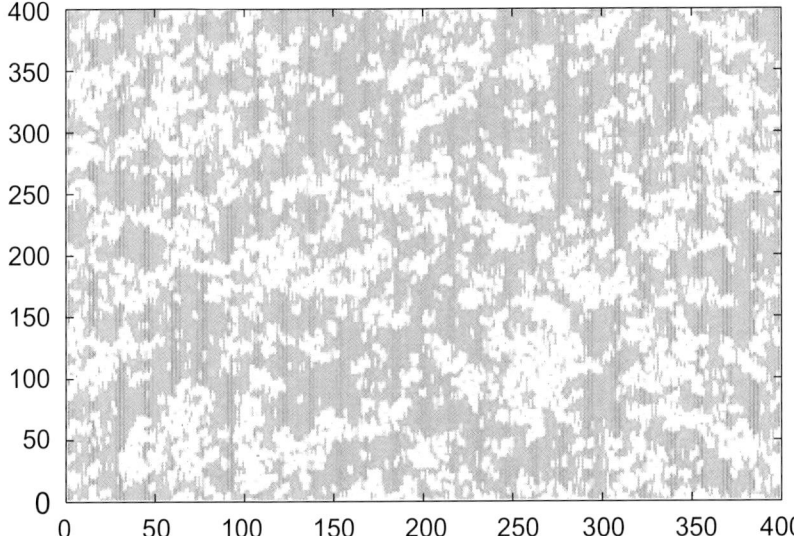

Fig. 18.1. Small clusters in modified Ising model: Cluster formation for slow forgetting in the Ising modification of Müller et al [2008]; no large ghetto is formed.

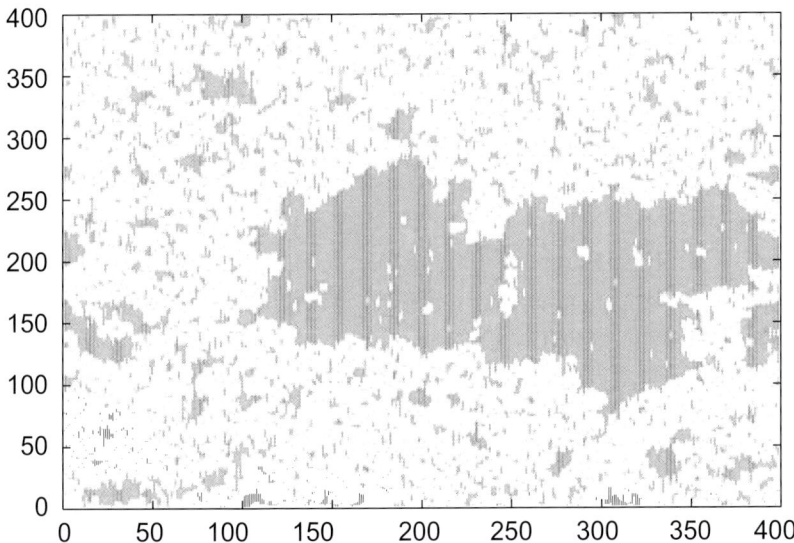

Fig. 18.2. Modified Ising model: feedback, memory loss. In the same model of the previous figure, a large ghetto is formed if people forget faster their learned tolerance.

18.5.2 Sociophysics and Networks

A good overall review of the sociophysics field (of which the Schelling model is just one example) was given by Castellano et al [2009]. Of particular interest is the reproduction of universal properties of election results with many candidates: The curves of how many candidates got n votes each are similar to each other [Chatterjee et al, 2013]. Car traffic [Chowdhury et al, 2000], economic markets [Bouchaud & Potters, 2009; Bonabeau, 2002], opinion dynamics [Malarz et al, 2011], stone-age culture [Shennan, 2001], histophysics [Lam, 2002], languages [Schulze et al, 2008], Napoleon's decision before the battle of Waterloo [Mongin, 2008], religion [Ausloos & Petroni, 2009; Ausloos, 2010], political secession from a state [Lustick, 2011], demography on social networks [Fent et al, 2011], insurgent wars in Iraq and Afghanistan [Johnson et al, 2011], and so on are other applications. For example, non-physicists Holman et al [2011] have acknowledged that physicists Serva and Petroni [2008] may have been the first to use Levinshtein distances to calculate the ages of language groups. (These distances are differences between words for the same meaning in different languages.)

Students in class may sit on a square lattice, but normally humans do not. (In universities, mostly only a part of the lattice is occupied, which physicists simulate as a "dilute" square lattice.) They may be connected by friendship or job not only with nearest neighbors but also with people further away. Such networks, investigated for a long time by sociologists [Stegbauer & Häussling, 2010], were studied by physicists intensively for a dozen years [Barabási, 2002; Albert & Barabási, 2002; Bornholdt & Schuster, 2003; Cohen & Havlin, 2010]. In the Watts-Strogatz (or "small world") network a random fraction of nearest-neighbor bonds is replaced by bonds with sites further away, selected randomly from the whole lattice. If that fraction approaches unity, one obtains the Erdös-Rényi networks, a limit of percolation theory [Flory, 1941]. More realistic are the scale-free Barabási-Albert networks, where the network starts with a small core and then each newly added site forms a bond with a randomly selected already existing member of the network. The selection probability is proportional to the number of bonds which the old member already has acquired: Famous people get more attention and more "friends" than others; no lattice is assumed here anymore. In all these networks, the average number of bonds needed to connect two randomly selected sites increases logarithmically with the number of sites in the network, whereas for d-dimensional lattices this average number of bonds increases stronger, with a power law exponent $1/d$.[1] Having simulated one network, one can also study connected sets of networks or other social networks [Watts et al, 2002], or demography on them [Fent et al, 2011].

The latest application is Statistical Justice: In May 2011, John Demjanjuk was sentenced for having helped in 1943 in the murder of more than 28,000 Dutch Jews in the Nazi concentration camp of Sobibór. One knows who was deported but not who survived the transport from the Netherlands to Poland. And one does not know which duties the accused had there on which day. Thus all acts of the camp guards were regarded as having helped in their murder and the number near 28,000 was estimated from the average death rate during the transports. The verdict thus gave neither the name of a murder victim nor the day of a murder, but was entirely based on statistical averages.[2]

Appendix 18.1: How to Program the Ising Model

The following Fortran manual and program are both short and should encourage the reader to learn this technique. A Fortran compiler may be available at http://gcc.gnu.org/fortran/ (Gnu compiler f77/f95).

1. Fortran manual

Fortran (= formula translator) was an early language (above machine code or assembler) for computer programming; many others followed and in particular C^{++} is widespread, but nevertheless this tutorial uses Fortran which is closer to plain English and allows to easily find a typical programming error (using an array outside its defined bounds). If your Fortran program is called name.f, it can be compiled with f95 -O name.f (or 77 instead of 95; O = optimization), and executed with

---

[1] If $y = \exp(x)$ then $x = \ln(y)$ is the natural logarithm of y.
[2] New York Times (nytimes.com), May 12, 2011, "Demjanjuk".

. /a.out (or just a.out). The just mentioned error message appears when using f95 -fbounds-check name.f ; ./a.out, but execution then is much slower and thus instead -O should be used after error correction.

Fortran commands usually start in column 7 and end before column 73. A C in column 1 signifies a comment for the reader, to be ignored by the computer. In column 6 we write a 1 if this line is a continuation of the previous line, while columns 2 to 5 are reserved for labels, i.e., numbers to control the flow of commands. For example, GOTO 7 means to jump to the line labeled by 7.

Variable names start with a letter; names starting with I, J, K, L, M, N signify integers without rounding errors; other names are real (floating-point) numbers and nearly always have rounding errors. The operations +, -, *, / and SQRT, COS, SIN, EXP, etc. have their usual meaning except that N/M is always rounded downwards to an integer value; e.g., 3/5 is zero. Also I = X means rounding downwards. Since the natural logarithm is not an integer it is denoted by ALOG instead of log. Random numbers between 0 and 1 can be produced by RAND().

Decisions are made automatically, e.g.,

IF (A.GT.0) B = SQRT(A)

where .GT. means greater than, with analogous meanings for .LT., .GE., .LE., .EQ., .NE., .NOT., .AND., .OR. .

A loop is executed by

DO 99 K=M, N

which means that all lines from this line down to and including the line with label 99 are executed for $k = m, m + 1, m + 2, ..., n$. One may put inner loops into outer loops, if needed.

Arrays need to be declared at the beginning of a program, for example through

DIMENSION A(100, 100), B(100), C(L), D(-4:4)

The allowed index for D varies between -4 and 4. Here, if C has an arbitrary dimension L, then L must be given a value before this dimension statement through

PARAMETER (L = 100)

and must not be changed throughout the program. Similarly, variables can be initialized via a data line like

DATA L/100/, B/100*1.0/

but only once at the beginning of the program, not later again.

Results are best printed out through

PRINT *, x, y, z

Thereafter execution should stop with a STOP line, followed by an END line.

The statement

n = n+1

is not an equality (which then could be simplified to the nonsensical 0 = 1) but a command to the computer: to find the place in the memory where the variable n is stored, to get the value of n from there, to add one to it, and to store the sum in that same memory place as the new value for n. Some computer languages therefore use := instead of the simpler but misleading = sign.

Normally it does not matter whether or not CAPITAL letters are used. The computer language Basic is rather similar to Fortran. As promised, here is a random walk program:

```
         iseed=1
         z=rand(iseed)
         ix=0
         iy=0
         do 1 istep=1, 100
           z=rand()
           zz=rand()
           if(z.lt.0.5) goto 2
           ix=ix-1
           if(zz.lt.0.5) ix=ix+2
           goto 1
2          iy=iy-1
           if(zz.lt.0.5) iy=iy+2
1          print *, istep, ix, iy
         stop
         end
```

If you show this to a computer scientist, he may teach you the construct:

```
        if(...)   then ...   else ...   end if
```
An old-fashioned economist may regard the resulting coordinates ix and iy as models for stock markets. Do you know a shorter Fortran manual?

Now we simulate the Ising model on the square lattice.

2. Ising model program

```
c        heat bath 2D Ising in a field
         parameter(L=1001, Lmax=(L+2)*L)
         dimension is(Lmax), ex(-4:4)
         data t, mcstep, iseed/0.90, 1000, 1/, h/+0.50/, ex/9*0.0/
         print *, ' #' , L, mcstep, iseed, t, h
         x=rand(iseed)
         Lp1=L+1
         LspL=L*L+L
         L2p1=2*L+1
         do 1 i=1, Lmax
1          is(i)=1
         do 2 ie=-4, 4, 2
           x=exp(-ie*2.0*0.4406868/t-h)
2          ex(ie)=x/(1.0+x)
         do 3 mc=1, mcstep
           mag=0
           do 4 i=Lp1, LspL
c            if(i.ne.L2p1) goto 6
c            do 5 j=1, L
c5             is(j+LspL)=is(j+L)
6            ie=is(i-1)+is(i+1)+is(i-L)+is(i+L)
             is(i)=1
             if(rand().lt.ex(ie)) is(i)=-1
4            mag=mag+is(i)
c          do 7 i=1, L
c7           is(i)=is(i+L*L)
3          if(mc.eq.(mc/100)*100) print *, mc, mag
         stop
         end
```

The parameter line fixes the size of the $L \times L$ square; the sites in it are numbered by one index, typewriter style. Thus the right neighbor of site $2L$ is $2L + 1$ and sits on the left end of the next line: Helical boundary conditions. The lower neighbor of site $2L$ is $2L + L$, the upper neighbor is $2L - L$, and the left neighbor is $2L - 1$. The spins in the top and bottom buffer lines ($1 \ldots L$ and $L^2 + L + 1 \ldots L^2 + 2L$) stay in their initial up orientation; if instead one wants periodic boundary conditions in the vertical direction (better to reduce boundary influence), one has to omit the five comment symbols C at and before loops 5 and 7.

The temperature enters the data line in units of T_c, i.e., $T = 0.9\ T_c$ in this example; the known value $J/T_c = 0.44068 \ldots$ is used nine lines later. In the same line the field, in units of $k_B T$, is given as 0.5. In the line after the first print statement, rand(iseed) initializes the random number generator (see next subsection for warning and improvement); a different seed integer gives different random numbers. Later rand() produces the next "random" number between 0 and 1 from the last one in a reproducible but hardly predictable way. Loop 2 determines the Boltzmann probabilities ex needed for Eq. (18.1) (in Appendix 18.2) and finishes the initialization. Loop 3 makes mcstep iterations (time steps = Monte Carlo steps per spin). Loop 4 runs over all spins except those in the two buffer lines and determines the local interaction energy ie as the sum over the four neighbor spins. Then we set the spin to +1, and if the conditions of Eq. (18.1) so require, instead it is set to −1. (Generally, a command is executed with a probability p if the random number rand() is smaller than p.) We print out the magnetization only every hundred time steps in order to avoid too much data on the computer screen. The results are plotted in Fig. 18.3(top) and show that the set temperature is actually very low: Even though it is only 10 percent below the critical temperature T_c, the magnetization barely changes from its initial value 1002001 and is after 100 iterations already in equilibrium, also due to the applied field. In three instead of two dimensions, without a field, and at temperatures closer to T_c longer times are needed for equilibration.

Perhaps you find it more interesting to simulate revolutions, as in Fig. 18.3(bottom) where the field h is equal to the fraction of overturned spins [Kindler et al, 2013]. So, what is difficult about computations?

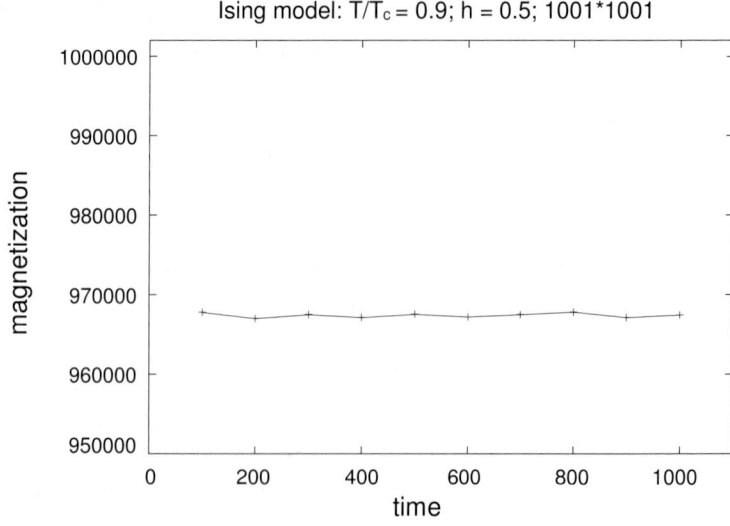

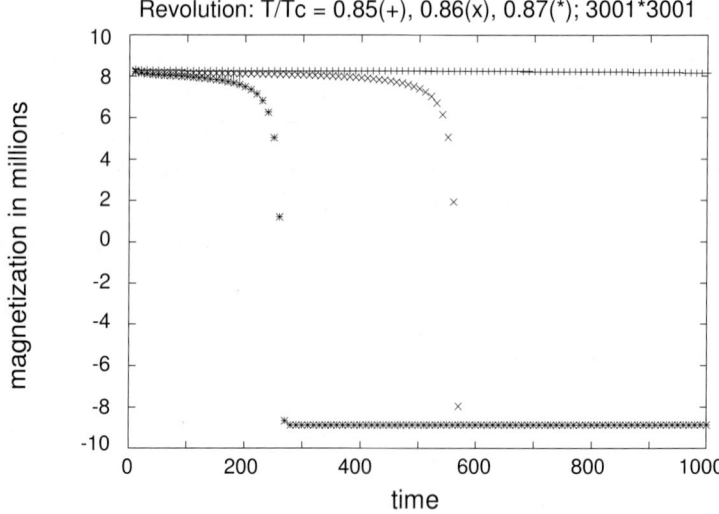

Fig. 18.3. *Top*: Results from the listed Ising program. *Bottom*: Revolutions.

3. Random numbers

The above rand produces random numbers in an easily programmed way, but often this may be slow and/or bad, or the used algorithm is

unknown to the user. It is better to program random number generation explicitly. If you multiply by hand two nine-digit integers, you may easily predict the first and the last digit of the product, but hardly the digits in the middle, except by tediously doing the whole multiplication correctly. Similarly, if ibm is a 32-bit odd integer, then the product
    ibm=ibm*16807
is again an odd integer, and normally requires 46 bits. (A bit = binary digit, which is a zero or one in a computer.) The computer throws away the leading bits and leaves the least significant 32 bits. The first bit gives the sign, thus plus times plus gives minus in about half the cases, in contrast to what you learned in elementary school. The last of these remaining bits is predictably always set to one (odd integers) but the leading (most significant) bits are quite random. (Actually, your computer may do something very similar when you call rand().) More precisely, they are pseudo-random; in order to search for errors one wants to get exactly the same random numbers when one repeats a simulation with the same seed.

These random 32-bit integers ibm between -2147483647 and $+2147483647 = 2^{31} - 1$ can be transformed into real numbers through ran=factor*ibm+0.5 where factor = 0.5/2147483647, but it is more efficient to normalize the probabilities p (here = $x/(1.0+x)$) to the full interval of 32-bit integers through (2.0*p-1.0)*2147483647, once at the beginning of the simulation. Then the next random integers ibm simply have to be compared with this normalized probability. In the above Ising program, one then stores the Boltzmann probabilities as dimension iex(-4:4) at the beginning, calculates them through
  2   iex(ie)=(2.0*x/(1.0+x) - 1.0)*2147483647
in the initialization, and later merely needs
    ibm=ibm*16807
    if(ibm.lt.iex(ie)) is(i)=-1
in the above Ising program.

With 32 bits the pseudo-random integers are repeated after 2^{29} such multiplications with 16807, which is a rather small number for today's personal computers. It is better to use 64 bits via
    integer*8 ibm, iex

at the beginning of the program in order to get many more different random numbers, using e.g.

    2   iex(ie)=2147483647*(4.0*ex/(1.0+ex) - 2.0)*2147483647

for the normalized 64-bit probabilities. Now the quality is much better without much loss in speed; unfortunately one now can make much more programming errors involving these random numbers.

Appendix 18.2: Some Formulas for Boltzmann Distributions

The Boltzmann principle applied to a binary choice, where a particle can be in two states A and B with energies E_A and E_B means that the two probabilities are

$$p_A = \frac{1}{Z}\exp(-E_A/T); \quad p_B = \frac{1}{Z}\exp(-E_B/T); \quad (18.1a)$$

$$Z = \exp(-E_A/T) + \exp(-E_B/T) \quad (18.1b)$$

since the sum over all probabilities must be unity. More generally, a configuration with energy E is in thermal equilibrium found with probability

$$p = \frac{1}{Z}\exp(-E/T); \quad Z = \sum \exp(-E/T) \quad (18.2)$$

where the sum runs over all possible states of the system. Z is called the gpartition function.

For the Ising model, the total energy is

$$E = -J\sum_{<ij>} S_i S_j - H\sum_i S_i \quad (18.3)$$

where the first sum goes over all ordered pairs of neighbor sites i and j. Thus the "bond" between sites A and B appears only once in this sum, and not twice (for i = A, j = B as well as for i = B, j = A). The second sum runs over all sites of the system. Thus $2J$ is the energy to break one bond, and $2H$ is the energy to flip a spin from the direction of the field into the opposite direction.

Appendix 18.3: History

Now comes a list of extended abstracts, a few pages each, about what this author finds interesting in recent history, available on request from dstauff@thp.uni-koeln.de:

1. Who is to blame for World War I
2. No miracle on the Marne, 9/9/1914
3. Lies and Art. 231 of Versailles Peace Treaty 1919
4. Was Hitler's 1941 attack against the Soviet Union a preemptive war?
5. Had Hitler nearly gotten Moscow in 1941?
6. Why was there no joint Japanese attack when Hitler attacked the Soviet Union?
7. The Sea Battle of Leyte, 25 October 1944
8. Did Soviet tanks approach Tehran in March 1946?
9. Missed chance for peace in Korea, October 1950?
10. Stalin's proposal of March 1952 for a united Germany
11. 1956: West German finger on the nuclear trigger?
12. Tank confrontation at Checkpoint Charlie 10/1961
13. The 1962 Cuban Missile crisis: security or prestige?
14. Lyndon B. Johnson (1908-1973) and the Dominican crisis (US Invasion 1965)
15. 1990: East Germany into NATO?
16. The non-conquest of Banja Luka (BiH) in 1995
17. Kosovo War 1999
18. The start of the Libyan war, March 2011

References

Albert, R. & Barabási, A.-L. [2002] "Statistical mechanics of complex networks," *Rev. Mod. Phys.* **74**, 47-97.

Ausloos, M. [2010] "On religion and language evolutions seen through mathematical and agent based models," in *Proceedings of the First Interdisciplinary CHESS Interactions Conference*, eds. Rangacharyulu, C. & Haven, E. (World Scientific, Singapore) pp. 157-182.

Ausloos, M. & Petroni, F. [2009] "Statistical dynamics of religion evolutions," *Physica A* **388**, 4438-4444.

Barabási, A.-L. [2002] *Linked* (Perseus, Cambridge, MA).
Billari, F. C., Fent, T., Prskawetz, A. & Scheffran, J. [2006] *Agent-Based Computational Modelling* (Physica-Verlag, Heidelberg).
Bonabeau, E. [2002] "Agent-based modelling: Methods and techniques for simulating human systems," *Proc. Natl. Acad. Sci. USA* **99**, 7280-7287.
Bornholdt, S. & Schuster, H. G. [2003] *Handbook of Graphs and Networks* (Wiley-VCH, Weinheim).
Bouchaud, J. P. & Potters, M. [2009] *Theory of Financial Risks and Derivative Pricing* (Cambridge University Press, Cambridge).
Bridson, R. & Batty, C. [2010] "Computational physics in film," *Science* **330**, 1756-1757.
Castellano, C., Fortunato, S. & Loreto, V. [2009] "Statistical physics of social dynamics," *Rev. Mod. Phys.* **81**, 591-646.
Chatterjee, A., Mitrovic, M. & Fortunato, S. [2013] "Universality in voting behavior: An empirical analysis," *Scientific Reports* **3**, article 1049.
Chowdhury, D., Santen, L. & Schadschneider, A. [2000] "Statistical physics of vehicular traffic and some related systems," *Physics Reports* **329**, 199-329.
Cohen, R. & Havlin, S. [2010] *Complex Networks* (Cambridge University Press, Cambridge).
de Vladar, H. P. & Barton, N. H. [2011] "The contribution of statistical physics to evolutionary biology," *Trends in Ecology and Evolution* **26**, 424-432.
Diamond, J. [1997] *Guns, Germs, and Steel* (Norton, New York).
Fent, T., Diaz, B. A. & Prskawetz, A. [2011] "Family policies in the context of low fertility and social structure," Vienna Inst. Demogr. Working Paper 2/2011 (www.oeaw.at/vid).
Flory, P. J. [1941] "Molecular size distribution in three-dimensional polymers: I, II, III," *J. Am. Chem. Soc.* **63**, 3083, 3091, 3096.
Fossett, M. [2011] "Generative models of segregation," *J. Math. Sociology* **35**, 114-145.
Galam, S. [2008] "Sociophysics: A review of Galam models," *Int. J. Mod. Phys. C* **19**, 409-440.
Geiss, I. [2008] *Geschichte im Überblick* (Anaconda, Köln).
Hadzibeganovic, T., Stauffer, D. & Schulze, C. [2008] *Physica A* **387**, 3242-3252.
Henry, A. D., Pralat, P. & Zhang, C.-Q. [2011] *Proc. Natl. Acad. Sci. USA* **108**, 8505-8610.
Holman, E. W., Brown, C. H., Wichmann, S., Muller, A., Velupillai, V., Hammarstrom, H., Sauppe, S., Jung, H. G., Bakker, D., Brown, P., Belyaev, O., Urban, M., Mailhammer, R, List, J. M. & Egorov, D. [2011] "Automated dating of the world's language families based on lexical similarity," *Current Anthropology* **52**, 841-875.

Johnson, N., Carran, S., Botner, J., Fontaine, K., Laxague, N., Nuetzel, P., Turnley, J. & Tivnan, B. [2011] "Pattern in escalations in insurgent and terrorist activity," *Science* **333**, 81-84.

Jones, F. L. [1985] "Simulation models of group segregation," *Aust. New Zeal. J. Sociol.* **21**, 431-444.

Kindler, A., Solomon, S. & Stauffer, D. [2013] "Peer-to-peer and mass communication effect on opinion shifts," *Physica A* **392**, 785-796.

Lam, L [2002] "Histophysics: A new discipline," *Mod. Phys. Lett. B* **16**, 1163-1176. (See also [Lam, 2008].)

Lam, L. [2008] "Human history: A Science Matter," in *Science Matters: Humanities as Complex Systems*, eds. Burguete, M. & Lam, L. (World Scientific, 2008) pp. 234-254.

Lim, M., Metzler, R. & Bar-Yam, Y. [2007] "Global pattern formation and ethnic/cultural violence," *Science* **317**, 1540-1544. (See also [Hadzibeganovic et al, 2008].)

Lustick, J. [2011] "Secession of the center: A virtual probe of the prospects for Punjabi secessionism in Pakistan and the secession of Punjabistan," *Journal of Artificial Societies and Social Simulation* **14**, issue 1, paper 7 (electronic only via jasss.soc.surrey.ac.uk).

Malarz, K., Gronek, P. & Kułakowksi, K. [2011] "Zaller-Deffuant model of mass opinion," *Journal of Artificial Societies and Social Simulation* **14**, issue 1, paper 2 (electronic only via jasss.soc.surrey.ac.uk).

Mongin, P. [2008] "Retour à Waterloo: Histoire militaire et théorie des jeux," *Annales. Histoire, Sciences Sociales* **63**, 39-69.

Müller, K., Schulze, C. & Stauffer, D. [2008] "Inhomogeneous and self-organized temperature in Schelling-Ising model," *Int. J. Mod. Phys. C* **19**, 385-391.

Prados, J. [2009] *Vietnam* (University Press of Kansas, Lawrence) p. xiii.

Richardson, L. F. [1935] "Mathematical psychology of war," *Nature* **135**, 830-831; **136**, 1025-1026.

Rogers, T. & McKane, A. J. [2011] "A unified framework for Schelling's model of segregation," *J. Stat. Mech. Theory Exp.*, article P07006 (electronic only).

Schelling, T. C. [1971] "Dynamic models of segregation," *J. Math. Sociol.* **1**, 143-186.

Schulze, C., Stauffer, D. & Wichmann, S. [2008] "Birth, survival and death of languages by Monte Carlo simulation," *Comm. Comput. Phys.* **3**, 271-294.

Shennan, S. [2001] "Demography and cultural innovation: A model and its implications for the emergence of modern human culture," *Cambridge Archeol. J.* **11**, 5-16.

Serva, M. & Petroni, F. [2008] "Indo-European language tree by Levinshtein distance," *Eur. Phys. Lett.* **81**, issue 6, article 68005.

Stauffer, D., Moss de Oliveira, S., de Oliveira P. M. C. & Sá Martins, J. S. [2006] *Biology, Sociology, Geology by Computational Physicists* (Elsevier, Amsterdam).

Stegbauer, C. & Häussling, R. (eds.) [2010] *Handbuch Netzwerkforschung* (VS-Verlag, Wiesbaden).

Sumour, M. A., El-Astal, A. H., Radwan, M. M. & Shabat, M. M. [2008] "Urban segregation with cheap and expensive residences," *Int. J. Mod. Phys. C* **19**, 637-645.

Sumour, M. A., Radwan, M. M. & Shabat, M. M. [2011a] "Highly nonlinear Ising model and social segregation," arXiv:1106.5574 (electronically only on arXiv.org section Physics).

Sumour, M. A., Radwan, M. M., Shabat, M. M. & El-Astal, A. H. [2011b], "Statistical physics applied to stone-age civilization," *Int. J. Mod. Phys. C* **22**, 1357-1360.

Vitanov, N. K. & Ausloos, M. R. [2012] "Knowledge epidemics and population dynamics models for describing idea diffusion," in *Models of Science Dynamics: Encounters between Complexity Theory and Information Sciences*, eds. Scharnhorst, A., Boerner, K. & van den Besselaar, P. (Springer, Berlin) pp. 69-125.

Watts, D. J., Dodds, P. S. & Newman, M. E. J. [2002] "Identity and search in social networks," *Science* **296**, 1302-1305.

Weidlich, W. [2000] *Sociodynamics; A Systematic Approach to Mathematical Modelling in the Social Sciences* (Harwood Academic Publishers, Newark, NJ); reprinted in 2006 by Dover, Mineola, NY.

Acknowledgments

All the chapters in this book are by invitation. But out of the 18 invited contributions, five (Chapters 5, 7, 9, 14 and 16) are expanded from talks presented at The Third International Conference on Science Matters, *All About Science: Philosophy, History, Sociology & Communication*, Lisbon, Portugal, November 21-23, 20011, which was under the auspices of the International Science Matters Committee. We thus want to thank the sponsors and people connected with this 3rd scimat conference:

Co-chairs: Maria Burguete and Lui Lam.

Sponsors: Fundação para a Ciência e Tecnologia, Fundação Calouste Gulbenkian, Instituto Rocha Cabral, Fundação Luso-Americana, Champalimaud Foundation, Câmara Municipal de Lisboa, Lisboa Convention Bureau, Tourismo de Portugal.

Local Scientific Committee: João Caraca, Roberto Carneiro, Claudina Rodrigues-Pousada, Fernando Ramôa Ribeiro.

International Advisors: Paul Caro (France), Bärbel Friedrich (Germany), Janos Frühling (Belgium), Dun Liu (China), Nigel Sanitt (UK), Brigitte Hoppe (Germany), Michael Shermer (USA), Edward O. Wilson (USA).

International Science Matters Committee: Manuel Bicho (Portugal), Peter Broks (UK), Maria Burguete (Portugal), João Caraça (Portugal), Paul Caro (France), Jean-Patrick Connerade (UK), Patrick Hogan (USA),

Brigitte Hoppe (Germany), Lui Lam (USA, *Coordinator*), Bing Liu (China), Dun Liu (China), John Onians (UK), David Papineau (UK), Nigel Sanitt (UK), Ivo Schneider (Germany), Michael Shermer (USA), Robin Warren (Australia).

    Finally, we are much grateful to the Fundação para a Ciência e Tecnologia for their unwavering support of the international scimat conference series since its 2007 beginning; Kim Tan, WS' editor overseeing the scimat series, for her superb assistance; Charlene Lam, the London-based artist, for her beautiful design of the section dividers in all the scimat books.

Contributors

Kajsa Berg finished her BA in Art History, Anthropology and Archaeology in 2004 at the University of East Anglia. There she continued with an MA in World Art Studies focusing on the phenomenology of collecting, and completed her PhD (with thesis "Caravaggio and a Neuroarthistory of Engagement") in 2010. She is currently teaching at the School of World Art Studies at the University of East Anglia, where she also continues her research on neuroarthistory and emotional responses to images. Email: k.berg@uea.ac.uk

Peter Broks obtained his BA and PhD from the University of Lancaster. In 1990 he was a specialist appointment at the University of the West of England, Bristol, to design, develop and teach a new undergraduate program in "Science, Society and the Media" jointly run by the Faculty of Humanities and the Faculty of Applied Sciences. He has published extensively in the history of science especially as it relates to popular culture and is the author of *Understanding Popular Science* (2006). In August 2011 he left UWE so that he could devote more time to exploring creative ways of presenting the history of science. He now teaches at the Herefordshire College of Art in the UK. Email: p.broks@hotmail.co.uk

Maria Burguete, a scientist at Bento da Rocha Cabral Institute in Portugal, received her PhD in History of Science (contemporary chemistry) from Ludwig Maximilians University, Munich, Germany

(2000). She graduated from the Faculty of Sciences in Lisbon (1982) after obtaining BS in Chemical Engineering (1979). She has research experience in a diversity of scientific fields, which enhanced the development of her interdisciplinarity and transdisciplinarity. She has published seven scientific books and seven poetry books, and over 25 scientific papers mostly in history and philosophy of science. Since 2010 she is a Corresponding Member of the European Academy of Sciences Arts and Letters (founded in Paris, 1980). *Email: mariamartins434@ gmail.com*

Harry Collins is Distinguished Research Professor of Sociology and Director of the Centre for the Study of Knowledge, Expertise and Science (KES) at Cardiff University and is a past President of the Society for Social Studies of Science. A Fellow of the British Academy, prizes include the 1997 Bernal prize for social studies of science and the 1995 ASA Merton book prize. He has published more than 150 papers and 16 books, most with major university presses, with a 17th in press and an 18th about to be submitted. The topics are sociology of scientific knowledge, artificial intelligence and, more recently, expertise and tacit knowledge, now supported by a major European Research Council Advanced Grant. *Email: CollinsHM@cf.ac.uk*

Lui Lam obtained his BS (First Class Honors) from the University of Hong Kong, MS from University of British Columbia, and PhD from Columbia University. He is Professor of Physics at San Jose State University, California and Adjunct Professor at both the Chinese Academy of Sciences and the China Association for Science and Technology. Lam invented Bowlics (1982), one of three existing types of liquid crystals in the world; Active Walks (1992), a new paradigm in complex systems; and two new disciplines: Histophysics (2002) and Scimat (Science Matters, 2008). He published 14 books and over 180 scientific papers. He is the founder of the International Liquid Crystal Society (1990); cofounder of the Chinese Liquid Crystal Society (1980); founder and editor of two book series, *Science Matters* and *Partially Ordered Systems*. His current research is in scimat, histophysics and complex systems. *Email: lui2002lam@yahoo.com*

Bing Liu obtained his BS in physics from Peking University and MS from the Graduate School of Chinese Academy of Sciences. He is professor of history of science at Tsinghua University, Beijing, vice director of the Center for Science Communication and Popularization of CAST and Tsinghua University, and Guest Professor at several universities in China including Shanghai Jiaotong University. He published 17 books (also translated 7 books and edited more than 30 books) and over 200 academic papers. Liu's research fields currently include history of physics, historiography of science, philosophy of science, and science communication. His blog: http://blog.sina.com.cn/liubing1958. Email: liubing@tsinghua.edu.cn

Dun Liu, former Director of the Institute for the History of Natural Science, Chinese Academy of Sciences (1997-2005) and past President of the International Union of History and Philosophy of Science (2009-2013), is currently Professor Emeritus of the Institute and Professor of Tsinghua University at Beijing. His main research field is Chinese mathematics/astronomy and its interaction with the social context, especially in the Ming-Qing transitional period (c. 17th century). Also serving as editor-in-chief of the bimonthly journal, *Science & Culture Review*, he currently focuses on such historiographic and cultural topics as the "Needham question" and the "C. P. Snow thesis". Email: liudun@ustc.edu.cn

Jin-Yang Liu obtained his BE and MA from Chengdu University of Technology, and PhD from Renmin University of China (RUC). He is associate professor of philosophy at RUC, member of the governing board of the Chinese Society for Dialectics of Nature, secretary general of Committee of History of Dialectics of Nature (i.e., philosophy of science and technology), and visiting scholar in Department of Philosophy, University of Pittsburg (2012-2013). He works on philosophy of science: scientific methodology, metaphysical problems in science, complexity and systems science. Liu has published more than 30 papers and several books; recent books are: *Complexity: A Philosophical View* (2008) and *Classic Readings in the Philosophy of Science and Technology* (2011). He is heading a National Research Project, "Whole and Parts: Interdisciplinary Study of Contemporary Holism" which is supported by China National Social Science Foundation. *Email: liujinyangchina@gmail.com*

Nigel Sanitt obtained his BS in Physics from Imperial College, London and Part III of the Mathematics Tripos and PhD from Cambridge University, where he trained as an astrophysicist at the Institute of Astronomy. He is founder and editor of *The Pantaneto Forum*, a journal which aims to promote debate on how scientists communicate, with particular emphasis on how such communication and research skills can be improved through a better philosophical understanding of science. *Email: nigel@pantaneto.co.uk*

Dietrich Stauffer, PhD 1970, Technical University of Munich, is retired professor of theoretical physics and studies history since retirement. Before that he worked on Monte Carlo simulations, like Ising models, percolation, ageing, opinion dynamics; and published 647 papers and six books of which *Introduction to Percolation Theory* (1994, 2nd printing of 2nd edition, with A. Aharony) is cited most. More relevant for his contributed chapter is the book *Biology, Sociology, Geology by Computational Physicists* (2006) by Stauffer et al. He edited nine volumes of Annual Reviews of Computational Physics (World Scientific, 1994-2001). *Email:dstauff@thp.uni-koeln.de*

Hong-Sheng Wang obtained BEgn from Northwestern Polytechnics University (Xian), Master of philosophy and PhD of political science from Renmin University of China. He is professor at Renmin University of China, and president of Beijing Society for Philosophy of Nature, Science and Technology. He was senior research fellow at Chung Ang University, Seoul (2004-2005) and had completed "A study on Nature and limits of Confucianism". He was visiting professor at Utrecht University (2012), working on the project "Human dignity in the context of Bioethics: China and the West". Wang had published several books, including *A World History of Science and Technology* (1996, 2001, 2008, 2011) and *The Waterfalls and Gorge of China's History* (2007). He teaches history of science and technology as well as methodology of academic research, while paying attention to Chinese culture and civilization with a scientific outlook. *Email: whs6558@ sina.com*

Ting-Ting Wang, a PhD student at Department of Chinese Language and Literature, Peking University, is China's foremost expert on *Tanbi* (danmei) novels. She obtained both her BA and MA in Chinese Literature from Renmin University of China. Wang has translated books such as *Seven Samurai* (Peking University Press, 2012) and *Mishima's Sword* (Jiangsu People's Press, 2012); wrote novels; and published research papers in journals, the latest: "Internal order: Resort to violence by members of vulnerable groups in Chinese independent films" (2014), which is based on her PhD thesis. Her current interest is in directing films. *Email: wtingting_fish @yahoo.com.cn*

Robin Warren was born in 1937, in Adelaide, South Australia. He graduated M.B., B.S. from the University of Adelaide in 1961. After training at the Royal Melbourne Hospital, he was admitted to the Royal College of Pathologists of Australasia in 1967. Since then, he was a senior consultant pathologist at the Royal Perth Hospital in Western Australia, becoming emeritus consultant pathologist in 1998. *Email: jrwarren@aapt.net.au*

Guo-Sheng Wu obtained his BS and MS from Peking University, and PhD from the Graduate School of Chinese Academy of Social Sciences. He is Professor at Department of Philosophy and Director of Center for History and Philosophy of Science, Peking University. He published more than 20 books and over 100 scientific papers. His current research is in phenomenological philosophy of science and technology, early modern history of mathematical physical sciences, and science communication. *Email: wugsh@pku.edu.cn*

Lin Yin is a PhD and associate professor of Division of Theoretical Studies on Science Popularization, China Research Institute for Science Popularization. She is also a member of Association of Chinese Science Writers. Her main research interests are: history of science popularization, science writing, and development of S&T communication around the world. She has completed several research programs including "Research on Science Popularization Studies in China", "Survey on Popular Science Publications in China", and "Looking into S&T Educational Activities for Youth and Adolescents in America". Her research has been published in the journal *Science Popularization* and in books such as *Science Communication in the World* (Springer, 2012). *Email: yinlin213@126.com*

Mei-Fang Zhang obtained her BBA from Anhui University, MS from University of Science and Technology of China, and PhD from Tsinghua University, Beijing. She is associated professor of history of science at University of Science and Technology Beijing. Zhang has published three books (including one translated book) and over 40 research papers. Her current research fields include historiography of science, science communication, and history of science in modern China, with particular focus on gender, science and technology. *Email: zhangmeifang@gmail.com*

Index

A

Academia Sinica, 258, 260
Academician, CAS, 260, 261, 263, 266, 273, 274
Acta Physica Sinica, 275
Active walk, 28, 232, 234
Actors' categories, 246
Aesthetics, 259
Aether, 141
Africa, 84
Against Method, 62
Agazzi, E., 150
Age of Insight, The, 32
Age of Revolution, 12, 52
Agency, The, 372
Aggregate collection, 155
Alchemy, 20, 112
Algebra, 189
Alias, 372, 377, 378
America. *See* United States
America's Army, 381
American Physical Society, 20, 78, 231, 232
  Focus, 78
  Outreach Program, 78
Anaheim, 232
Anatomy, 200
Anatomy Theatre, 195
Anaximander, 10
Anaximenes, 10
Anthropology, 16, 246
Antimarriage, 39
Antiscience, 36, 39
Anti-scientism, 302
Antiscientist, 40
Antwerp, 256
Apheliotes, 8
Apollo, 8
Apple, Inc., 7, 214
Approximation, 23, 29, 34, 35, 40, 64, 65, 70, 79, 83, 87, 91
Arab, 86
Arch of Knowledge, The, 54
Archaeology, 16, 41
Archimedes, 12, 85
Archimedes' Principle, 12, 22, 42
Aristotle, 9, 11, 12, 22, 32, 41, 42, 60, 63, 66, 85, 86, 119, 148, 150, 164
  tradition, 23, 95
Art, 349 352, 403
Art and Illusion, 357
Artificial intelligence, 72
Artist, 21, 76
Arts, 1, 14, 16, 94
  performing, 16
  visual, 16
Arts, 38
Astrology, 80
Astronomy, 16, 22, 27, 41
Astrophysics, 138, 145
Athias, Marck, 197, 204, 207, 208
Atom, 3, 24, 41, 84, 150, 387
Atomism, 152

416 Index

Autumn and Spring period, 26
Auyang, Sunny Y., 148, 150, 157

B

Ba Tiao Mu, 127
Babbage, Charles, 189
Babylon, 118
Bacon, Francis, 187
Bahm, Archie John, 155
Ballet, 224
Baltimore, 256, 257
Bangalore, 216, 217, 219, 220, 274
Baodiao Movement, 254-257, 282
Bardeen, John, 24
Bata, Lajos, 223
Baxandall, Michael, 357, 367
Bednorz, Johannes, 24
Bee, 5, 6, 87, 92
Beijing, 134, 215, 217, 218, 221, 234, 237, 256-260, 265, 266, 268-270, 273, 276, 281, 284
Beijing Normal University, 267
Beijing Zoo, 263
Being-in-the-world, 105
Belgium, 215, 256
Bell, Johan Adam Schall von, 333
Bell Labs, 71, 82, 213, 217, 256, 267, 276
Ben Cao Gang Mu, 123
Ben-Ari, Moti, 55
Benevolence (ren) system, 126, 134
Bergersen, Birger, 225
Bergson, Henri-Louis, 157, 159
Berkeley, 59, 62, 88, 89, 91, 216, 274, 276
Berlin, 61, 257
Berlin University
  Berlin Histology Laboratory, 198
  Berlin School of Medicine, 199
  hospital, 200
  medical faculty, 200
Berlin Wall, 80
Bern, 265
Bernal, John, 188
Bertalanffy, L. v., 148, 152, 153, 155, 168
Bethe, Hans, 268
Big Bang, 83, 84, 94

theory, 13
Biochemistry, 200, 203, 204
Biology, 16, 23, 28, 85, 150, 158
Black-body radiation, 23
Bloor, David, 69
Blue Brain Project, 169
Boddaert, Richard, 198
Bodmer, Walter, 316
Bohm, David, 151, 159
Bohr, Niels, 90
Boltzmann distribution, 389, 402
Boltzmann, Ludwig, 383, 389
Bombs, atomic and hydrogen, 257, 258, 261, 268
Bonaparte, Napoleon, 6
Book of Changes, The, 163, 164
Book Drop Test, 67
Bordeaux, 222, 224
Boreas, 8
Born, Max, 268
Borne Conspiracy, The, 372
Botany, 200
Bottom-up approach, 78. *See also* Research level
Boulder, 224
Boundary, 151
Bowles, Samuel, 171
Bowlic. *See* Liquid crystal
Boxer Indemnity, 285
Bridges of Madison County, The, 89
Brief History of Time, A, 89
Broks, Peter, 77
Brown, Glenn, 210, 219, 224
Brownian motion, 4, 56, 87
Brun, Charles Le, 349, 352, 354, 355, 360, 366
Bruno, Giordano, 39
Brussels, 256
Bryn Mawr College, 139
Bryson, Norman, 356, 357, 367
Budapest, 223, 224
Buddhism, 131
Butterfly effect, 80, 82
Byrne, Peter, 43

C

Cai, Jun-Dao, 262, 263
Cai, Lun, 163

Cai, Shi-Dong, 274
Cai, Yuan-Pei, 259
Calculator, HP, 259
Calculus, 189
California, 80, 222, 232, 278
Cambridge University, 93
Canada, 210, 213, 254, 269
Cao, Cao, 32
Caravaggio, Michelangelo Merisi da, 349, 350, 363
Carnegie Institution, 59
Caro, Paul, 294
Castaing, Raimond, 217
Categories
  actor's, 92, 246
  technical, 92, 246
Ce Yuan Hai Jing, 121
Cell Phone Test, 36
Cellular automaton (CA), 159-163, 165, 168
Centenary Conference of Liquid Crystal Discovery (1988, Beijing), 223
CERN, 71
Chance, 31
Chandrasekhar, Sivaramakrishna (Chandra), 216-220, 224, 227, 231, 232, 234, 237
Changan Avenue, 282
Chaos, 26, 32, 33, 80
Charvolin, Jean, 217, 222
Chemical Abstracts, 215
Chemical analysis, 203
  course, 203
Chemistry, 16, 63, 196, 200
  organic, 204
Chen, Jie, 276
Chen, Ruo-Xi, 282
Chen, Yi, 133
Chen,Yin-Ke, 133, 134
Cheng, Bing-Ying, 270
Chiang, Kai-Shek, 269
Chimp, 7, 84, 92
China, 50, 74, 76-78, 80, 103, 118, 185, 213-217, 219, 221, 222, 225, 253, 270, 276, 281, 282, 284, 286, 379
  ancient, 85, 116, 124, 131, 134, 330, 333
  contemporary, 290

  mainland, 213, 216, 254, 271, 274, 285, 286, 290
  modern, 330, 344
  National Day, 258, 265
  People's Republic of, 78, 133, 254, 260, 286
  Republic of, 77, 133, 218, 259, 269
China Art Gallery, 263
China Association for Science and Technology (CAST), 78
China Construct, 272
China Daily News, 255, 270, 271
China Knowledge Resource Integrated Database (CNKI), 297
China Research Institute on Science Popularization (CRISP), 78
Chinatown, 254, 255
Chinatown Food Co-op, 213, 256
Chinese, ancient, 95
Chinese Academy of Sciences (CAS), 134, 215, 258, 261, 267, 283, 296
Chinese classic, 332
Chinese Communist Party (CCP), 133
Chinese Dialectics of Nature Research Society, 292
Chinese Liquid Crystal Society (CLCS), 218, 219, 222, 223, 237, 273
Chinese Physical Society, 260, 266, 269
Chinese physics, Father, 260
Chinese University of Hong Kong, 281
Cholos, Alan, 73
Chomsky, N., 159
Christian, 86
Chuang-tse, 332
Church News, The, 338
City University of New York (CUNY), 213, 214, 221
  City College (CCNY), 213, 256
Civilization, 1, 125
  Chinese, 117, 118, 127, 131, 134
  Greece, 132
Cladis, Patricia, 217, 234
Clark, Noel, 220, 224
Class F person, 68
Climate, 82
  change, 82, 94
  system, 26
Climatology, 22

Cliodynamics, 32
CNRS, 217
Cognitive science, 148
Coimbra University, 194, 196, 197
  Botanical Garden, 195
  Chemical Laboratory, 195
  Experimental Physiology Laboratory, 199
  history of medicine, 202
  Laboratories of Experimental Physics, 195
  Laboratory of Experimental Physiology, 201
  Laboratory of Histology, 201
  medical curriculum, 200
  medical faculty, 199, 201, 202, 203
  microscopes collection, 202
  Pharmaceutical Dispensary, 196
  Science Museum, 195
  Teaching Hospital, 195
Cold fusion, 80
Cold War, 313, 377, 381
Collingwood, Robin, 9, 12
Collins, Harry, 53, 68, 72, 243
Columbia University, 71, 195, 213, 254, 260
Communism, 58
Complex, 152
  three types, 153
Complexity, 27, 154, 155
  science, 148
Computer simulation, 29, 66, 71, 82
Computing Center, CAS, 266
Compton profile, 213, 214
Comte, Auguste, 15, 86
Condorcet, 253, 254
Confucianism, 26, 85, 125, 128, 131, 132
Confucianism and Taoism, 125
Confucius, 85, 94, 127
Consciousness, 12
Consilience, 33
Constructivity, 104
Contextual brain, 349, 350, 363
Cooper, Leon, 24
Copernicus, Nicolaus, 66, 79, 336
Cornell University, 78
Costa Simões, António Augusto da, 196-199, 201-203, 207, 208

Counter-Revolution of Science, The, 31
Counter-Strike, 374
Crick, Francis, 29
Crowd control, 29
Crystallography, 63
Cultural Revolution, The Great Proletarian, 254, 262, 265, 268, 272-274, 282, 284, 286
Cultural Spirit of Western Modern Science, The, 293
Cultural study, 6
Cunningham, Andrew, 52
Curie
  Madame (Irène), 257, 260
  Madame (Marie), 260
Czechoslovakia, 223

D

Da Ming Li, 120
Da Xue, 127
Dao, 163
Daoism, 131
Daqing, 257
DARWARS Ambush, 381
Darwin, Charles, 3, 6, 11, 15, 17, 87, 186, 191
Darwin's evolutionary theory, 15, 23, 29, 58, 63, 84, 93, 186, 187
Dasein, 113
Death to Spies, 372, 375
Debate, 85
  between Science and Metaphysics, 291
Decision making, 82, 83, 91, 248, 249
Deduction, 88
Demarcation, 40, 57, 112
Democracy, 77, 132, 249, 251
Democritus, 24
Demus, Dietrich, 217, 220
Deng, Xiao-Ping, 133, 265, 266, 268, 270, 274, 285
Department of Energy, 81
Destrade, Christian, 217, 221
Dialectics of Nature, 293
Dialectics of Nature, 185, 186, 188, 191
Diamond, Jared, 125, 126
Diaoyudao Islands, 254, 255
Disciplinary matrix, 60

Discipline
  historical, 25, 41
  research, 25, 29, 50, 54, 75, 94, 148
Dissipation function, 215
DNA, 66, 169
Doane, J. William, 210, 220, 224, 225, 227, 234
Dodgson, Charles Lutwidge, 189
Dong, Dong, 216
Doufu, 262
Double helix, 29
Dowell, Flonnie, 231
Drummond, Henry, 13
Du, Ya-Quan, 342
Duan, Shi-Yao, 282
Duhem, Pierre, 57, 148
Dunmur, David, 212, 234
Durand, Georges, 217, 220, 223
Dynamical Theory of Crystal Lattices, 268
Dynamical Theory of the Electromagnetic Field, A, 187
Dynasty
  Han, 26, 120, 123
  Jin, 121
  Ming, 122-124, 129, 330, 331, 334
  Northern, 120
  Northern Song, 332
  Qin, 129, 130
  Qing, 77, 123, 124, 129, 133, 259, 285, 290, 330, 331, 334, 337
  Shang, 120
  Song, 121, 129, 332
  Southern, 120
  Tang, 120
  West Han, 129
  Xie, 120
  Yuan, 121, 123
  Zhou, 119, 120

E

Earth, 26, 58, 70, 76, 79, 82, 120
Earth sciences, 16, 22
Earthquate, Loma Prieta, 80
Eclipse, 9, 69, 120
Ecole Polytechnique Fédérale de Lausanne, 169
Economic meltdown, 2008, 30

Economist, The, 30, 254
Economics, 14, 16, 30, 86, 158, 245, 382, 392
Economy, 7, 29
Econophysics, 30
Eddington, Arthur, 51
Edinburgh, 219
Edkins, Joseph, 124
Educational
  reform, 302
  system, 51, 62, 94, 203
Egypt, 118
Eigen, M., 165
Eight-Power Allied Forces, 77
Einstein, Albert, 4, 6, 17, 23, 35, 42, 56, 60, 65, 68, 72, 87, 90, 139, 188
  mistake, 72
Ekman, Paul, 356
Elective modernism, 251
Electricity, 187, 188
Electroconvection, 219
Electromagnetism, 23, 187, 189
Elements, The, 124
Emerging property, 86
Emotion, 349
Empathy, 349, 352, 355, 363
Empedokles, 383
Emperor
  Han Wu, 131
  Kuang-Xu, 285
  Kublai, 122
  Qin, First, 129, 130
  Taizhong, 120
Empirical approach. *See* Research level
Energy conservation law, 186
Engel, Friedrich, 133, 185, 186, 188, 191
England, 189, 191, 268
Enlightenment, 13, 15, 31, 44, 86, 87, 188
Enrico Fermi Summer School, 222
Entertainer, 21
Epistemology, 103
Ericksen-Leslie equation, 215
Eros, 8
ETH Zürich, 62
Ethics, 1, 27, 85, 382
Ethnography, 246
Eudemus, 63

European
  country, 194
  laboratory, 199
  university, 195
Euscience, 42
Evanston, 216, 270
Evolution paradigm, 169, 170
Evolutionary behavior, 158
Evolutionary theory. *See* Darwin's evolutionary theory
Exemplar, 60
Exogenesis, 169
Experiment, 34, 71, 72, 74, 81, 198, 201, 245
  control, 25, 41, 42, 74
Experimental histology and physiology, 199
Experimentation, 23, 196
Experimenter's regress, 72
Expert, 72, 250, 269, 310
Expertise, 248
  Contributory, 250
  Interactional, 250, 251

F

Faculty of
  Mathematics, 204
  Medicine, 195-197, 203, 204
  Medicine Hospital, 196
  Philosophy, 196, 202, 203, 204
Falsification, 56-58
Falsificationism, 57
Fang, Fu-Kang, 267
Fang, Yi, 267
Faraday, Michael, 141, 187, 188
Federal Institute of Technology Zürich, 31
Feminist, 65
Feng, Duan, 266
Feng, Huan, 266
Feng, Hui, 266
Feng, Kang, 266, 275, 277, 278
Feng, Ke-An, 263
Fergason, James, 231
Fermi, Enrico, 90
Fermi Lab, 71
Feyerabend, Paul, 51, 55, 61-62, 88, 108
  on Kuhn, 88

Feynman Lectures on Physics, The, 264
Finance, 30
Fish, 84
Fortran, 384, 395
Foucault, Michel, 375
France, 6, 40, 89, 222, 257-260, 265, 285
Frank free energy, 215
Fractal, 28
Free will, 12, 32
Free University of Brussels, 138
Freedberg, David, 349. 350, 361-363, 367
Freiburg, 227, 229
French Revolution, 12, 13, 188, 253
Freud, Sigmund, 57
Fribourg, 262
Friedel, Jacques, 217
From the Earth to the Moon, 344
Frontier, 95
Fukuda, Atsuo, 211, 220, 237
Full Spectrum Warrior, 381
Fung, You-Lan, 163

G

Galaxy, 141
Galileo Galilei, 2, 11, 13, 18, 22, 23, 42, 61, 85, 86, 94, 188, 336
Game, 371, 379, 382
Gao, Xing-Jian, 283
Ge Zhi Hui Bian, 338, 339
Gene, 51, 169
General education, 43, 59
Generalization trap, 68, 93
Generation, 163
Generation of 1911, 207
Generative grammar, 159
Generativism, 163, 164
Geneva, 139
Gennes, Pierre-Gilles de, 214, 217, 220, 222, 223
Geology, 16
Geometry, 124
Geometry, 334
German Democratic Republic, 221
German model, 198
Germany, 189, 192, 215, 221, 256, 259, 264, 278
Global positioning system (GPS), 69

Global warming, 82
God, 3, 8-10, 12, 14, 17, 37, 40, 52, 54,
    84, 123
  existence of, 2, 10, 15
God of the gaps, 13, 14
Gödel, Kurt, 87
Godfrey-Smith, Peter, 55
Gods hypothesis, 9
Golem, The, 53, 68, 69
Gombrich, Ernst, 352, 356, 366
Gong, Zi-Zhen, 129
Goodby, John, 237
Gordon, Martin, 219
Gordon & Breach, 219
Government, 40
Gravitational wave, 72, 73
Gravity, 67, 80, 140
  Law of, 16, 58, 70
Gray, George, 220, 227
Great Hall of the People, 258
Great Leap Forward, 269
Greece, ancient, 20, 54, 63, 85, 352
Greek, ancient, 8, 11, 54, 86, 95, 138
Gross, David, 43
Gu, Shi-Jie, 270
Guan, Wei-Yan, 261, 270, 282, 283, 285
Guangdong Province, 266
Guangzhou, 134, 336
*Guide to the Culture of Science,
    Technology, and Medicine, A*, 292
Guinier, André, 217
Guo, Mo-Ruo, 134
Guo, Shou-Jing, 122
Guomingdang (KMT), 133

H

Hai Guo Tu Zhi, 336-338
Haigui, 259, 286
Halle, 221
Han, Fei, 130
Han, Min-Qing, 292
Handbook of Liquid Crystals, 223, 279
Hangzhou, 255, 259
Hao, Bai-Lin, 262, 263, 285
Harry, Rome, 111
Hayek, Friedrich, 31
Harvard University, 59, 63, 91, 256, 257
Hawkings, Stephen, 89

Healey, Richard, 148
Heidegger, Martin, 104
Hei Longjiang Province, 292
Helbing, Dirk, 31
Helicobacter, 177
  discovery, 178-183
Helmholtz, Hermann von, 198
Helfrich, Wolfgang, 217, 222, 234
Hepatitis, 267, 274
Heriot-Watt University, 219
Hermeneutics, 109
Herodotus, 63
Higgs, Peter, 142
Higgs boson/particle, 142
Hillary, Edmund, 64
Histological preparation, 198
Histology, 197, 198
  human, 199
Histophysics, 234
History, 14, 16, 31, 32, 62, 63, 148, 246,
    403
  Father of, 63
  narrative, 65
  Natural, 195, 203
  temporal, 166
History of science, 50, 62-66, 84, 91-94,
    110, 185, 246, 290, 296
  a discipline, 63
  Father of, 63
  institute, 64
  open problem, 66, 88
  professionalization, 63
  scope, 63
*History of Science and the New
    Humanism, The*, 292
History of Science Society, 63
Hitchcock, Alfred, 372
Hitler, Adolf, 68, 403
Hobson, Benjamin, 336
Höglund, Johan, 380
Holism, 147, 171
  absolute, 151, 171
  conformation, 148
  constitutive, 147, 149, 155, 166, 171,
    172
  generative, 147, 149, 158, 159, 168,
    171, 172
  meaning, 148

422 Index

methodological shift, 149
omological, 148
ontological, 148
property/relational, 148
relative, 152, 171
strong, 155
substantial, 163
system, 163
Holism and Evolution, 147
Homo
  erectus, 7
  sapiens, 8, 53, 84
Honesty, 18, 57
Hong Kong, 213, 216, 254, 259, 263, 267, 270, 285, 335
Hongqi, 268
Hospital 301, Beijing, 267
Hu, Shi, 291
Hua, Guo-Feng, 265, 267, 269
Huang, Kun, 268, 282, 285
Huang, Yun, 278
Huazhong Institute of Technology, 278
Hufbauer, Karl, 59
Hunger, 6
Human Brian Project (HBP), 169
Humanism, New, 63
Humanist, 7, 21, 28, 31, 51, 383
Humanities, 5, 6, 14, 16, 18, 31, 41, 43, 50, 54, 64, 87, 91-95, 247
  importance of, 6, 37
  revolt against, 91
Humans, 5, 28, 31, 35, 37, 53, 84, 246, 383, 387
Humboldt, Wilhelm von, 200
Humboldt University, 200
Hundred Schools of Thought period, 131, 332
Husserl, Edmund, 103

I

IBM, 231, 284
Idea of Nature, The, 9
Idea of a Social Science and Its Relation to Philosophy, The, 31
Imitation Games, 250
Impact 100, 256
Impact of Science on Society, 300
Incommensurability, 60, 88, 91

India, 118, 216, 217, 221, 229
Induction, 88
Industrial Revolution, 12, 188
Information communication technology (ICT), 169
Insider, 51, 55, 91
Institute of Applied Physics, 260
Institute for the History of Natural Science, CAS, 296, 301
Institute of Mathematics, CAS, 266
Institute of Physics, NAB, 260
Institute of Physics, NAC, 260
Institute of Physics (IoP), CAS, 215, 258-267, 269, 270, 272, 274, 276, 279-282
  Department of Magnetism, 262, 263, 273
  theoretical physics group, 262
  Theoretical Physics Section, 273
  Theory Group, 262, 263, 273
Institute of Semiconductors, CAS, 258, 261, 268
Institute of Theoretical Physics (ITP), CAS, 268, 269, 273, 278
Instruction
  anatomical, 200
  surgical, 200
Integration of Science and Humanities in Modern Society, 293
ntegrationism, 137, 138
Intelligence (zhi) system, 127
Intelligent design, 40
Intentionality, 103
International Liquid Crystal Conference (ILCC), 210, 211, 219, 232, 234, 274
  8th (1980, Kyoto), 217, 218, 237, 274
  9th (1982, Bangalore), 219
  11th (1986, Berkeley), 210, 222
  12th (1988, Freiburg), 227, 229
  13th (1990, Vancouver), 210, 232, 235, 236
  16th (1996, Kent), 234
International Liquid Crystal Society (ILCS), 209, 212, 216, 224-227, 230, 232, 233, 237, 274
  birth, 232, 234, 235
  Board member, 230, 234

Bylaws, 231, 233, 234, 236
Conference Committee, 210, 211, 234
founding, 209, 224-234
   Act I (March 2, 1988), 225-227
   Act II (May 25, 1988), 227, 228
   Act III (September 20, 1988), 230, 231
   Act IV (1988-1990), 231, 233
   Finale (July 22-27, 1990), 232, 235, 236
   Prologue (July 1987-March 1988), 224
Officers Meeting, first, 234
Petition, The, 229-231
pre-history, 210
President, 231, 232, 237
Vice President, 234
International Union of History and Philosophy of Science, 63
   Division of History and Technology, 63
Introduction to Chinese Classic, 129, 130
Introduction of Nonlinear Physics, 281
Intuition, 88
Ionian school, 11
Ireland, 189
Iron Man, 376
Irreproducibility
   historical, 27, 32
Irreversibility, 166
Irreversible thermodynamics, 214
Ising, Ernst, 389
Ising model, 383, 389, 395, 398
Isis, 63
Israel, 90
Italy, 121, 188, 222, 349

J

Jacob, Margaret, 294
Jacobsohn, Kurtz, 197, 204, 207, 208
Janik, Jerzy, 217, 221
Japan, 77, 222, 254
JCOM, 74
Ji, Meng-Liang, 123
Jia, Bin-Xiang, 292
Jiang, Qing, 273
Jiang, Xiao-Yuan, 293, 301
Jin, Wu-Lun, 163

Jobs, Steven, 7
Johns Hopkins University, 257
Journal of Dialectics of Nature, 292, 300
Journey to the Center of the Earth, A, 344
Just a Theory, 55
Justification, 109

K

Kandel, Eric, 32, 363, 367
Kannimei, 380
Kant, Immanuel, 10, 113
Kao, Kuen, 285
Kent, 210, 218, 224, 234, 237
Kent State University, 210
Kepler, Johannes, 79
Killing Time, 62
Kléman, Maurice, 217
Knowledge, 12, 16, 21, 35, 52, 86, 95, 124, 127, 155, 244
   absolute, 85
   human, 3
   human-dependent, 4
   human-independent, 4, 25, 68
   Primary Source, 250
   scientific, 2, 8, 13, 35, 42, 76, 118, 125
   tacit, 71, 250
Knowledge and Social Imagery, 69
Knowscape, 2-5, 25, 61, 64, 65, 73, 74
Kobayashi, Shunshuke, 217, 218, 219, 221
Kobe University, 188
Koyré, Alexandre, 110
Kraft Circle, 61
Kraków, 219
KTSF, 276
Kuhn, Thomas, 51, 55, 59-61, 65, 88, 91, 108, 244, 246, 315
   career plan, 91
   on Sarton, 59
Kumar, Satyendra, 237
Kunming, 257, 262
Kwong, Peter, 213
Kyoto, 217, 218, 220, 237

L

Laboratory, 193, 195, 199
   course, 203

Laboratory of Experimental Histology and Physiology, 198
Laboratory of Paris University, 199
Laboratory visit, 71
Lady Gaga, 91
Lagerwall, Sven, 217, 218, 221
Lagranian, 214, 215
Lam, Lui, 117, 211, 213, 216, 221, 222, 232, 234, 244, 264, 278
Lam-Platzman Correction Term, 213
Lam-Platzman Theorem, 213
Landau, Lev Davidovich, 191
Laudau-de Gennes mean-field model, 217
Lang, Graeme, 125, 126
Language, 16, 40
Lao-tse, 85
Lao-tse, 163
Laplace, Pierre-Simon, 15
Lanting Xu, 120
Large Hadron Collider (LHC), 139, 142
Lax, Melvin, 213-215
Law, 16
  exponential, 153
  allometric growth, 153
Lawrence Radiation Laboratory, 276
LCer, 217, 224, 237
Leadbetter, Alan, 217, 221
Lee, Tsung-Dao, 27, 285
Lee, Yuan-Tseh, 285
Lee-Yang dispute, 27
Legalist, 130
Lei Feng hat, 263
Leibniz, Gottfried, 13, 93
Les Houches Summer School, 214
Levelut, Anne, 217, 221
Li, Dao-Yuan, 332
Li Ji, 127
Li, Jie-Li, 123
Li, Shan-Lan, 122, 124, 336
Li, Shi-Min, 120
Li, Shi-Zhen, 122, 123
Li, Shu-Hua, 163
Li, Si, 130
Li, Tie-Cheng, 262, 263, 270
Li, Ye, 121
Li, Yin-Yuan, 263, 286
Liao, Qiu-Zhong, 274

Life
  artificial, 4, 168
  origin of, 84
  synthetic, 29
Life of Pie, 20
Lin Lei, 216, 221, 270, 271, 279
Lin, Pei-Wen, 278
Lin, Ze-Xu, 337
Linguistics, 16
Linyi, 380
Liquid crystal (LC), 210, 213, 215, 223, 269, 270, 273, 274
  biaxial nematic, 215
  bowlic, 209, 216, 217, 219, 221-223, 225, 279
  cholesteric, 215
  discotic, 209, 217
  history, 210
  nematic, 215
  polar, 222
  polymer, 222
  rodic, 209
  superconductor, 223
Liquid crystal cell, 276
Liquid Crystal Conference of Socialist Countries
  6th (1985, Halle), 221
  7th (1987, Pardubice), 223
Liquid crystal display (LCD), 209, 273
Liquid Crystal Institute (LCI), Kent, 210, 219, 223
Liquid Crystal Polymer Conference (1987, Bordeaux), 222, 224
Liquid crystal thermometer, 223, 224
Liquid crystalist, 217, 222, 224, 225
Liquid Crystals, 222, 231
Liquid Crystals Today (LCT), 211, 212, 231
Liquid Crystals West, 231
Lister, J. David, 221, 222
Literature, 16, 283
  English, 32
Liu, Bing, 301
Liu, Dun, 301
Liu, Hua-Jie, 293
Liu, Hui, 120, 121
Liu, Stanley, 281
Logic, 85, 134

Logical positivism-empiricism, 56, 107
London School of Economics, 56, 61
Looijen, Rick C., 150
Lorentz, Edward, 33
Los Alamos, 231, 232
Losee, John, 107
Love Story, 89
Lü, Bu-Wei, 129, 130
Lu, Fu, 123
Lü Shi Chun Qiu, 129
Lu, Xun, 343
Luckhurst, Geoffrey, 237
Lushan conference (1978), 269

M

Ma, Lai-Ping, 292
Macau, 278, 285
Mach, Ernst, 51, 55-56, 87
Macroprocess, 154
Macroproperty, 154
Magnet, 90
Magnetism, 187, 188
Mallamace, Francesco, 222
Manhattan Project, 90
Mannheim, Karl, 66
Mao, Ze-Dong, 133, 134, 217, 224, 256, 265, 273, 274
Marginer, 51, 54, 91
Markram, Henry, 169
Marshall, Barry, 177, 181, 184
Martin, Paul, 214
Marx, Karl, 66, 133
Marxism, 56, 58, 134
Mass, 60, 141
 point, 79
Mass communication, 75
Massachusetts Institute of Technology (MIT), 59, 60, 256
Mathematical Principles of Natural Philosophy, 15, 79, 86, 87
Mathematical Theory of Communication, The, 314
Mathematics, 63, 64, 122, 134, 141, 148, 189
Mathematization, 23
Maxwell, James Clerk, 186-188, 191

Maxwell's
 equations, 36, 188
 kinetic gas theory, 15
May Fourth Movement, 77, 259, 298
Mbeki, Thabo, 247, 249, 250
Mead, Margaret, 92
Meaning, 144
Mechanics
 Aristotle's, 60, 61, 73, 88
 continuum, 214
 Newton's, 13, 15, 20, 23, 31, 36, 58, 63, 68, 86, 88, 190
Medal of Honor: Rising Sun, 375, 377
Medhurst, Walter Henry, 336
Medical
 course, 201
 curriculum, 200
 student, 200
Medicine, 196, 204
 Internal, 200
 experimental, 208
Mendel, Gregor, 187
Mendeleyev, Dmitry Ivanonich, 188
Meng, Jian-Wei, 293, 296
Meng Xi Bi Tan, 121, 332
Merleau-Ponty, Maurice, 104
Merton, Robert, 251
Metaphysics, 9
Metal Gear Solid, 372, 377, 378
Metascience, 54
Meteorology, 22
Methodology, 103, 147-149, 172, 247
Meyer, Mihel, 138
Miao, Li-Tian, 127
Mickey Mouse, 260
Micro mechanism, 158
Microentity, 154
Microreduction, 154
Microscopy, general, 199
Milesian school, 10
Millikan, Robert, 90
Mind-body problem, 12
Mireille, 224, 233
Mirkovitch, Michel, 214
Mirksy, Steve, 70
Mirror neuron, 349, 360
Mo, Yan, 283
Mo-tse, 332

Model building, 386
Modernity, 111
Mohist, 119
Mohist, 332
Molecular Crystals and Liquid Crystals (MCLC), 219, 222, 231
Molecular fluid, 215
Molecule, 150
Monadism, 152
Moniz, Egas, 197, 204, 205, 207, 208
Morin, Edgar, 159
Morris, Arthur, 182
Moscow, 266
Moscow State University, 261
Mott, Nevill, 268
Mou, Zong-San, 117, 126
Movement, 350
Movie, 372, 373, 380
Müller, Johannes, 198
Müller, Karl, 24
Multi-narrative, 377
Myth, 34
Mythology, 8

N

Nanjing, 260
Nanjing Normal University, 221
Nanjing University, 266
Nanning, 282
Napoleon. *See* Bonaparte, Napoleon
Narrative, 371, 372, 376
National
　Academia Sinica (NAS), 259, 260
　Academy of Art, 260
　Academy of Beiping (NAB), 260
　Academy of Sciences (USA), 257
　Central University, 260, 266, 273
　Conference on Science (Beijing), 265
　Educational Technology Standards (NETS), 293
　Endowment for the Humanities, 44
　Institute of Health (NIH), 28
　Science Foundation (NSF), 18, 28, 44, 313
　Southwest Associated University, 256, 257, 262
Nature, 3, 5, 11, 15, 33, 36, 53, 85, 94, 144, 166
　idea of, 11
Nature, 82, 243
Natural science, 15, 51, 124, 195, 196, 200, 227, 245
"Natural science", 14-17, 41, 51, 53, 54, 63, 64, 85, 89-91, 93, 95, 126
　Room, 21
Needham, Joseph, 116, 125
Needham Question, 85, 116, 125, 134
Nelson, David, 222
Nematic-isotropic transition, 217, 270
Nematodynamics, 214
Netease, 379
Netherlands, 395
Network, 143, 168, 394
New Thesis of Human Body, 336
New York City (NYC), 213, 214, 221, 224, 255, 257, 282
New Yorker, 234
New Zealand, 62
Newton, Issac, 13, 15, 18, 35, 42, 52, 86, 87, 93, 188, 189, 390
　dark secrets, 15, 35
　laws of motion, 79
Neuro plasticity, 349
Neuroarthistory, 350, 361, 367
Neurohumanities, 32
Neuroscience, 351
Neutron, 81
Nevis Labs, 71
Nie, Rong-Zhen, 133
Nietzsche, Friedrich, 66
Nickles, Thomas, 59
9/11, 372, 376, 377, 381
Nobel Prize, 24, 27, 30-33, 43, 60, 72, 81, 90, 91, 142, 184, 214, 217, 220, 260, 268, 281, 283, 285, 392
Noether, Emmy, 139
Noether's theorem, 139, 140, 142
Non-science, 112
Nonseparability, 164
Norgay, Tenzing, 64
Northwestern University, 216, 269, 270
Novak, Igor, 42
Notus, 8

O

Occam's Razor, 9, 58

Oldroyd, David, 54
Olympic Games, 85
On Astronomy, 336
Oneness, Heaven and Man, 85
Ong, Hiap Liew, 231
On Humanistic Value of Science, 293
Onians, John, 350, 362, 363, 367
Onnes, Heike, 24
Onsager reciprocal relations, 215
On Scientific Spirit, 293
On the Origin of Species, 87
Ontology, 103, 148
Opium War, 335, 337
Oppenheimer, J. Robert, 258
Organism, 150, 166
Organizing principle, 164
Orsay, 214, 217, 219, 221, 224, 281
   "gang of four", 217
Ortega y Gasset, José, 105
Osiris, 63
Our Scientific Culture, 294, 301
Outsider, 51, 54, 69, 70, 73, 92
Ou-Yang, Zhong-Can, 273
Overseas Chinese, 255, 271
Overseas Chinese Hotel, 262

P

Paradigm shift, 60, 91
Parascience, 42
Pardubice, 223, 224
Paris, 258, 260, 274
Parity, 24
   nonconservation, 24, 25, 73, 81, 281
Parodi relation, 215
Partially Ordered Systems, 222
Pathological Anatomy, 197
Pathology, 200, 204
PBS (Public Broadcasting System), 15, 35, 84
Peacock, George, 189
Pedestrian modeling, 29
Peking University, 216, 259, 278, 297, 377
Peng, Huan-Wu, 267, 268, 281, 282, 284, 285
People's Daily, 264
Periodic Table of Expertises (PTE), 250
Perrin, Jean, 56

Phaedrus, 166
Pharmacy, 200
Phase transition, 215, 216, 269, 275, 276
Phenomenological approach. *See* Research level
Phenomenology, 103
Philosophy, 6, 7, 9, 10, 12, 14, 17, 84, 91, 148, 259
   Natural, 14, 17, 41
   'Philosophy', 13, 14, 17
   "Philosophy", 14, 16, 32, 39, 54
Philosophy of science, 50, 54-62, 84, 91, 93, 94, 103, 247, 290, 296
   Second, 109
Phoedo, 151
Physical Review, 72
Physical Review A, 214
Physical Review Letters (PRL), 24, 216, 219, 259, 270, 276, 278-281
Physicist, 19, 23, 30, 51, 55, 89
Physics, 16, 19, 23, 55, 59, 72, 75, 79, 85-88, 90, 91, 93, 157, 158, 196, 200, 260, 270
   classical, 82, 188
   mathematical, 139
   messy, 82
   microscopic picture, 25
   neat, 82
   Newtonian, 157
   nonlinear, 224, 231, 232
   particle, 140, 145
   quantum, 82, 188
   solid state, 59, 268
   very messy, 82
Physics (Wuli), 274
Physics Today, 224
Physiognomic Perception, 357
Physiology, 55, 197, 198, 200, 204
   experimental, 198
   general, 197, 201
Piaget, Jean, 154
Pinch, Trevor, 53, 68
Planck, Max, 23
Planning and Steering Committee (PSC), 210, 218, 219, 223, 225-227, 229, 230, 234, 274
Plato, 3, 6, 10, 32, 61, 63, 66, 85, 92, 148, 151, 166

Platzman, Philip, 256
Pleasure, 371, 372
Pogson's ratio, 191
Poland, 219, 220, 395
Politician, 21
Pombalin Reform, 195
Pope, 11
Popper, Karl, 51, 55, 56-58, 61, 87
Popular science (popsci), 71, 270, 272, 307
  book, 89, 217, 316
  magazine, 341
  newspaper, 341
  origin, 307
  publishing, 331
  redefined, 311
  writing, 330, 337, 341, 343, 345
Portugal, 193, 208
Positive thinking, 106
Postmodernism, 89, 247, 248
Pre-givenness, 105
Pre-Qing period, 332
Prediction, 25, 30, 43, 57, 58, 74, 82, 86, 144, 222, 245, 256, 279
Present-at-hand, 105
Priest, 21
Princeton University, 59, 139, 275
Probability, 15, 31, 82, 86, 90, 389
Problematology, 137, 138, 143
Projectile path, 66
Promeheus, 118
Prost, Jacques, 217, 221, 222
Protestant, 335
Protestant Ethic and the Spirit of Capitalism, The, 125
Provo, Utah, 80, 81
Pseudoscience, 36, 40, 57, 112
Psychology, 16, 55, 56, 86, 148, 259
  Gestalt, 148
  in history, 148
Psychology of science, 110
Pu, Fu-Ke, 263, 274, 278, 285
Public Engagement with Science and Technology (PEST), 77, 321, 324
Public Understanding of Science (PUS), 77, 315, 324
Public Understanding of Science, 74, 317
Pythagoras, 9, 10, 70

Q

Qian, Mu, 129, 130
Qian, San-Qiang, 257, 258, 260, 267, 285
Qian, Xue-Sen, 286
Qian, Yong-Jia, 270
Qu, Yuan, 119, 332
Quality-oriented education, 293
Quan, Jean, 213
Quant, 30
Quantum mechanics, 25, 36, 82
Quellette, Marc, 376
Quine, Willard Van Orman, 148

R

Rabi, Isidor Isaac, 260
Raman Research Institute, 216, 219
Random number, 400
Random walk, 58, 384, 385
Rauschning, Hermann, 68
Read, Rupert, 59
Reality, 40, 379, 382
Reality Check (RC), 35, 44, 52, 60, 61, 65, 69, 70, 74, 88, 91
Reductionism, 148, 150
  micro-, 154, 157
  strong, 154, 155
Reform-and-opening up revolution, 270, 291
Reinitzer, Frederick, 209
Relativity theory. *See* Theory
Religion, 1, 7, 14-17, 37, 40, 76, 194
  organized, 9
Remembering Cai Shi-Dong, 275
Ren, Zhi-Gong, 256, 257, 285
Renaissance, 188, 349, 353
Reocentrism, 140
Republic of Science, 77, 308, 328
Rescher, Nicholas, 172
Research, 193, 200, 265
  Dialectics of Nature, 293
  life science, 200
  school, 193, 207
Research level, 53
  bottom-up, 25, 29, 32, 53, 66
  empirical, 18, 25, 29, 32, 40, 53, 86, 89
  phenomenological, 22, 25, 29, 32, 53, 65, 86

Resistance-against-Japan War, 286
Returnee, Chinese, 254, 272, 275, 284, 286
Reverse thinking, 106
Revolution, 253, 256, 270, 274, 284
Reymond, Emil Du-Bois, 198
Ribotta, Roland, 214, 217, 219
Ricci, Matteo, 124, 333, 334
ROC Taiwan Liquid Crystal Society, 237
Rockefeller University, 257
Romania, 257
Rome, 349, 352
Rong, Hong, 285
Royal Society, 188, 316
Ruan, Liang, 218
Ruan, Wei, 132
Russia, 211

S

Saarbrücken, 256, 262
Sackmann, Horst, 221
Safina, Carl, 76
Salt Lake City, 80, 81
San Diego State University, 278
San Francisco, 276
San Jose, 223, 224
San Jose Mercury News, 41
San Jose State University (SJSU), 71, 224, 231, 232
   Nonlinear Physics Laboratory, 232
Santa Fe Institute, 168, 169
Sarton, George, 59, 63-65, 93, 292
Saupe, Alfred, 217, 221, 224
Schadt, Martin, 222, 223, 234
Scheler, Max, 104
Schelling model, 392
Schelling, Thomas C., 382
Schleiden, Matthias, 187
Scholastic culture, 128
Schrieffer, Robert, 24
Schrödinger, Erwin, 168
Schuster, P., 165
Scicomm. *See* Science communication
Science, 1, 3, 5, 11, 14, 16, 37, 42, 50-52, 62, 73, 77, 87, 94, 129, 137, 148, 185, 191, 244, 248
   aim of, 2, 18, 53
   applied, 21
   atmospheric, 16
   bad, 144
   characteristics of, 41, 44
   Chinese word, 342
   cognitive, 75
   definition, 16, 40, 62
   education, 31, 76
   essence of, 34
   "exact", 34
   experimental, 204
   Father of, 2, 9, 12
   good, 88, 144
   historical development, 63
   historical/natural, 203
   human, 44, 83, 94, 243
   idea of, 12
   medical, 5, 23, 28, 64, 68, 94, 193, 194, 196, 202
   metaphor for, 53
   modern, 13, 42, 95, 126, 148, 290
   nonhuman, 91, 94
   physical/chemical, 196, 203
   physiological, 198
   political, 85
   revolt against, 89, 91, 94
   writing, 331, 335, 336
"Science", 6, 14, 43, 53, 94
Science, 82
Science appreciation, 95
Science and Civilization in China, 116
Science communication (scicomm), 31, 50, 74-83, 84, 93, 94, 307, 330, 338
   degree, 78
   difficulty, 79
   history, 76, 307
   Law, 78
   magazine, 338
   meanings, 324
   newspaper, 338
   review, 307
Science Communication, 74
Science & Culture Review, 294
Science fiction, 343
   with illustrations, 344
Science Matters. *See* Scimat
Science Matters, 38
Science Popularization, 78, 301

Science Review, 293, 294
Science Room, 16, 19, 20
Science and Society, 300
Science Society of China, 291
Science Spring 1978, 261, 265
Science Squirrels Club, 78
Science Studies, 243
　First Wave, 244
　Second Wave, 243, 245, 247-249
　Third Wave, 243, 248-251
Science as a Cultural Process, 292
Science and Technology Chronological Table, 188
Science and technology (S&T), 330, 331, 338
　book, translation and publication, 340
　study, 296
Science and technology studies (STS), 292-294
Scientific
　apparatus, 193
　center, 188
　discovery, 188
　instrument, 198, 201
　instrumentation, 198
　literacy, 93, 325
　literature, foreign, 333
　method, 14, 33, 69, 245, 251
　procedure, 150
　process, 17, 20, 22, 52, 66, 67, 88
　questioning, 143
　research, 204, 207
　result, 52, 62, 65, 67, 88
　Revolution, 13, 29, 195
　society, 207
　spirit, 75, 124, 133, 134
　theory, 143-145
　thinking, 76
Scientific culture (SC), 51, 116, 132, 134, 290, 331
　Chinese, 292
　definitions, 294
　group, 301
Scientific culture study (SCS), 296, 291, 299
　National Symposium (2002, Beijing), 293
　School, 302

Scientific Spirit and Humanistic Spirit, 293
Scientific World, 342
Scientificity, 54, 87
Scientism, 42, 90, 291
Scientist, 14, 16, 35, 41, 43, 65, 77, 78, 93, 137, 142, 144, 189, 225, 244, 247, 248, 284
　Chinese, 254, 333, 336
　definition, 18
　social, 21, 28
　natural, 21
"Scientist", 36, 63, 64, 80
Scientometrics, 188
Scimat, 2, 5, 14-16, 32, 43, 51, 53, 64, 95, 117, 234, 244, 251
　action plan, 38
　Center, 38
　conference, 38
　Program, 7, 38
　Proposal, 38
　Q & A, 36-39
　ramification, 36, 39
　Standard, 18, 43
Scimatist, 38
Scotsman, The, 219
Scott Russell, John, 219, 275
Second Asia Pacific Physics Conference, 220
Secularity, 44, 52
Segal, Eric, 89
Self-organizing mechanism, 165
Semantics, 89
Semiconductor, 4, 36
Sendai, 234
Sex, 8, 65
Shangdong Provience, 380
Shanghai, 217, 336, 342
Shanhaiching, 332
Shantou, 213
Sharrock, Wes, 59
Shen, Jue-Lian, 263, 270, 278
Shen, Kuo, 121, 332
Shen, Nong, 123
Shen Nong Ben Cao Jing, 123
Shen, Yuen-Ron, 216, 231
Shi, Ru-Wei, 260-262, 270, 285
Shou Shi Li, 122

Shu, Chang-Qing, 273, 276-278, 281, 282
Shui Jing Zhu, 332
Si Zhou Zhi, 336
Sichuan Province, 276
Sign, 373, 382
Sketch for a Historical Picture of the Progress of the Human Mind, 253
Skirrow, Martin, 182
Slave, 85
Sluckin, Tim, 212
Smith, Adam, 13, 15, 86
Smuts, Jan Christiaan, 147-149, 152
Snow, Charles Percy, 247, 292
Social physics, 86
Social science, 5, 14-17, 29, 51, 54, 63, 64, 87, 92, 124, 246, 247
  computational, 31
Social segregation, 392
Social Studies of Science, 243
Social Studies of Science, 243
Sociedade Portuguesa de Biologia, 207
Sociedade Portuguesa de Ciências Naturais, 207
Sociology, 14-16, 31, 36, 86, 246
Sociology of knowledge, 66
  Strong program, 67, 69
Sociology of science, 50, 66-74, 84, 93, 94, 110, 246, 290, 296
Sociology of scientific knowledge (SSK), 67, 69, 108, 112, 245
  intrinsic limitation, 67, 93
Sociophysics, 394
Socrates, 10, 26, 39, 85, 349, 350, 352, 360
  Method, 26, 85
Soft matter, 232, 274
Sokal hoax, 81-82
Solidariity, 219, 220
Soliton, 216, 219, 221, 222, 253, 266, 272, 274, 284
  definition, 275
  in China, 275-282
  in condensed matter, 278, 279
  in shearing liquid crystal, 277
  kink, 277
Solitons in Liquid Crystals, 222, 279, 281
Sommerville, Johann, 79

Song, Ci, 121
Song, Ying-Xing, 123, 333
Soul, 9, 10, 12
South Africa, 247
Southeast University, 260
Soviet Union. *See* USSR
Soy sauce, 265
Special Economic Zone, Shenzhen, 266
Spin, 387
Splinter Cell, 372, 377, 378
Spoon bending, 70
Spring and Autumn period, 119, 129-131
Springer, 222, 223
Sputnik, 77, 313
Spy, 371
Spyhunter, 372
Stalin, Joseph, 133
Standard Model, 43, 139, 140
Stanford University, 41
Stanley, Eugene, 222
Star, 84
Star War, 20
Statistical Physics, 383, 389
  application, 391
Statistical Physics and Condensed Matter Theory, National Conference on
  First, 278
  Second, 282
Strange attractor, 272
Structure, 152
  complex, 160
  stable, 156
  temporal-spatial, 188
Structure of Materials, 217
Structure of Scientific Revolutions (*SSR*), *The*, 59, 61, 89, 244, 315
Student movement, 89
Studies
  medical, 194, 196, 197, 202
  physical/chemical, 195
  philosophical/natural, 195
  theological, 195
Studies in Dialectics of Nature, 292, 300
Studies in Philosophy of Science and Technology, 300
Study on Science Popularization, 78
Studies in Science of Science, 300
Sun, 8

Sun, Bai, 377
Sun, De-Zhong, 295
Supercomputer, 30
Superconductivity, 24
Superconductor
  high-T_c, 18, 223
Superconductor Week, 223
Superman, 91
Supernatural, 3, 8, 37, 52
Surgery, 200
Switzerland, 262
Synthetic microanalysis, 157, 158
System, 152, 155
  closed, 26
  complex, 21, 26, 51, 63, 65, 75, 78, 82, 83, 85, 86, 90, 95, 154, 222, 231, 234
  complex adaptive, 168
  deterministic, 15, 58, 82, 90, 388
  history-dependent, 68, 90
  human, 52, 53, 85, 86
  messy, 82
  muscular, 201
  nearly decomposable, 156
  nervous, 201
  nonhuman, 52, 53, 85, 86, 94
  nonlinear, 231
  open, 26, 165
  probabilistic, 15, 33, 43
  science, 148
  simple, 21, 51, 58, 83, 85, 86, 90, 95, 234
  statistical, 388
  stochastic, 31, 58
  time-evolving, 68

T

Tacit
  knowledge, 71, 250
  writing, 371
Tagore, Rabindranath, 217
Taiwan, 218, 237, 254, 261, 285
Tang, Ming-Zhao, 255
Technology, 21
Television (TV), 209, 265, 276
Teller, Edward, 72
Temporal-spatial localization, 28
Tenure system, 284

Thales, 2, 9, 10, 12, 18, 22, 42, 52, 54, 84
Thermodynamics, 23
  Second Law, 28
Theology, 13, 14, 17, 40, 54
  Natural, 13, 14
Theory, 36, 69, 71-73, 266
  BCS, 24
  catastrophe, 168
  Confucian, 120
  dissipative structure, 168
  electromagnetic, 36
  general relativity, 36, 68, 69, 72, 82, 87
  heliocentric, 79
  hypercycle, 168
  mature, 57
  special relativity, 23, 36, 60, 82, 87
  synergetic, 168
Theory and Reality, 55
Theory-ladenness, 109
Therapy, 200
Thermodynamics, 189
Thermoviscous solid, 215
Third Plenum, CCP (1978), 270
Three-Dimensional Morphology of Systems Engineering, 167
Three Kingdoms period, 32, 120
Tian Gong Kai Wu, 123, 333
Tian, Song, 293
Tian Wen, 119, 332
Tian Yuan Shu, 121
Tiananmen Gate, 258
Tiananmen "incident", 80, 237
Time
  real, 157
  spatial, 157
Ting, Chao-Chung, 285
Titov, Viktor, 211
Top-down approach, 78
Toulmin, Stephen, 107
Transformer, 376
Travel Diary of Xu Xia-Ke, The, 333
Truffaut, François, 372
Truth, 143, 248
Tsinghua University, 256, 257, 266, 268, 273-275, 279, 283, 284
Tsui, Chee, 285
Turney, Jon, 331
Two-culture problem, 64, 247

The Two Cultures, 292
2061 Project, 293

U

Uncertainty, 32, 33, 35, 43, 82
Undergrad, 256
United Kingdom (UK), 76, 77, 219, 229, 247, 264
United Nation, 255
United States (US), 40, 44, 78, 82, 89, 222, 261, 269, 270, 273, 274, 276, 278, 285, 286
Universal Gazette, The, 338
Universe, 3, 53, 73, 79, 138
Universitaire Instelling Antwerpen, 215
Universität des Saarlandes, 215
Université de Paris-Sud
   Laboratoire de Physique des Solidides, 217
University, 193, 194, 207
   studies, 194
University of
   Auckland, 62
   British Columbia (UBC), 213, 225, 234
   California, 32, 59, 274
   Edinburg, 69
   Ghent, 63, 198
   Hong Kong, 213, 256
   Paris, 198, 260
   Science and Technology of China, 261
   Vienna, 55
University College London, 61
University of Wisconsin-Madison, 78
   Department of Life Sciences Communication, 78
USA. *See* United States
USSR, 77, 185, 186, 211, 261, 274, 285
Utah, 80

V

Vancouver, 210, 225, 232, 235, 254
Vanderbilt University, 273
Varrena, 222
Vatican, 11, 13
Venn, John, 189
Verne, Jules, 343, 344

Vesalius, Andreas, 121
Verbiest, Ferlinand, 333
Vertebrate, 74
Vibrational Spectra and Structure of Polyatomic Molecules, 262
Video game, 371
   first-person shooter (FPS), 374, 375
   online, 371
   personal computer (PC), 371
   role-playing (RPG), 371, 373, 381
   spy, 371, 376, 381
Video movie, 380
Vienna, 32, 56, 61, 87
Vienna Circle, 56, 61, 87
Vietnam War, 89
Victorian era, 185, 187, 189, 191
Vinci, Leonardo da, 19, 349, 353, 360, 365, 366
Virchow, Rudolf, 198
Viscous fingering, 221
Vries, Adrian de, 217-219, 221, 225

W

Wałęsa, Lech, 220
Wall Street, 30
Waller, Robert, 89
Walrasian paradigm, 170
Wang, Da-Yan, 293
Wang, Ding-Sheng, 262, 263, 270
Wang, Hao, 257, 285
Wang, Rong-Jiang, 290
Wang, Tao, 336
Wang, Xi-Zhi, 120
War and Peace, 20
Warcraft, 379, 382
Warren, Robin, 184
Warring States period, 85, 123, 129, 130, 332
Washington University, 278
Watson, James, 29
Watts, Duncan, 31
Wealth of Nations, The, 14
Weather, 26, 82
   forecasting, 82
Web Addiction War, 380
Weber, Max, 125
Weinberg, Steven, 60, 61
Weizmann, Chaim, 90

Wen Wei Po, 270, 271
West Lake, 255
West Learning Spreading to the East Movement, 334, 336
Western missionary, 335
Western Returned Scholars Association, 284
Westernization Movement, 124, 331, 337, 341
Whewell, William, 14, 311
Whitehead, Alfred North, 157, 159, 189
Whole, 147, 148, 150, 152, 155, 158, 164, 171
  basic conditions, 151
  generative, 164
  mechanical, 155
  organic, 155
  six phases, 150
  three kinds, 155
Williams, Perry, 52
Winch, Peter, 31
Winter School in Nonlinear Physics, 232
Wölfflin, Friedrich, 349, 356, 360, 366
Wolfram, Stephen, 161
Woo, Chia-Wei, 216, 218, 269, 276
Woodward Conference, 231
World Journal, 33
World War I, 189, 403
World War II (WWII), 56, 61, 89, 90, 244, 291
World Wide Web, 214
Worldview, 244
  clockwork, 86
Wright, Talmadge, 381
Wu, Chien-Shiung, 24, 73, 81, 285
Wu, Da-You, 262, 285
Wu, Guo-Sheng, 293, 297
Wu, Ling-An, 264
Wuhan, 217
Wylie, Alexander, 124, 336

X

Xi, Jin-Ping, 266, 270
Xi, Zhong-Xun, 266
Xi Yuan Ji Lu, 121
Xia Er Guan Zhen, 335
Xiao, Feng, 293

Xie, Yu-Zhang, 218, 222, 273, 275, 286
Xinhua News Agency, 272
Xu, Guang-Qi, 124, 334
Xu, Hong-Zu, 333
Xu, Nian-Ci, 343
Xu, Xia-Ke, 123
Xu Xia KeY ou Ji, 123
Xun-tse, 130

Y

Yaquan Magazine, 342
Yale University, 62, 261
Yan, Ji-Ci, 258-260, 267, 285
Yang, Chen-Ning, 27, 72, 269, 281, 285
Yang, Yong-Xin, 380
Yang Wu Movement, 337
Yong Shi, 129
Yu, Guang-Yuan, 293
Yu, Lu, 263, 278, 285
Yuan, Jiang-Yang, 296
Yuan, Wei, 336
Yuasa, Mitsutomo, 185, 186, 188, 191, 192

Z

Zeng, Er-Man, 222
Zephyrus, 8
Zeus, 118
Zhang, Heng, 120
Zhao, Jiang-An, 218
Zhao, Lei-Jin, 292
Zheng, Jia-Qi, 270
Zhongguancun, 261, 262, 265
Zhongwen Yundong, 256
Zhou, Chang-Zong, 293
Zhou, En-Lai, 133, 256, 282, 285
Zhou, Guang-Zhao, 446
Zhu, De, 133
Zhu, Guo-Zhen, 275, 276, 278, 279
Ziegler, Jean, 6
Ziman, John, 51
Zipf plot, 65, 384
Zu, Chong-Zhi, 120, 121
Zu, Heng, 120
Zürich, 62